VOLUME 2

Environmental Effects on Composite Materials

EDITED BY

George S. Springer

TECHNOMIC
PUBLISHING CO., INC

VOLUME 2

Environmental Effects on Composite Materials

WITHDRAWN

EDITED BY

George S. Springer

Department of Aeronautics and Astronautics
Stanford University, Stanford, California 94305

Technomic Publishing Company, Inc.
851 New Holland Avenue
Box 3535
Lancaster, Pennsylvania 17604

Printed in the United States of America
10 9 8 7 6 5 4 3 2 1

Main entry under title:
  Environmental Effects on Composite Materials, Volume 2

A Technomic Publishing Company book
Bibliography: p. 413
Includes index p. 437

Library of Congress Card No. 81-50309
ISBN No. 87762-348-1

Table of Contents

Preface

This book is a sequel to the first volume on *Environmental Effects on Composite Materials*. The first volume summarized work published prior to 1980. This second volume takes up where the first left off; it covers experimental and analytical results generated through 1983 by industry, government, and university laboratories both in the United States and abroad.

Chapters 3-31 were published either in the *Journal of Composite Materials* or in the *Journal of Reinforced Plastics and Composites*. The first two and the last chapters were prepared specifically for this volume. Chapter 1 provides a summary and overview. Chapter 2 presents data not found in the open literature. Chapter 32 lists all the references cited both in Volume 1 and Volume 2, and serves as a guide to other articles and reports of interest.

Together, the two volumes of this book provide a comprehensive summary and should be of practical value to engineers and specialists interested in the effects of moisture and temperature on composite materials.

GEORGE S. SPRINGER
Stanford

List of Contributors

D. F. ADAMS
Composite Materials Research Group
University of Wyoming
Laramie, Wyoming

R. E. ALLRED
Sandia National Laboratories
Albuquerque, New Mexico

K. H. G. ASHBEE
University of Bristol
H.H. Wills Physics Laboratory
Tyndall Avenue
Bristol BS8 1TL, England

N. BALASUBRAMANIAN
R&D Centre
Asbestos Cements Limited
Bangalore-560 058, India

CH. S. BARRETT
University of Denver Research Institute
Denver, Colorado

C. A. BIGELOW
NASA Langley Research Center
Hampton, Virginia

P. BONNIAU
Centre des Materiaux
Ecole des Mines
BP 87-91003 EVRY Cedex, France

R. J. BRADSHAW
Department of Mechanical Engineering
The University of Alabama in Huntsville
Huntsville, Alabama

A. R. BUNSELL
Centre des Materiaux
Ecole des Mines
BP 87-91003 EVRY Cedex, France

D. S. CAIRNS
Hercules Aerospace Division
Bacchus Works
Magna, Utah

M. CHANDA
Department of Chemical Engineering
Indian Institute of Science
Bangalore-560 012, India

C. I. CHANG
Materials Science and Technology Division
Naval Research Laboratory
Washington, D.C.

T. W. CHOU
Mechanical and Aerospace
Engineering Department
University of Delaware
Newark, Delaware

R. M. CHRISTENSEN
Lawrence Livermore National Laboratory
University of California
Livermore, California

T. J. CHUNG
Department of Mechanical Engineering
The University of Alabama in Huntsville
Huntsville, Alabama

W. J. CRAFT
North Carolina Agricultural and
State University
Greensboro, North Carolina

F. W. CROSSMAN
Lockheed Palo Alto Research Laboratory
Palo Alto, California

R. J. DEIASI
Research Department
Grumman Aerospace Corporation
Bethpage, New York

F. DELALE
Lehigh University
Bethlehem, Pennsylvania

J. A. DICARLO
National Aeronautics
and Space Administration
Lewis Research Center
Cleveland, Ohio

E. C. EDGE
Principal Stress Engineer
British Aerospace, Aircraft Group
Warton Division, Warton Aerodome
Preston PR 4 1AX
United Kingdom

F. ERDOGAN
Lehigh University
Bethlehem, Pennsylvania

D. L. FLAGGS
Lockheed Palo Alto Research Laboratory
Palo Alto, California

P. FURMAŃSKI
Heat Conduction Laboratory
Institute of Heat Engineering
Warsaw Technical University
Warsaw, Poland

R. F. GIBSON
Mechanical Engineering Department
University of Idaho
Moscow, Idaho

W. GOGÓŁ
Heat Conduction Laboratory
Institute of Heat Engineering
Warsaw Technical University
Warsaw, Poland

C. A. GRIFFIS
Material Science and Technology Division
Naval Research Laboratory
Washington, D.C.

K. HARAKAWA
Department of Fiber and Polymer Engineering
Nagoya Institute of Technology
Gokiso-cho, Showa-ku, Nagoya 466, Japan

A. HOENIG
Department of Mathematics
John Jay College of Criminal Justice
New York, New York

T. ISHIKAWA
Mechanical and Aerospace
Engineering Department
University of Delaware
Newark, Delaware

K. KONDO
Department of Aeronautics
University of Tokyo
Bunkyo-ku, Tokyo 113, Japan

E. R. LONG, JR.
NASA, Langley Research Center
Hampton, Virginia

R. A. MASUMURA
Material Science and Technology Division
Naval Research Laboratory
Washington, D.C.

E. W. MENDE
Mechanical Engineering Department
University of Idaho
Moscow, Idaho

R. K. MILLER
General Dynamics Convair Division
San Diego, California

K. MORIMOTO
Nisshin Spinning Co., Ltd.
Nishiarai Laboratory
1-81-1, Nishiarai Sakaecho
Adachi-ku, Tokyo, 123 Japan

S. NOMURA
Mechanical and Aerospace
Engineering Department
University of Delaware
Newark, Delaware

W. E. OSBORN
Mechanical Engineering Department
University of Idaho
Moscow, Idaho

H. R. PHELPS
Hercules Corporation
Magna, Utah

P. PREDECKI
University of Denver Research Institute
Denver, Colorado

R. M. V. G. K. RAO
Materials Science Division
National Aeronautical Laboratory
Bangalore-560 017, India

D. A. RIEGNER
General Motors Manufacturing Development
General Motors Technical Center
Warren, Michigan

T. SAKAI
Department of Textile and Polymeric Materials
Faculty of Engineering
Tokyo Institute of Technology
O-okayama, Meguro-ku, Tokyo 152, Japan

J. P. SARGENT
University of Bristol
H.H. Wills Physics Laboratory

Tyndall Avenue
Bristol BS8 1TL, England

R. L. SCHULTE
Research Department
Grumman Aerospace Corporation
Bethpage, New York

G. S. SPRINGER
Department of Aeronautics and Astronautics
Stanford University
Stanford, California

T. SUZUKI
Nisshin Spinning Co., Ltd.
Nishiarai Laboratory
1-81-1, Nishiarai Sakaecho
Adachi-ku, Tokyo, 123 Japan

K. TAKAHASHI
Department of Fiber and Polymer Engineering
Nagoya Institute of Technology
Gokiso-cho, Showa-ku, Nagoya 466, Japan

T. TAKI
Aircraft Manufacturing Division
Kawasaki Heavy Industries Ltd.
Kagamihara, Gifu 504, Japan

Y. WEITSMAN
Civil Engineering Department
Texas A&M University
College Station, Texas

A. YAU
Mechanical Engineering Department
University of Idaho
Moscow, Idaho

R. YOSOMIYA
Department of Industrial Chemistry
Chiba Institute of Technology
2-17-1, Tsudanuma
Narashino-Shi, Chiba, 275 Japan

Table of Contents for Volume 1

(Provided as a reference for those who have not read Volume 1)

1

Introduction

G. S. SPRINGER

IN ORDER TO UTILIZE FIBER REINFORCED COMPOSITE MATERIALS TO THEIR full advantage, the performance characteristics of such materials during their entire service life must be known. This requires analytical and experimental methods which predict changes in the mechanical, physical, thermal, and chemical properties of the material caused by exposure to moisture and elevated temperature. Specifically, methods are required which provide the following information:

a) temperature inside the material as a function of position and time,
b) moisture concentration inside the material as a function of position and time,
c) total amount (mass) of moisture inside the material as a function of time,
d) dimensional changes of the material as functions of position and time,
e) changes in the mechanical properties of the material as functions of position and time.

A survey of the material in this book is given below. Details are provided in the articles which follow and in the cited references.

DATA

The response of composites to moisture and temperature cannot be assessed without extensive data. Data are required

a) to evaluate the accuracies of analytical models,
b) to provide the material properties and constants needed in analytical and numerical calculations, and

1

c) to establish those performance characteristics which cannot be obtained by analytical means.

Data on different types of continuous graphite fiber reinforced epoxy matrix composites and short glass fiber reinforced polyester and vinylester composites (sheet molding compounds) were given in the first volume on *Environmental Effects of Composite Materials*. Since the publication of the first volume, large amounts of data became available both for graphite epoxy composites and for sheet molding compounds. Much of these data are summarized in Chapters 2 and 6.

Properties of several commonly used graphite epoxy composites (manufactured by Fiberite, Hercules, and Narmco) are given in Chapter 2. The data in this chapter include the following information: a) maximum moisture content; b) diffusivity; c) moisture and thermal expansion coefficients; d) glass transition temperature; e) tensile, shear, flexural, compressive, and buckling properties. In most cases the data are of sufficient detail to provide an appreciation of the detrimental effects of moisture and temperature on these properties.

Additional data on moisture and thermal expansion coefficients of some graphite and glass fiber reinforced composites can be found in Chapter 25. Data on the effects of space environments (including the effects of ultraviolet light and vacuum) on graphite epoxy composites are given in Chapter 3.

Changes in the flexural behavior of Kevlar epoxy laminates due to absorbed moisture and elevated temperature are examined in Chapters 4 and 5. A comprehensive review of the effects of moisture and temperature on chopped and continuous glass fiber reinforced sheet molding compounds is presented in Chapter 6. Data in this chapter include tensile, compressive, shear, flexural, fatigue, and creep properties. Information is also provided regarding moisture absorption, thermal expansion, and vibration damping. Additional data on changes in the vibration damping characteristics of sheet molding compounds are given in Chapter 7.

As this survey indicates, most of the available data are for epoxy and polyester matrix composites. Relatively little information exists on composites with other types of matrices. There are some data for glass fiber reinforced polyurethane foam (Chapter 8), graphite-polyimide (Chapter 9), and metal matrix (Chapter 10) composites. Nonetheless, the data are quite meager for composites with matrices other than epoxy or polyester. More data are needed for nonepoxy and nonpolyester matrix composites already in use, as well as for new types of thermosetting, thermoplastic, and metal matrix composites which are now becoming commercially available.

MEASUREMENT TECHNIQUES

Most of the measurement techniques necessary for determining the properties of interest in environmental studies are well established. However, until recently, moisture distribution inside the material could not be measured without destroying the material. The x-ray diffraction and nuclear resonance analyses described in Chapters 11 and 12 offer the possibility of determining the moisture distribution by nondestructive means.

PREDICTION OF MECHANICAL PROPERTIES AND THE FAILURE TIME

Data are necessary for evaluating the changes in material behavior during exposure to moist, high temperature environments. Unfortunately, in many cases the data are inadequate to predict the material performance beyond the conditions of the experiments. This may pose problems if it is required to know changes in the material properties over long periods of time, since testing for several years is often impractical and unfeasible. This difficulty could be overcome by the use of analytical methods which can predict changes in material properties. A model is described in Chapter 13 for estimating changes in the strengths and moduli and the time of failure of loaded composite structures exposed to elevated temperatures.

ANALYSES OF MOISTURE AND TEMPERATURE DISTRIBUTIONS

Analyses for calculating temperature and moisture distributions inside composites exposed to moisture and elevated temperature were presented in the first volume on *Environmental Effects on Composite Materials*. With the computer code (designated as W8GAIN) given in the first volume the moisture and temperature distributions can be calculated over wide ranges of conditions. The calculations can be performed using mainframe computers or, for some conditions, personal computers or programmable pocket calculators.

Alternate methods of solutions (applicable under certain restrictions on the material and on the environmental conditions) are presented in Chapters 14 and 15. The user should select the calculation procedure most suitable for his or her needs. In selecting a computational procedure, its range of applicability, accuracy, and ease of use should be taken into account. In the analyses of moisture distribution, the fiber is generally assumed to be impermeable to moisture and viscoelastic effects are neglected. Permeabilities of the fiber and viscoelastic effects during moisture absorption are considered in Chapters 16 and 17, respectively.

It is noted that most of the analyses currently in use are based on the assumption that the moisture absorption can be described by Fick's law. The validity of this assumption for glass epoxy composites is examined in Chapter 18.

The analytical procedures referred to above can be used to calculate moisture and temperature distributions under different environmental conditions. Examples of moisture gradients in graphite epoxy composites during natural weathering are shown in Chapter 19. The transient thermal response of fiber reinforced organic matrix composites subjected to intense surface heating is discussed in Chapter 20. The results in this chapter shed light on the thermal processes occurring during laser irradiation.

ANALYSES OF THERMAL AND TRANSPORT COEFFICIENTS

In order to calculate the moisture and temperature distributions inside the material and the dimensional changes of the material, the thermal and transport coefficients must be known. These parameters can be determined experimentally. There are also analytical methods available for estimating the overall or effective thermal conductivity, diffusivity, and thermal expansion coefficient from the known properties of the resin and the fiber. Approximate expressions for thermal conductivities and diffusivities of continuous fiber composites were given in the first volume of this book. Further approximations for the effective thermal conductivities and diffusivities of unidirectional composites are presented in Chapters 21 and 24, respectively. Information regarding the effective thermal conductivities of chopped fiber composites and of solids containing cracks are given in Chapters 22 and 23.

Procedures for estimating moisture and thermal expansion coefficients are developed in Chapters 25–28. In Chapter 25 a finite element, numerical procedure is discussed for calculating moisture and thermal expansion coefficients of unidirectional, continuous fiber reinforced composites. Models for the thermal expansion coefficients of fabrics, of composites with randomly oriented fibers, and of particle filled polymers are presented in Chapters 26–28.

VISCOELASTIC ANALYSES OF LAMINATES AND ADHESIVE JOINTS

The exposure of organic matrix composites and adhesives to hygrothermal exposure alters the mechanical behavior of the material. The linearly viscoelastic response of polymeric matrix, laminated composites is analyzed in Chapter 29. The results of this analysis are applicable to problems involving a) dimensional stability and alteration of residual stresses, b) strain rate dependence of tensile response, and c) thermal stresses and creep.

The responses of adhesive joints to temperature and moisture exposure are discussed in Chapters 30 and 31. In Chapter 31 the viscoelastic behavior of bonded lap joints at elevated temperatures is considered. In Chapter 31 experimental results are presented showing moisture induced changes in the geometries of adhesive joints involving composites.

CONCLUDING REMARKS

Volumes 1 and 2 of this book provide a comprehensive summary of environmental effects on composite materials. The contents of these books show that our knowledge of the influence of moisture and temperature on the properties of organic matrix composites has improved greatly during the past decade. In particular, considerable progress has been made towards understanding the behavior of those types of graphite fiber reinforced epoxy and glass fiber reinforced polyester matrix composites which are already being used by the aerospace and automotive industries. Data are scarce, at least in the open literature, on other types of composites. Data are badly needed for the newly developed thermosetting, thermoplastic, and metal matrix composites. Additional data which might extend the applicabilities of graphite epoxy and glass polyester composites would also be useful and would be welcome. Hopefully, data will be forthcoming which will facilitate the use of new materials, and extend the range of application of existing, commercially available material systems.

2

Moisture and Temperature Induced Degradation of Graphite Epoxy Composites

G. S. SPRINGER

INTRODUCTION

LARGE NUMBERS OF TESTS HAVE BEEN PERFORMED IN THE PAST evaluating the effects of the environment on organic matrix composites. The results of many of these tests were summarized in reference 1 and in the subsequent chapters of this volume. Recently, extensive tests have been performed by Laurities and Sandorff at Lockheed California Company [2], by Kibler at General Dynamics Corporation [3], and by Gibbins and Hoffman at Boeing Commercial Aircraft Company [4]. The important data generated by these investigators are summarized in this chapter.

The data presented in the following provide information on the response of graphite epoxy composites to moisture and elevated temperature. These data should be useful in the design of composite elements and structures. However, the data must be used with caution. Some important considerations, which must be borne in mind in using the data, are discussed below.

Most of the data is represented only by a single value. In reality, there is always a considerable spread in the data. Twenty to thirty percent variations in the data are quite common. This is illustrated in Figures 1 and 2. A comparison of the data in these figures show that, for T300/5208 laminates, the maximum (equilibrium) moisture contents measured by Kibler [3] and by Gibbins and Hoffman [4] differ by 30 to 40 percent. These differences may have been caused by a) variations in the material (prepreg) properties, b) differences in the manufacturing and cure processes, or c) differences in the test procedures. Even with stringent controls, variations in prepreg properties and in processing variables seem to be unavoidable, often giving rise to undesirable variations in the material properties.

Test conditions (e.g. specimen geometry, fixture geometry) may also affect the experimental results. Data obtained under one set of conditions may be

6

invalid or in error when applied under a different set of conditions. The applicability of the data in a given problem should be carefully evaluated.

DATA

Data are presented for five different types of graphite-epoxy composites commonly used in the aerospace industry: T300/5208, T300/5209, T300/934, AS/3501-A, and AS/3502. The data given for each type of material are summarized in Table 1. In this table the figures and tables containing the relevant data are also identified.

No attempt was made to draw general conclusions regarding the data. In those cases where the data are sufficiently detailed the trend is generally obvious and needs no further elaboration. Generalizations and extrapolations cannot and must not be made in the case of limited data.

Table 1. Summary of data presented in this chapter. The letters F and T refer to the relevant figure and table numbers.

	T300/5208	T300/5209	T300/934	AS/3501-5A	AS/3502
Max moisture content	F1	F2	F2	F2	F1
Diffusivity	T3				T3
Moisture expansion coeff.	T3				T3
Thermal expansion coeff.	F3-5, T3			F6-8	T3
Glass transition temp.	F9, T4	T4	T4		F9
Tensile properties	T4-8	T4	T4	T9	T5
Shear properties	T4, T10	T4, T10	T4, T10	T11	
Flexure properties	T10	T10	T10		
Compressive properties	T4, T12-14	T4	T4	T15	
Buckling properties	T16			T16	

Table 2. Laminate configurations and conditions used in specimen preparations for the results given in ref. 2 (Figures 3-8, Tables 8-9 and 11-16).

Test Laminates

Designation	Type	No. of Plies	Layup
U1L	100% – 0° unidirectional	16	$(0)_{16}$
U1T	100% – 90° unidirectional	16	$(90)_{16}$
U2	100% – ±45°	16	$(\pm 45)_{4s}$
L1L	25% – 0°, Quasi-isotropic	16	$(0/45/90/-45_2/90/45/0)_s$
L1T	25% – 0°, Quasi-isotropic	16	$(90/-45/0/45_2/0/-45/90)_s$
L2L	67% – 0°,/33% – ±45°	24	$(0/45/0_2/-45/0_2/45/0_2/-45/0)_s$
L2T	67 – 90/33% – ±45°	24	$(90/-45/90_2/45/90_2/-45/90_2/45/90)_s$

Moisture Content

"Dry" Laminate: Moisturized at 72°F-40% rh until stable moisture content (of about 0.3%) was reached.

"Wet" Laminate: Moisturized at 180°F-90% rh until stable moisture content was reached.

Table 3. *The moisture induced swelling coefficients (β), the thermal expansion coefficients (α), and the diffusivity parallel (11 directions) and normal (22 directions) to the fibers of T300/5208 and AS/3502 laminates (ref. 3).*

Coefficient	Environment	T300/5208	AS/3502
β_{11} ($\mu\varepsilon/°K$)		0	0
β_{22} ($\mu\varepsilon/°K$)	M(%) = 0.5	1700	
	= 1.0	2440	
	= 1.4	2800	
α_{11} ($\mu\varepsilon/°K$)	Dry, 300°K	−0.56	−0.27
	400°K	−0.61	−0.20
α_{11} ($\mu\varepsilon/°K$)	95% RH	−1.0	
α_{22} ($\mu\varepsilon/°K$)	Dry, 300°K	23.0	21.8
	400°K	26.7	25.9
	450°K	28.6	28.0
α_{22} ($\mu\varepsilon/°K$)	95% RH, 300°K	28.2	26.4
	370°K	34.0	30.8
D (cm²/sec.)	RT	17.0	12.0

Table 4. *The strengths and glass transition temperatures of T300/934, T300/5208 and T300/5209 laminates (ref. 4).*

T300/934	Strength, ksi		
	Room Temp	120°F	180°F
0-deg short beam shear	15.39	14.38	12.51
Flexure	256.78	250.94	235.85
±45-deg tension	23.23	22.09	23.06
0-deg compression	252.08	235.60	225.42
Quasi-isotropic tension	56.11	53.86	47.13
90-deg compression	27.68	28.01	25.17
Quasi-isotropic compression	130.56	124.22	118.41

Tg, 401°F

T300/5208	Strength, ksi		
	Room Temp	120°F	180°F
0-deg short beam shear	15.70	14.44	12.33
Flexure	243.63	239.17	226.16
±45-deg tension	22.98	21.43	19.46
0-deg compression	247.44	226.49	174.01
0-deg tension	210.02	—	223.91
Quasi-isotropic tension	48.68	47.09	49.39
90-deg compression	28.63	29.73	27.04
Quasi-isotropic compression	152.14	133.37	125.84

Tg, 417°F

(continued)

Table 4. (continued).

T300/5209	Strength, ksi		
	Room Temp	120°F	180°F
0-deg short beam shear	13.22	11.74	9.22
Flexure	246.48	232.97	209.30
±45-deg tension	25.10	26.21	25.83
0-deg compression	240.35	225.07	174.94
0-deg tension	249.94	—	223.91
Quasi-isotropic tension	51.45	47.91	49.93
90-deg compression	30.40	26.05	23.00
Quasi-isotropic compression	83.19	78.16	68.97
Tg. 262°F			

Table 5. Retention of tensile strength of T300/5208 and AS/3502 laminates (relative to room temperature dry) (ref. 3).

		Room Temp		325°F	212°F	212°F
		75%RH	95%	Dry	75%RH	95%RH
T300/5208	$(0)_6$	0.90	0.88	0.78	0.87	1.05
	$(90)_{15}$	0.56	0.58	0.71	0.42	0.33
	$(\pm45)_{2S}$	0.99	0.99	0.70	0.73	0.74
	$(0/\pm45)_S$	0.81	0.78	1.01	0.76	0.88
	$(90/\pm45)_S$	1.57	1.20	0.98	0.96	0.89
	$(0/90/\pm45)_S$	0.94	0.90	0.93	0.88	1.02
AS/3502	$(0)_6$	1.06	1.14	1.02	1.04	1.03
	$(90)_{15}$	0.61	0.64	0.82	0.45	0.36
	$(\pm45)_{2S}$	0.93	0.95	0.77	0.79	0.87
	$(0/\pm45)_S$	1.00	0.87	0.94	1.01	1.03
	$(90/\pm45)_S$	1.13	1.13	0.91	0.85	0.83
	$(0/90/\pm45)_S$	0.91	0.90	0.95	0.92	0.92

Table 6. Tensile properties of dry and wet T300/5208 laminates. Wet laminates moisturized in humid air at 75 or 95 percent relative humidity (ref. 3).

Lay up	Test Temp °F	Ultimate stress, ksi			Ultimate strain, %			Initial modulus, psi × 10⁻⁶		
		Dry	75%	95%	Dry	75%	95%	Dry	75%	95%
$(0)_6$	RT	219	231	248	1.09	1.23	1.24	18.1	17.1	18.5
	210[a]	223	227	225	1.24	1.19	1.19	18.0	18.2	18.0
$(90)_{15}$	RT	6	4	4	0.41	0.24	0.26	1.6	1.6	1.6
	210[a]	5	3	2	0.39	0.23	0.17	1.4	1.5	2.2
$(\pm45)_{2S}$	RT	24	22	23	1.33	1.96	3.08	2.9	2.6	2.5
	210[a]	18	19	21	2.65	3.75	4.78	2.3	1.5	1.5
$(\pm45)_S$	RT	85	85	73	1.04	1.10	0.96	8.1	7.9	8.0
	210[a]	80	86	87	1.15	1.28	1.27	7.6	7.1	7.2
$(90/\pm45)_S$	RT	27	31	31	0.82	0.99	0.99	3.7	3.8	3.8
	210[a]	25	23	23	0.82	0.79	0.82	3.4	3.4	3.4
$(0/90/\pm45)_S$	RT	78	71	70	1.10	0.96	0.94	7.1	7.5	7.5
	210[a]	74	71	72	1.04	1.14	1.05	7.1	6.7	6.9

[a]For dry specimens the test temperature was 325°F.

9

Table 7. Tensile properties of dry and wet T300/5208 laminates. Wet laminates moisturized in humid air at either 75 or 95 percent relative humidity (ref. 3).

Lay up	Test Temp °F	Ultimate stress, ksi Dry	75%	95%	Ultimate strain, % Dry	75%	95%	Initial modulus, psi × 10⁻⁶ Dry	75%	95%
$(0)_6$	RT	217	195	191	1.03	0.94	0.92	19.2	20.0	19.3
	210[a]	168	189	228	0.92	0.93	1.19	20.2	18.7	18.2
$(90)_{15}$	RT	7	4	4	0.50	0.24	0.26	1.4	1.6	1.6
	210[a]	5	3	2	0.39	0.23	0.17	1.3	1.5	2.2
$(\pm45)_{2S}$	RT	23	23	23	1.17	1.69	1.92	3.0	2.5	2.5
	210[a]	16	17	17	2.30	2.85	3.12	2.4	1.4	1.3
$(\pm45)_S$	RT	86	70	67	1.03	0.92	0.85	8.3	7.6	7.9
	210[a]	87	66	76	1.17	0.93	1.07	7.9	7.1	7.2
$(90/\pm45)_S$	RT	21	34	26	0.63	1.07	0.80	3.7	3.7	3.7
	210[a]	21	21	19	0.70	0.72	0.63	3.3	3.2	3.3
$(0/90/\pm45)_S$	RT	75	70	67	1.02	0.92	0.85	7.4	7.6	7.9
	210[a]	69	66	76	0.94	0.93	1.07	7.4	7.1	7.2

[a]For dry specimens the test temperature was 325°F.

Table 8. Tensile strengths and moduli of dry and wet T300/5208 laminates (see Table 2) (ref. 2).

Laminate Type	Test Temp °F	Average Ultimate Stress, ksi Dry	Wet	Average Ultimate Strain in/in Dry	Wet	Average Initial Apparent Modulus psi × 10⁻⁶ Dry	Wet	Average Final Apparent Modulus psi × 10⁻⁶ Dry	Wet	Poisson's Ratio Dry	Wet
L1	−65	67.8	74.4	0.0089	0.0099	7.69	7.78	6.74	5.97	—	—
Longitudinal	72	76.4	79.7	0.0105	0.0100	7.59	7.84	6.47	—	0.273	0.220
	200	72.5	78.3	0.0099	0.0102	7.51	7.85	—	—	—	—
	275	76.3	74.3	0.0102	0.0114	7.48	6.83	—	5.17	—	—
L1	−65	68.7	74.6	0.0087	0.0101	7.86	7.45	—	—	—	—
Transverse	72	77.8	79.7	0.0106	0.0105	7.54	7.62	7.06	—	0.273	0.300
	200	73.9	74.3	0.0108	0.0102	7.26	7.57	—	—	—	—
	275	69.1	71.9	0.0108	0.0102	7.47	7.23	—	—	—	—
L2	−65	152.8	149.1	0.0097	0.0371	15.10	14.64	—	—	—	—
Longitudinal	72	149.4	163.0	0.0100	0.0100	14.75	15.75	—	—	0.569	—
	200	162.5	159.7	0.0110	0.0101	14.69	15.36	—	—	—	—
	275	158.1	158.3	0.0105	0.0111	13.83	14.92	—	—	—	—
L2	−65	21.0	23.0	0.0123	0.0110	2.66	2.73	0.63	1.24	—	—
Transverse	72	20.3	21.8	0.0102	0.0108	2.47	2.79	0.91	1.21	0.035	0.158
	200	21.0	22.5	0.0117	0.0105	2.51	2.70	0.99	0.91	—	—
	275	21.4	20.1	0.0104	0.0102	2.58	2.54	1.23	1.31	—	—
U1	−65	173.8	219.9	0.0081	0.0090	19.5	21.7	20.6	—	—	—
Longitudinal	72	196.7	238.0	0.0091	0.0107	19.7	21.0	21.7	—	0.181	0.118
	200	235.5	230.1	0.0112	0.0098	19.8	21.7	20.8	—	—	—
	275	229.7	242.2	0.0122	0.0100	20.0	23.7	19.4	—	—	—

(continued)

Table 8. (continued).

Laminate Type	Test Temp. °F	Average Ultimate Stress, ksi		Average Ultimate Strain in/in		Average Initial Apparent Modulus psi × 10⁻⁶		Average Final Apparent Modulus psi × 10⁻⁶		Poisson's Ratio	
		Dry	Wet	Dry	Wet	Dry	Wet	Dry	Wet	Dry	Wet
U1	−65	4.3	2.1	0.0026	0.0012	1.57	1.6	—	—	—	—
Transverse	72	4.0	2.8	0.0030	0.0021	1.35	1.3	—	—	—	—
	200	4.1	1.8	0.0035	0.0022	1.21	1.1	—	—	—	—
	275	4.2	1.5	0.0037	0.0017	1.11	0.8	—	—	—	—

Table 9. Tensile strengths and moduli of dry and wet AS/3501-5A laminates (see Table 2) (ref. 2).

Laminate Type	Test Temp. °F	Moisture Level	Average Ultimate Stress, ksi	Average Ultimate Strain in/in	Average Initial Apparent Modulus psi × 10⁻⁶	Average Final Apparent Modulus psi × 10⁻⁶	Poisson's Ratio
L1 Longitudinal	72	Dry	83.0	0.0120	7.33	6.51	0.153
	275	Wet	76.9	0.0121	6.11	—	—
L1 Transverse	72	Dry	84.9	0.0121	7.42	6.95	0.153
	275	Wet	68.3	0.0145	5.75	5.61	—
L2 Longitudinal	72	Dry	142.5	0.0103	13.5	—	—
	275	Wet	152.0	0.0173	13.1	—	—
L2 Transverse	72	Dry	23.3	0.0146	2.6	0.79	—
	275	Wet	16.2	0.0141	1.4	—	—
U1 Longitudinal	72	Dry	214.2	0.0101	21.1	—	0.109
	275	Wet	228.2	0.0111	22.2	—	—
U1 Transverse	72	Dry	3.7	0.0026	1.5	—	—
	275	Wet	0.4	0.0026	0.5	—	—

Table 10. Short beam shear and flexure strengths of T300/934, T300/5208, and T300/5209 laminates. 180°F values ratioed against 180°F dry baseline (ref. 4).

Property	Exposure Humidity, %	Percentage of Baseline Strengths					
		5208		5209		934	
		Room Temperature	180°F	Room Temperature	180°F	Room Temperature	180°F
Short beam shear strength	40	94	95	93	86	96	90
	60	88	85	83	72	91	80
	75	88	81	84	68	89	75
	95	80	66	—	30	80	57

(continued)

11

Table 10. (continued).

		Percentage of Baseline Strengths					
		5208		5209		934	
Property	Exposure Humidity, %	Room Temperature	180°F	Room Temperature	180°F	Room Temperature	180°F
Flexure strength	40	98	96	102	96	101	97
	60	99	92	105	88	100	91
	75	97	91	101	77	101	86
	95	94	84	86	41	94	76

Table 11. In plane shear (±45° tension) test results for dry and wet AS/3501-5A laminates (see Table 2) (ref. 2).

Temp °F	Max shear stress ksi		Initial in plane shear modulus, psi × 10⁻⁶		In plane shear modulus at 20 ksi, psi × 10⁻⁶	
	Dry	Wet	Dry	Wet	Dry	Wet
−65	14.3	14.3	0.80	0.89	0.33	0.35
72	12.5	11.7	0.77	0.73	0.23	0.20
200	10.2	10.1	0.70	0.62	0.12	0.70
275	9.2	9.0	0.73	0.64	0.07	0.60

Table 12. Fully supported compression test results for dry and wet T300/5208 laminates. Gauge length: 6 in, specimen width: 1 in (see Table 2) (ref. 2).

	Temp °F	Ultimate Stress ksi		Ultimate Strain in/in		Secant modulus at failure, psi × 10⁻⁶		Secant modulus at 35 ksi, psi × 10⁻⁶	
		Dry	Wet	Dry	Wet	Dry	Wet	Dry	Wet
L1 Longitudinal	−65	84.0	81.7	0.0136	0.0136	6.2	6.0	7.0	7.0
	72	70.6	70.7	0.0112	0.0108	6.3	6.6	7.0	7.3
	200	60.7	57.5	0.0094	0.0091	6.5	6.2	7.0	6.9
	275	63.5	60.0	0.0100	0.0092	6.2	6.5	6.8	7.2
L1 Transverse	−65	81.1	70.6	0.0125	0.0107	6.4	6.8	7.1	7.2
	72	58.2	64.1	0.0087	0.0098	6.7	6.5	7.2	7.3
	200	61.0	51.6	0.0095	0.0082	6.4	6.2	6.7	6.8
	275	56.3	51.8	0.0088	0.0087	6.4	6.2	7.0	6.9

12

Table 13. Fully supported compression test results for dry and wet T300/5208 laminates. Gauge length: 6 in, specimen width: 1 in (see Table 2) (ref. 2).

	Temp °F	Ultimate Stress ksi		Ultimate Strain in/in		Secant modulus at failure, psi × 10⁻⁶		Secant modulus at 70 ksi, psi × 10⁻⁶	
		Dry	Wet	Dry	Wet	Dry	Wet	Dry	Wet
L2 Longitudinal	−65	144.3	135.6	0.0133	0.0120	11.5	11.4	13.5	13.4
	72	130.6	140.0	0.0118	0.0124	11.8	11.3	13.0	12.8
	200	118.6	114.7	0.0103	0.0134	11.8	11.1	13.1	12.6
	275	112.9	109.2	0.0095	0.0090	11.9	12.2	12.9	13.0
								(at 17 ksi)	
L2 Transverse	−65	39.5	38.5	0.0147	0.0147	2.6	2.6	2.8	2.8
	72	33.8	35.9	0.0136	0.0152	2.5	2.4	2.6	2.7
	200	32.9	31.0	0.0161	0.0140	2.1	2.2	2.6	2.5
	275	30.1	25.6	0.0137	0.0137	2.2	1.9	2.5	2.2

Table 14. Fully supported compression test results for dry and wet T300/5208 laminates. Gauge length: 6 in, specimen width: 1 in. (see Table 2) (ref. 2).

	Temp °F	Ultimate Stress ksi		Ultimate Strain in/in		Secant modulus at failure, psi × 10⁻⁶		Secant modulus at 70 ksi, psi × 10⁻⁶	
		Dry	Wet	Dry	Wet	Dry	Wet	Dry	Wet
U1 Longitudinal	−65	173.6	—	0.0095	—	17.9	—	19.9	—
	72	143.4	152.8	0.0077	—	17.8	—	19.0	—
	200	136.4	—	0.0075	—	17.6	—	18.9	—
	275	111.0	82.3	0.0063	0.0055	17.9	17.2	18.9	18.6
								(at 17 ksi)	
U1 Transverse	−65	33.0	24.6	0.0239	0.0173	1.4	1.4	1.5	1.5
	72	30.0	29.8	0.0238	0.0273	1.2	1.1	1.4	1.3
	200	24.7	21.1	0.0234	0.0276	1.1	0.7	1.3	0.8
	275	22.3	18.1	0.0248	0.0331	0.9	0.6	1.2	0.6

Table 15. Fully supported compression test results for dry and wet AS/3501-5A laminates. Gauge length: 6 in, specimen width: 1 in (see Table 2) (ref. 2).

	Temp °F	Ultimate Stress ksi		Ultimate Strain in/in		Secant modulus at failure, psi × 10⁻⁶		Secant modulus at 35 ksi, psi × 10⁻⁶	
		Dry	Wet	Dry	Wet	Dry	Wet	Dry	Wet
L1 Longitudinal	72	83.0	—	0.0148	—	5.8	—	6.9	—
	275	—	60.5	—	0.0103	—	5.9	—	6.1
L1 Transverse	72	68.7	—	0.0114	—	6.1	—	6.8	—
	275	—	51.1	—	0.0093	—	5.5	—	6.0

(continued)

13

Table 15. (continued).

	Temp °F	Ultimate Stress ksi		Ultimate Strain in/in		Secant modulus at failure, psi $\times$ 10^{-6}		Secant modulus at 35 ksi, psi $\times$ 10^{-6}	
		Dry	Wet	Dry	Wet	Dry	Wet	Dry	Wet
								(at 70 ksi)	
L2	72	138.2	—	0.0134	—	10.9	—	12.7	—
Longitudinal	272	—	72.5	—	0.0066	—	11.1	—	12.3
								(at 17 ksi)	
L2	72	43.8	—	0.0182	—	2.4	—	2.7	—
Transverse	275	—	26.2	—	0.0176	—	1.5	—	1.8
								(at 70 ksi)	
U1-L 0°	72	155.6	96.6	0.0095	0.0052	16.5	19.1	18.1	18.6
Unidirectional	275	—	67.6	—	00.40	—	17.0	—	16.6
								(at 17 ksi)	
U1-T 90°	72	34.8	—	0.0969	—	—	—	—	—
Unidirectional	275	—	15.6	—	0.0308	—	0.5	—	0.5

Table 16. Average buckling strength (ksi) of dry and wet T300/5208 and AS/3501-5A laminates (see Table 2) (ref. 2).

Laminate Type	Column Length in.	Longitudinal				Transverse			
		72°F Dry		275°F Wet		72°F Dry		275°F Wet	
		T300/ 5208	A-S/ 3501-5A	T300/ 5208	A-S/ 3501-5A	T300/ 5208	A-S/ 3501-5A	T300/ 5208	A-S/ 3501-5A
	0[a]	70.7	83.0	63.3	60.5	58.2	68.7	53.4	51.1
	0.347	58.4	70.4	52.3	53.9	56.1	64.3	49.0	44.4
Laminate	0.571	54.8	68.8	44.1	49.7	54.0	58.6	42.6	39.6
L1	0.789	50.1	51.1	44.8	43.1	41.3	42.1	39.7	35.9
Quasi-Isotropic	1.058	32.4	33.4	34.0	28.8	26.2	26.2	26.7	22.2
	1.276	24.0	22.7	23.0	21.6	18.7	17.5	18.9	17.0
	1.606	15.8	15.7	15.6	15.4	12.5	11.8	12.5	11.7
	0[a]	130.6	138.2	119.0	72.5	33.8	43.8	27.6	26.2
Laminate	0.347	144.8	125.6	96.8	81.5	31.6	35.8	27.7	24.2
L2	0.571	121.9	117.3	93.9	74.5	29.7	36.2	24.7	23.6
67% − 0°	0.789	106.7	121.8	103.8	80.1	27.6	34.3	24.1	21.8
33% − ±45°	1.058	93.9	85.8	80.2	66.7	23.4	24.1	20.7	17.4
	1.276	73.1	68.2	67.8	61.6	17.2	17.3	15.2	13.3
	1.606	47.6	47.5	48.6	43.7	11.1	11.3	10.7	9.3
	0	143.3	155.6	72.4	67.6	30.0	34.8	19.1	15.6
	0.347	140.0	153.0	102.5	—	29.4	29.1	16.2	12.6
U1	0.571	134.3	—	81.2	68.0	19.0	18.9	10.3	7.1
0° −	0.789	105.1	—	78.1	64.0	11.2	11.3	7.5	6.0
Unidirectional	1.058	67.9	47.3	60.9	—	6.8	6.8	—	4.8
	1.276	48.1	—	45.0	34.8	4.6	4.7	3.4	3.4
	1.606	32.0	—	31.4	26.6	3.0	3.1	2.1	1.8

[a]Fully supported.

14

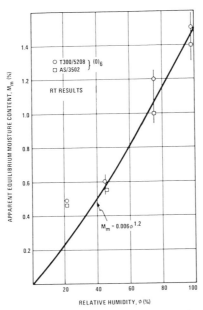

Figure 1. *Maximum moisture content versus relative humidity for T300/5208 and AS/3502 laminates (ref. 3).*

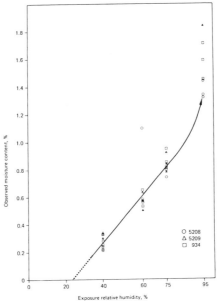

Figure 2. *Maximum moisture content versus relative humidity for T300/934, T300/5208, and T300/5209 laminates (ref. 4).*

15

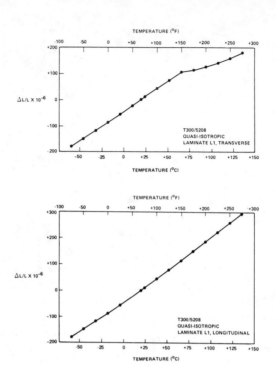

Figure 3. Thermal expansion coefficients of T300/5208 laminates (see Table 2) (ref. 2).

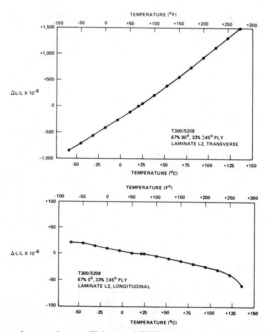

Figure 4. Thermal expansion coefficients of T300/5208 laminates (see Table 2) (ref. 2).

16

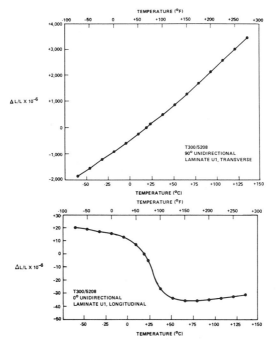

Figure 5. Thermal expansion coefficients of T300/5208 laminates (see Table 2) (ref. 2).

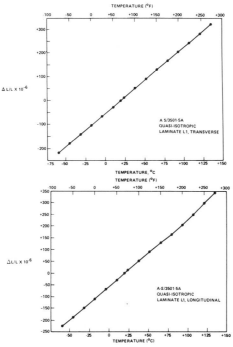

Figure 6. Thermal expansion coefficients of AS/3501-5A laminates (see Table 2) (ref. 2).

17

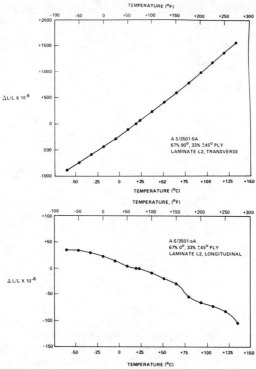

Figure 7. Thermal expansion coefficients of AS/3501-5A laminates (see Table 2) (ref. 2).

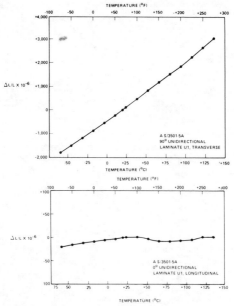

Figure 8. Thermal expansion coefficients of AS/3501-5A laminates (see Table 2) (ref. 2).

18

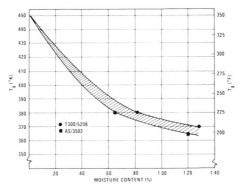

Figure 9. *Glass transition temperatures as functions of moisture content of T300/5208 and AS/3502 laminates (ref. 3).*

REFERENCES

1. Springer, G. S. *Environmental Effects on Composite Materials.* Lancaster, PA: Technomic Publishing Company (1980).
2. Laurities, K. N. and Sandorff, P. E., "The Effect of Environment on the Compressive Strengths of Laminated Epoxy Matrix Composites," Technical Report AFML-TR-79-4179, Materials Laboratory, Wright Patterson Air Force Base, Dayton, Ohio (December 1979).
3. Kibler, K. G., "The Time Dependent Environmental Behavior of Graphite/Epoxy Laminates," Technical Report AFWAL-TR-80-4082, Materials Laboratory, Wright Patterson Air Force Base, Dayton, Ohio (May 1980).
4. Gibbins, M. N. and Hoffman, D. J., "Environmental Exposure Effects on Composite Materials for Commercial Aircraft," NASA Contractor Report 3502 (January 1982).

3

Property Changes of a Graphite/Epoxy Composite Exposed to Nonionizing Space Parameters

H. R. PHELPS AND E. R. LONG, JR.

LARGE ANTENNAS FOR COMMUNICATION SATELLITES, STRUCTURAL frameworks for earth monitoring and solar energy conversion systems, and platforms for space observation telescopes are examples of large space structure concepts that have been presented in aerospace publications [1]. Each of these structures will require materials that have high stiffness, small coefficients of thermal expansion, and light weight, all of which must be retained in a space environment for mission durations of 30 or more years. Thus, the selection of the structural materials will be based upon the effects that each orbital environment will have on the properties of the materials and upon the structural requirements in each of these environments. Graphite/epoxy composites are prime candidates for structural materials. However, their ability to maintain their properties for long times in a space environment such as low earth orbit (LEO) remain to be proved.

The LEO environmental parameters consist of ultraviolet (UV) radiation, vacuum, and thermal conditions. The latter two parameters have often been used to screen out materials that are definitely not usable in a space environment. Epoxy materials generally pass these screening tests, although some researchers have reported significant weight losses in vacuum at temperatures in excess of 310 K [2-4]. UV radiation may cause additional effects. A considerable amount of study has been done of UV effects [2,3,5,6,7]; however, the effects of UV in combination with some of the other environmental parameters has not been reported.

This paper reports a study of the effects of individual and combined nonionizing space parameter exposures on the mass, thickness, and flexural properties of a graphite/epoxy composite material. The exposures were conducted to determine if combined space environmental parameters have synergistic effects. Both wet and dry specimens were exposed to determine if moisture content altered the effects of the LEO type environment.

EXPERIMENTAL PROCEDURE

The test specimens were four-ply, unidirectional composites with nominal dimensions of 25.4 mm by 12.7 mm by 0.5 mm. The fibers were in the direction of the specimens' lengths. The specimens were machined from 330 mm by 76 mm autoclaved panels from which a 6 mm border had been removed. The panels were fabricated from T300/5208 prepreg, using procedures recommended by the prepreg supplier. The specimens were preconditioned for two weeks at 344 K in a dessicated nitrogen environment. At the end of the drying period, each specimen's mass, width, and thickness were recorded. The specimens to be exposed while wet were then placed in distilled water at 355 K for two weeks. The gain in mass due to moisture absorption was approximately 1.8 percent with respect to the original mass. Immediately after preconditioning (which included moisture absorption if specimens were to be exposed while wet), each group of five specimens was placed in an environmental chamber for exposure to nonionizing space parameters. The exposure facility consisted of a solar simulator, the environmental chambers containing specimens, a thermal bath for controlling specimen temperature, and support equipment to monitor UV intensity and temperature at the specimens' front surfaces. The temperature was monitored by three thermocouples attached to the front surface of a sixth speciment in each chamber. Each thermocouple was attached with a mixture of graphite powder and epoxy and a thin coat of this mixture covered the thermocouple. The solar simulator used xenon lamps to provide two equivalent suns in the near and middle spectral regions, 200–400 nm, of UV. The beam diameter from the simulator's collimating lens was 254 nm. As many as four environmental chambers could be placed in the beam at any one time. Each chamber had either a quartz window if UV irradiance was used or a pyrex window if no UV irradiance was used. The sample mounting block in each chamber contained passages for fluid from the thermal bath, which controlled the specimens' temperatures. The specimens' atmosphere consisted of either a static gas (air or argon) or a vacuum at 7×10^{-3} Pa. Each chamber could be removed from the exposure location without interrupting the exposure conditions for the other chambers.

After completion of an exposure for the specimens in a chamber, the chamber containing the specimens was placed in a glove box which contained a microbalance and bench top micrometer in dry argon gas. The specimens were removed from the chamber and their masses and thicknesses were recorded. The specimens were then transferred in dry argon to a second glove box which contained a table top universal tester for flexural property measurements. Flexural properties were also measured in a dry argon environment. The dry argon environment was used to minimize any post exposure effects of oxygen. The exposure conditions for each group of specimens are given in Table 1.

Table 1. *Summary of four week exposure environments for study of the effects of nonionizing space parameters.*

Group Number	UV[1] Intensity	Type of Atmosphere[2] Temperature[3]	Sample Moisture Percent Content[4]
1	None	Air/450	0.0
2	None	Air/450	1.8
3	None	Argon/450	0.0
4	None	Argon/450	1.8
5	None	Vacuum/450	0.0
6	None	Vacuum/450	1.8
7	2 Suns	Vacuum/400	0.0
8	2 Suns	Vacuum/420	0.0
9	2 Suns	Vacuum/420	1.8

[1]The 2 suns intensity refers to the near and middle ultraviolet spectra.
[2]The argon was 99.998 percent pure. Some undetermined amount of oxygen was present.
[3]Temperature is in Kelvin.
[4]The moisture content is given in percent with respect to the specimen dry weight.

RESULTS AND DISCUSSION

All of the exposures reported in this paper were for four weeks. The data for all the exposure durations, except two sun exposures of dry specimens at 400 K for four weeks, may be found in [8]. The additional two sun exposures at 400 K was done after the completion of [8].

Ultraviolet Exposures

Mass Loss

The percent changes in mass for the specimens not exposed to UV are shown in Figure 1a as the average and range of five specimens per exposure. Each specimen's change is the difference in predried and post-exposure mass divided by the predried mass and multiplied by 100, regardless of the specimen's moisture content at the beginning of exposure.

The effects of initial moisture content can be observed by comparing initially dry (D) and wet (W) specimen data for air, argon, and vacuum atmospheres. The moisture content at the beginning of exposure had little or no effect on mass loss at elevated temperatures.

The effects of atmosphere on mass loss at elevated temperatures may also be seen in Figure 1a. The greatest mass losses occurred in air. Mass loss in vacuum was approximately half of that which occurred in air. Oxidation may explain the largest mass loss for the air environment and intermediate loss for the argon environment. (The argon also contained some oxygen, although some orders of magnitude less than for air.) At 450 K the thermal energy was sufficient to break bonds in the cured epoxy [9]. The breakage of bonds may have created small molecules which were capped by the oxygen from the air.

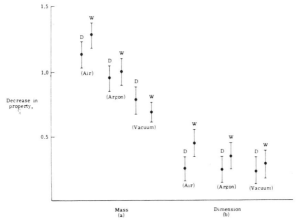

Figure 1. *Decreases in mass and dimensions of composite specimens at 450 K and no ultraviolet after four weeks for several gaseous environments.*

These capped groups might then be free to migrate out, causing mass loss. For a vacuum, the broken bonds would not have been capped and therefore were able to rebond.

Changes in Thickness

The percent changes in thickness of the graphite/epoxy composites are shown in Figure 1b. As in the case of mass loss, the percent changes after exposure were with respect to preconditioned dry values. In general, the changes in dimension correlated with loss of mass.

Ultraviolet Exposures

Mass Loss

For exposures at temperature of 400 K the UV did not cause additional loss of mass. Above 400 K mass loss was partly due to UV. In fact, the losses were in excess of those for still higher temperature exposures without UV. This may be seen in Figure 2a. The loss of mass for UV at 420 K, for both wet and dry specimens, was larger than the loss of mass at 450 K without UV. Indeed, the comparison shows that UV affected the loss of mass at temperatures above 400 K and that the effects were in excess of those due to thermal conditions alone. This suggests a synergism between UV and temperature.

Change in Thickness

Figure 2b shows the data for thickness changes for the UV exposures at 400 and 420 K and no UV exposures at 450 K. The relative changes are similar to

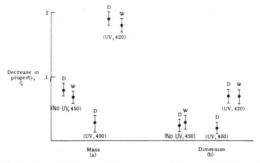

Figure 2. The effects of ultraviolet on the decrease of mass and dimensions of composite specimens at elevated temperatures and in a vacuum after four weeks.

the mass loss data, and also suggest that a synergism exists for combined UV and temperature exposures above 420 K.

Flexural Properties and Surface Morphology

The flexural properties of composite specimens tested with their UV-exposed sides in tension were not altered by any of the conditions for exposure groups. The data are included in [8]. This result agrees with Mauri [5] and is in contrast with Bassewitz [6] and Hancox and Minty [7]. However, comparison to these reports are not valid because the authors did not report thermal conditions. The studies reported in [6] and [7] used highly accelerated UV exposures which may have caused thermal degradation. If thermal degradation occurred, then incorrect conclusions for effects of UV on mechanical properties could have been made.

Changes in surface morphology were observed which suggest that mechanical properties would be changed by long duration exposure. The most extensive form of damage for high temperatures and no UV occurred in air. The damage, which may be seen in Figure 3, was in the form of cracking at the fiber/resin interface. The same form of damage, but to a lesser extent, was observed for high temperature exposures in argon. The vacuum exposures at high temperatures did not show this damage. For UV exposures, a loss in surface material was observed (Figure 4). Small chunks of resin material spalled from the surface. Surface erosion would explain why UV exposures caused larger mass loss than exposures without UV. Eventually, this erosion may cause changes in mechanical properties.

SUMMARY

The following general statements can be made from studying the effects of four week exposures to nonionizing space environmental parameters on mass, thickness, flexural properties, and surface morphology of graphite/epoxy composites:

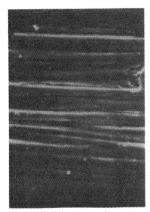

Figure 3. *Electron microscope photograph, magnification factor of 500, of a composite specimen exposed at 450 K in air without UV. Cracking occurred at fiber/resin interface.*

Figure 4. *Electron microscope photograph, magnification factor of 500, of a composite specimen exposed in vacuum to UV at 420 K. Resin spalled from surface.*

Whether wet or dry at the beginning of exposure, specimens had the same changes in physical and mechanical properties after exposure to nonionizing space parameters, relative to their initial dry properties.

A vacuum environment at high temperatures had less effect on the composite specimens' physical properties than did an oxygen environment.

For exposure of four weeks, UV in a vacuum at temperatures below 400 K had little effect on the mass and dimensions of composite specimens. Above 400 K, the loss of mass and shrinkage due to UV was greater than for just thermal exposure at still higher temperatures. Thus, UV and high temperature synergistically affected changes in physical properties.

An examination of the surface morphology of specimens after UV exposure showed that the additional mass loss was probably due to spalling and subsequent loss of surface resin.

No change of flexural properties was found for nonionizing space parameter exposures. However, the physical property and surface morphology changes that did occur suggest that years of exposure may cause significant changes in mechanical properties.

REFERENCES

1. *Astronautics and Aeronautics, 16*(10) (Oct. 1978).
2. Broadway, N. J., King, R. W., and Palinchak, S., "Space Environmental Effects on Materials and Components," *Elastomeric and Plastic Materials, 1,* Appendix A-1, Battelle Memorial Institute (April 1964).
3. Garzek, L. E. and Isenberg, L., "The Effects of Space Environment on Reinforced Plastics,"

National Symposium on the Effects of Space Environment on Materials, St. Louis, MO (May 7-9, 1962).

4. Van Der Waal, P. C. *Effects of Space Environment on Materials*. First Edition, RV-22, Fokker, Amsterdam: Royal Netherlands Aircraft Factories (1968).

5. Mauri, R. E., "Organic Materials for Structural Applications," in *Space Materials Handbook,* eds. J.B. Rittenhouse and J.B. Singletary. Third edition, SP-3051, pp. 355-381 (1969).

6. Bassewitz, H. V., "Behavior of Carbon Fiber Composites Under Simulated Space Environment," Non Semiconducting Materials, Session II, International Conference on Evaluation of the Effect of the Space Environment on Materials, Toulouse, France, 525-537 (June 17-21, 1974).

7. Hancox, N. L. and Minty, D. C., "Materials Qualification and Property Measurements of Carbon Fiber Reinforced Composites for Space Use," *British Interplanetary Society Journal, 30,* 391-399 (Oct. 1977).

8. Phelps, H. R., "Effects of Nonionizing Space Environmental Parameters on Graphite/Epoxy Composites," Masters Thesis, George Washington University (August 1979).

9. Charlesby, A. *Atomic Radiation and Polymers*. New York: Pergamon Press (1960).

4

The Effect of Temperature and Moisture Content on the Flexural Response of Kevlar/Epoxy Laminates: Part I. [0/90] Filament Orientation

R. E. ALLRED

INTRODUCTION

THE UNIQUE THERMOPHYSICAL PROPERTIES OF KEVLAR/EPOXY COM-posites make them attractive candidates for use in many applications. They are particularly suitable for load-bearing, thermal and/or electrical insulators. Compatibility with the environment is a primary consideration prior to extensive use of Kevlar/epoxy composites in fielded systems. Of particular concern is their aging behavior in moist environments. Absorbed atmospheric moisture followed by rapid heating has been shown to severely degrade the performance of other high-performance composites such as graphite and boron-reinforced epoxy [1–4]. Depression of the matrix glass transition temperature, T_g, below the service temperature is believed to be the primary degradation in those composite systems [5–7].

The influence of moisture on Kevlar/epoxy composites is potentially greater than in other composite systems because the filament also absorbs moisture [8–11]. One study has indicated that moisture alone degrades Kevlar 49 filaments at room temperature and that the degradation is further enhanced by elevated temperatures [12]. The present study was undertaken to determine the magnitude of the effects of moisture and temperature on the flexural response of Kevlar 49/epoxy laminates. A test temperature of 150°C was chosen as representing an upper bound for current applications of these materials. Such data will serve as a baseline for design and will define further work necessary to characterize Kevlar 49/epoxy composite aging behavior.

EXPERIMENT

Two commercially available 175°C cure epoxy resins reinforced with Kevlar 49 181 style fabric (50 x 50 picks/in. (2 x 2 picks/mm) 380 denier yarn) were selected for evaluation in this study. The epoxide resins chosen were 5208 (Narmco Materials, Inc.) and CE-9000 (Ferro Corp). The 5208 resin system appears to be primarily composed of the epoxides tetraglycidyl methylene dianiline (Ciba MY-720) and glycidyl ether of bisphenol A novolac (Celanese SU-8) [13]. The CE-9000 system appears to contain the epoxides glycidyl ether, digylcidyl amine of p-amino phenol (Ciba 0510) and a glycidyl ether of bisphenol A novolac (Celanese SU series) [14]. Both are cured with diamino diphenyl sulfone (Ciba Eporal). Flat plates 12 in. (305 mm) square by 0.094 in. (2.4 mm) thick were prepared from each system by vacuum bag autoclave molding 12 plies of fabric prepreg stacked with the warp direction aligned. After molding to a final temperature of 175 °C for two hours, the plates were subjected to a four hour post cure in an air circulating oven at 190 °C. The resultant laminates contained a nominal filament volume fraction of 57 percent and a void content of less than 1 percent as determined by chemical dissolution [15].

A three-point bend geometry was chosen for initial testing because it is more matrix sensitive than a tensile test, and also because in most plate applications failure due to bending appears more critical. A span-to-depth ratio of 27.5 was maintained throughout the test matrix. Specimen dimensions were 5.0 in. × 1.0 in. × 0.094 in. (127 mm × 25.4 mm × 2.4 mm). Moisture was introduced into the specimens by immersion in 90 °C distilled water until a specified weight gain occurred. The specimens were then inserted into a thermally stabilized test chamber and allowed to sit four minutes before the application of load. An additional four minutes was nominally required to load the samples to their ultimate load carrying capacity. A three dimensional diffusion analysis for the heating times used in these experiments shows that only a small quantity of the outer ply thickness is subjected to drying during the duration of the test [16]. Deflections were monitored with a linear variable differential transformer (LVDT). After flexural testing, representative specimens were sectioned and examined metallographically to determine failure modes resulting from the various test conditions.

THREE-POINT BEND TEST

Prior to discussing the results of this study, it is appropriate to discuss the limitations, significance, and unknowns of the three-point bend test for testing Kevlar/epoxy composites as a function of temperature and moisture content. The primary advantage of the three-point bend test for determining environmental effects is that it is a simple test to run and to instrument. A further possible advantage of the flexure test is that in the temperature ranges of interest, it has shown that significant failure mode changes can occur in graphite/epoxy with moisture present [17].

When Kevlar filaments are used as the reinforcing phase, flexural results become less certain. Flexure testing of Kevlar-reinforced composites has been examined extensively by Zweben [18–20]. Complexity in the analysis of flexural loading of Kevlar-reinforced composites is introduced by material nonlinearity. Non-linearity in the load-deflection traces is believed to be caused by buckling failure or kinking of the filaments on the compressive side of the specimen prior to failure of the filaments subjected to tension [20]. The mechanism of compressive failure in Kevlar 49/epoxy composites has been attributed to internal compressive failure in the filaments themselves [21], or to a combination of local failures within the filaments and at the filament-matrix interface [22]. Simple elastic beam theory cannot be used to calculate the stress distribution or maximum stress at failure for a non-linear load-deflection response [23]. Therefore, the flexural strengths reported and discussed hereafter should be considered to be apparent flexural strengths and actually correspond to critical moments [18].

The flexural stiffness data to be presented later are calculated from the slope of the linear elastic portion of the load-deflection curves after correction for shear deflection effects [17,18]. The shear deflection correction is not large for the Kevlar fabric/epoxy laminates examined in this study because the in-plane modulus is low compared to most unidirectional composites. At room temperature, the shear correction adds 2.5 percent to the calculated stiffness; when the matrix is elastomeric (above T_g), the shear correction amounts to a 6.0 percent increase. Since Kevlar fabric/epoxy laminates have equal elastic moduli in tension and compression [24], measured stiffness in flexure should be representative of the material.

The strength data have been corrected for horizontal thrust as described in ASTM D-790-71 [25]. The horizontal thrust or end thrust correction must be applied to specimens tested at large span-to-depth ratios (>16:1) that exhibit deflections greater than ten percent of the span distance [25]. The correction is necessary because, at large deflections, the reaction vector at the end supports is not parallel to the loading ram; this effectively increases the moment on the specimen. Correcting for horizontal thrust is important in flexure tests of Kevlar/epoxy because of the large deflections which occur. Many of the specimens examined in this paper deflected 15–18 percent of the span distance. In those instances, the correction amounts to an increase of 20 percent in calculated stress.

RESULTS AND DISCUSSION

Mechanical Response

Load versus deflection responses of Kevlar 49 fabric/CE-9000 epoxy laminates tested at the extremes in this study are plotted in Figure 1. It should be noted that the curves shown in Figure 1 are plotted only a little beyond the maximum load. Catastrophic failure in the form of specimen separation does not take place; rather, either the load slowly dropped off (e.g., at 150°C with moisture), or the load dropped sharply and then built up to a lower maximum followed by another drop, etc. In the dry condition, the load-deflection curves over the temperature range from –55°C to 150°C appear quite similar in shape and amount of deflection at maximum load. In addition, all the

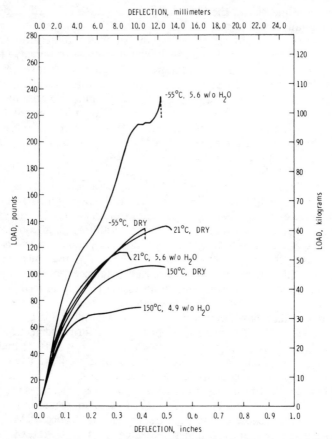

Figure 1. *Load versus deflection behavior for [0/90] Kevlar 49 181 style fabric/CE-9000 epoxy laminates as a function of temperature and moisture content.*

curves display a significant amount of nonlinearity. Some nonlinearity is due to horizontal thrust effects; however, substantial nonlinearity is exhibited by the material itself. At −55°C, Kevlar/epoxy appears somewhat more brittle compared to room temperature. At 150°C, a reduced load carrying capacity is seen.

The addition of moisture reduces the maximum loads and the deflections attained at 21°C and 150°C compared to like specimens in the dry condition, but has very little effect in the linear region. A substantial deviation in behavior is seen with moisture present at low temperature (Figure 1). The initial deviation from linearity (proportional limit) occurs at nearly double the loads characteristic of the other conditions tested, and a second linear region appears after the initial yielding. At high loads, another nonlinear region is evident which is followed by another region of rapidly increasing load that terminates in brittle failure. The cause of the behavior at 210 lbs (95 Kg) of load is unknown, but may have been caused by slipping of the specimen at the end supports. Similar behavior is seen with the 5208 epoxy resin matrix at all conditions examined with the exception that a somewhat higher proportional limit load is seen with the 5208, especially at -55 °C and moisture present.

The macroscopic failure appearance of the laminates plotted in Figure 1 took the form of one of two modes. At -55 °C with moisture and at all temperatures in the dry condition, failure corresponds to the formation of a tensile crack across the specimen surface as shown in Figure 2. A large stress whitened zone around the crack is also seen in Figure 2. Close examination of the tensile surface showed the crack ran preferentially in a fill yarn bundle (transverse tensile failure of the fill yarn bundle) and failed the warp yarns in axial tension only where the fill yarns traveled under the warp yarns (Figure 3). Thus, in the eight harness satin 181 style fabric, tensile failure corresponds

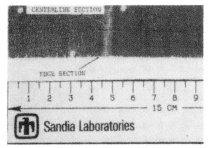

Figure 2. *Macroscopic tensile surface appearance of [0/90] Kevlar 49 181 style fabric/CE-9000 epoxy laminate tested in bending at 21°C.*

Figure 3. *Tensile crack in Kevlar 49 181 style fabric/epoxy laminate tested in bending at 21°C. 4X.*

to longitudinal tensile fracture of only one eighth of the Kevlar 49 filaments on the surface. The compression surface of laminates exhibiting tensile failure revealed the presence of small ridges which formed on either side of the loading ram at the specimen edges and propagated towards the specimen centerline. The laminates tested at 21 °C and 150 °C with near saturation moisture content exhibited only the compression ridges with no tensile crack formed. In those cases, the ridges were more abundant, higher, and completely transversed the width of the specimen.

Figures 4 and 5 give an overview of moisture effects on the apparent flexural strength and stiffness of [0/90] Kevlar 49, 181 style fabric/epoxy laminates as a function of temperature. As discussed in the previous section on the three-point bend test, the strengths presented in Figure 4 were calculated from elastic beam theory which is not strictly valid for these materials because of the nonlinearity in the load-deflection curves and are, therefore, considered to be apparent strengths. Each data point in Figures 4 and 5 is for an individual specimen. The curves drawn in Figures 4 and 5 are visual best fits to the data. For ease of presentation, property changes with temperature and moisture will be discussed in terms of property retention compared to the room temperature, dry condition.

The strength data for the dry condition given in Figure 4 show that Kevlar 49/epoxy composites have a strong temperature dependence. A 30 percent

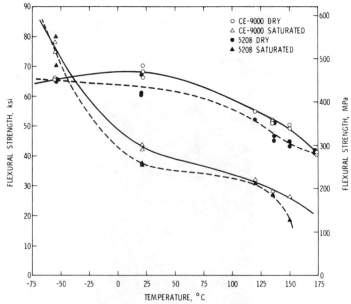

Figure 4. Flexural strength versus test temperature for [0/90] Kevlar 49 fabric/epoxy laminates dry and with near saturation moisture contents.

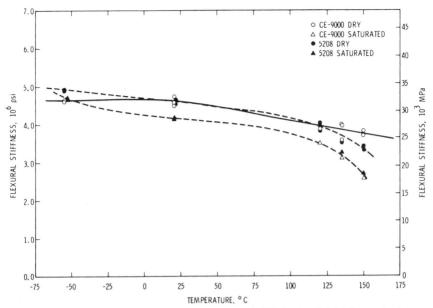

Figure 5. *Flexural stiffness versus test temperature for [0/90] Kevlar 49 fabric/epoxy laminates dry and with near saturation moisture contents.*

loss in strength is seen at 150 °C with both epoxide matrices. Strength data for neat epoxides with similar chemical composition show little temperature dependence below 150 °C [5,26]. The temperature dependence of the composite, is, therefore, most likely due to the thermal sensitivity of the Kevlar 49 filaments. At low temperature (-55 °C), strengths are observed to be nearly equal to room temperature data. The proportional limit or yield strength data (not plotted) also show a temperature dependence similar to, but not as pronounced as the ultimate strength data. Yield strengths for dry laminates varied from 20 ksi (138 MPa) at -55° to 18 ksi (124 MPa) at 21 °C to 14 ksi (97 MPa) at 150 °C.

As shown in Figure 4, the addition of moisture causes substantial changes in the apparent strength exhibited by [0/90] Kevlar 49 fabric/epoxy laminates over the entire temperature range examined in this study. At 21 °C and near saturation moisture content, a 35-40 percent loss of dry strength is evident. Examination of the load-deflection curves in Figure 1 reveals that only a 15 percent loss in load carrying capacity has occurred with the addition of moisture at 21 °C. The larger difference in strength is due to the substantial moisture induced swelling which takes place in the laminate thickness. Through-the-thickness swelling strains in the specimens used for this study averaged 0.50 percent strain/percent moisture absorbed with the CE-9000 matrix, and 0.67 with the 5208 resin. Because specimen thickness enters the equations for strength and stiffness as the reciprocal squared and cubed, respectively, the reduction in apparent strength and stiffness is greater than the reduction of load observed in the load-deflection curves.

High moisture content and 150°C temperature results in a 60 percent strength loss with the CE-9000 matrix and a 70 percent loss with the 5208 matrix. The reverse behavior is seen at low tempratures (-55°C) with moisture present where the strength increases 15–20 percent (Figure 3). Yield strengths (not plotted) show a 35–40 percent decrease at 150°C with moisture. Like ultimate strength, they also show substantial increases at low temperatures. Yield strength at -55°C and 5 percent moisture content doubles with the 5208 matrix to 36 ksi (248 MPa). Under the same conditions yield increases 40 percent with the CE-9000 matrix to 25 ksi (172 MPa). Little effect of moisture is seen on yield strengths at 21°C.

Laminate stiffness is also affected by temperature (Figure 5), but not as greatly as the strength. Dry CE-9000 matrix laminates exhibit a nearly linear dependence of stiffness on temperature at elevated temperatures and stabilize at low temperatures to a value near that at 21°C. At 150°C, a 19 percent loss in stiffness is seen with the CE-9000 matrix. The dry 5208 matrix laminates also have a nearly linear temperature dependence that shows a slight increase in stiffness at low temperatures but begins to diminish at high tempratures. A 27 percent loss in stiffness is seen in the 5208 matrix laminates at 150°C.

The effect of the addition of moisture on laminate stiffness is also shown in Figure 5. Accurate stiffness data with the CE-9000 matrix material could not be obtained at 21°C and -55°C becuase of noise in the load-deflection curves. In the 5208 matrix composites at room temperature, moisture causes a 10 percent loss in stiffness. The 150°C data show that a 40 percent loss in stiffness has occurred. The moduli of specimens with high moisture content at -55°C are equivalent to those of room temperature, dry specimens.

Failure Modes

Comparison of the strength and stiffness data in Figures 4 and 5 implies that the larger percentage changes in composite strength may be due to failure mode changes with increased severity of aging conditions. The failure zones of the flexural specimens were sectioned at the midpoint of the specimen width and at the edge, and then polished to a 1 micron diamond finish for microscopic examination of composite failure modes. The locations of the sections prepared for microscopic examination are indicated by the arrows in Figure 2.

At 21°C dry, filament compressive buckling (kinking)* is exhibited at the specimen edges (Figure 6b) with a slight tearing on the tensile surface as the only evidence of failure (Figure 6a). As seen in Figure 6b, buckling occurs

---

* Because the mechanism responsible for the type of permanent deformation seen in Figure 6b is unknown, buckling and kinking cannot be distinguished. For this disussion, that failure mode will be referred to as buckling.

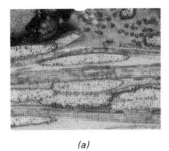
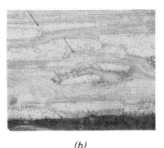

(a) (b)

Figure 6. *Edge cross sections of [0/90] Kevlar 49 fabric/CE-9000 laminate tested at 21°C dry. (a) Tension side; (b) Compression side. 25X.*

predominately where the fill yarn bundles cross the warp yarn bundles (see arrows in Figure 6b). This result is not unexpected because the transition region is where the warp yarns are at the largest angle to the compressive force and, thus, is more susceptible to buckling. As also seen in Figure 6b, buckling is evident into the third ply from the compressive surface for specimens tested in the 21°C, dry condition. The area shown in Figure 6b was adjacent to the loading ram at a distance of 0.075 in. (1.9 mm). The direct influence of the loading ram on the observed failure modes is unknown. Four-point bend tests may be used to determine failure modes without the presence of a center loading ram.

While filament buckling was observed at the specimen edge (Figure 6b), the opposite behavior is seen at the center of the specimen width (Figure 7) where no filament buckling is evident on the compression side, but extensive matrix and interfacial cracking is seen in tension. These results indicate that the nonlinearity seen in the load-deflection curve (Figure 1) is due to filament buckling and also, perhaps, to matrix cracking. The different modes seen at the specimen center and edge would indicate an unexpected edge effect is also present.

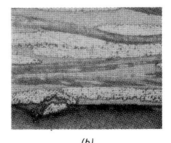

(a) (b)

Figure 7. *Center cross sections of [0/90] Kevlar 49 fabric/CE-9000 laminate tested at 21°C dry. (a) Tension side; (b) Compression side. 25X.*

Tests run at -55 °C and 150 °C in the dry condition exhibited failure modes identical to those tested at 21 °C. At -55 °C, buckling was observed in the first two plies at the specimen edge with no buckling observed on the compression side in the center of the specimen. Substantial matrix cracking and interfacial delamination were evident on the tensile surface. The specimens tested at 150 °C also revealed compressive buckling only at the specimen edge. At 150 °C, however, filament buckling was not localized to yarn cross-over regions but was found throughout the first four plies. On the tensile surface, a reduced amount of matrix cracking was observed along with some evidence of delamination at the center of the specimen width.

The extent of filament buckling becomes more pronounced at all temperatures with saturation moisture content. Whereas filament buckling was only observed at the specimen edges in the dry condition, buckling was observed all across the specimen width in the wet specimens. Buckling extended into the eighth ply from the compressive surface in the specimens tested at 150 °C with moisture. Eight plies represent two-thirds of the specimen cross section. Some surface tensile damage in the form of interfacial separations could be seen in the center of the specimen tested at 150 °C with near saturation moisture content; however, it is likely that the tensile damage did not occur until large deflections had been obtained. The massive filament compressive buckling seen at 150 °C in the presence of high moisture concentrations would indicate that the matrix is above its glass transition temperature at the test temperature, and, in its elastomeric condition, allows the filament buckling mechanism to operate at low loads.

The specimens tested at room temperature with high moisture content exhibited some matrix cracking on the tensile surface and compressive buckling into the first two plies. Testing of the 21 °C, wet specimens was terminated after the large drop in load seen in the load-deflection curve of Figure 1; otherwise, buckling would have likely been seen extending further into the specimen. It appears that the large drop in strength at 21 °C with the addition of saturation levels of moisture (Figure 4) is due to the filament buckling mechanism being allowed to operate with increased ease. Such behavior may be explained by the matrix becoming more compliant as it is plasticized by moisture.

Conversely, the specimen tested at -55 °C with saturation moisture content show only limited buckling in the first ply. On the tensile surface, extensive matrix cracking is seen which occurs into the specimen thickness between the fifth and sixth plies (Figure 8). In addition, large cavitations and some interfacial delaminations are seen on the tensile surface (Figure 8). One may speculate that the initial yielding seen in the -55 °C, 5.6 w/o H_2O load-deflection curve (Figure 1) corresponds to the buckling of the first ply in compression. The unbuckled portion of the specimen then begins to load elastically with a shifted neutral axis as discussed by Zweben [19] until the unusual behavior above 200 lb (90 kg) occurs which may be due to slippage on the end supports or the formation of delaminations near the tensile surface. The load then again begins to increase until the outer ply fails in tension at 236 lb (107

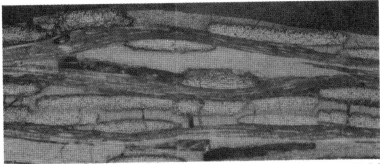

Figure 8. *Cross section of tensile side of [0/90] Kevlar 49 fabric/CE-9000 laminate with 5.6 per moisture content tested at −55°C. 40X.*

kg). A series of experiments with loading to different levels, unloading, and examining the specimen cross section would be required to verify this thesis. However, it does appear that the filament buckling mechanism is suppressed with the combination of moisture and low temperature except in the unconstrained surface layer. Moisture at low temperature may either stiffen the filaments and/or matrix or, by collecting at the filament-matrix interface, form a rigid layer around the filaments.

Intermediate Moisture Levels

To determine the effect of lower moisture concentrations, specimens were tested at elevated temperatures with moisture contents varying between dry and saturation. Figure 9 presents a plot of apparent flexural strength versus moisture content at the test temperatures 21 °C, 120 °C, 135 °C, and 150 °C for [0/90] laminates with the CE-9000 matrix resin. An identical plot for laminates with a 5208 matrix is given in Figure 10. Each data point in Figures 9 and 10 is for an individual specimen. The curves shown in Figures 9 and 10 are visual best fits to the data.

At 21 °C, an initial drop in apparent flexural strength of about 8 percent is seen with the addition of moisture for both resin matrices. Strength then stabilizes with increasing moisture content up to about 4 weight percent. At that point, strength again begins to drop, and falls to a final value of 42.9 ksi (295 MPa) with the CE-9000 resin and 37.2 ksi (256 MPa) with the 5208 matrix at saturation. The shape of the 21 °C curve is governed somewhat by the method of introducing moisture to the specimens. The initial drop in apparent strengths seen in Figures 9 and 10 is probably more severe than would be expected because the outer, most highly stressed plies are saturated. That condition would result, however, from storage at high relative humidity. At each elevated temperature, a linear hygrothermal degradation is seen with increasing moisture content until a moisture concentration is reached which depresses the matrix T_g below the test temperature. At that point, a

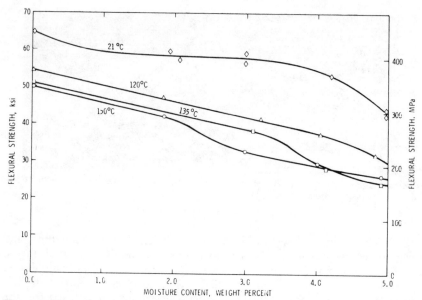

Figure 9. Effect of temperature and moisture on the flexural strength of [9/90] Kevlar 49 fabric/5208 laminates.

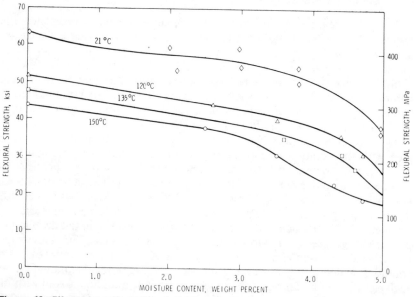

Figure 10. Effect of temperature and moisture on the flexural strength of [0/90] Kevlar 49 fabric/5208 laminates.

38

downward inflection is seen in the strength versus moisture curve which is followed by a region of relatively constant strength. The flexural data in Figure 9 indicate that a moisture content of approximately 3.0 weight percent corresponds to a T_g of 135 °C and 2.0 weight percent to 150 °C. Glass transitions for Kevlar 49, 181 style fabric/CE-9000 epoxy laminates have been determined in Reference [27] by thermomechanical analysis. Those results show a dry T_g of 195 °C which decreases to 135 °C with 2.0 weight percent moisture and to 125 °C with a 3.0 weight percent moisture.

Microscopic examination of the specimens plotted in Figure 9 showed that, as moisture is introduced at a given temperature, the filament buckling mode becomes more pronounced. Buckling occurs more generally along the filaments and further below the specimen surface (Figure 11). Although buckling becomes more predominant with increasing moisture content, tensile failure continues to occur in the midpoint of the width of the specimen. Once the test temperature exceeds the matrix T_g, only massive filament buckling is seen with isolated areas of matrix cavitation in evidence on the tensile side.

Similar behavior is seen in the flexural strength versus moisture content curves at various temperatures for laminates with a 5208 matrix (Figure 10). The same failure modes previously discussed for [0/90] laminates with a CE-9000 matrix were observed with the 5208 binder. A comparison of the data in Figures 9 and 10 provides some useful insight into the effect of matrix chemistry on the properties of aged Kevlar-reinforced composites. The matrix glass transition temperature of a dry Kevlar 181/5208 laminate post-cured at 190 °C is 207 °C [27]. The higher initial T_g of the 5208 matrix laminates is seen as a shift to higher moisture contents required to depress T_g to the test temperature (Figure 10). For example, a 4.0 percent moisture content lowers the 5208 T_g to 135 °C, whereas only 3.0 percent reduces the CE-9000 T_g to 135 °C.

Once T_g is exceeded by the test temperature, however, the 5208 laminates exhibit a sharp drop in strength and then stabilize at a strength of 13 ksi which

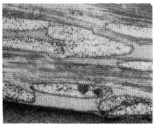 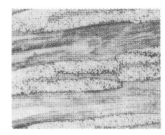

(a)   (b)

Figure 11. *Edge cross section of compression side of [0/90] Kevlar 49 fabric/CE-9000 laminate tested at 120°C with 3.2 weight percent moisture. (a) Surface plies 1 and 2; (b) Internal plies 5 and 6. 25X.*

represents a 77 percent loss from the 21 °C dry condition. The glycidyl ether, diglycidylamine of p-amino phenol (0510) is a small, highly reactive molecule with a 25 percent lower epoxide equivalent weight than the tetraglycidyl methylene dianaline (MY-720) [28]. Apparently, the higher cross-link density of the 0510 based system (CE-9000) results in a structure with a higher rubbery modulus than the MY-720 based system (5208); this in turn provides the Kevlar filaments with a more rigid lateral support and results in higher composite flexural strengths.

The data in Figures 9 and 10 show that hygrothermal effects result in a 40 percent loss of the room temprature, dry strength of Kevlar/epoxy laminates at temperatures approaching the matrix T_g. As T_g is exceeded, 40–50 percent of the strength remaining below T_g is lost (60–70 percent total loss from room temperature, dry). For purposes of design, it would, therefore, be desirable in applications of Kevlar/epoxy that experience long term exposure to atmospheric moisture followed by rapid heating to limit service temperatures to those below the reduced matrix T_g. Laminate moisture contents for Kevlar 49, 181 style fabric/5208 resin of 3–4 percent would be attained upon long term exposure to atmospheres of 55–75 percent relative humidity [9]. Similar moisture sorption behavior is seen with CE-9000 matrix laminates [27]. Based upon the data presented in Figures 9 and 10, and the moisture sorption data given in References [9] and [27], strength critical designs should be limited to temperatures below 120 °C. Stiffness critical designs would not be as sensitive to absorbed moisture, but are more influenced by the thermal sensitivity of the Kevlar 49 filaments themselves.

An alternative to the 120 °C temperature limitation may be to protect the laminates from atmospheric moisture by the addition of moisture barrier coatings. Thin electroplates Cu and Ni surface layers are currently being evaluated as potential moisture barriers on Kevlar/epoxy composites by the author.

SUMMARY AND CONCLUSIONS

This study has examined the flexural properties of Kevlar 49 fabric/epoxy laminates as a function of temperature and moisture content. Two 190 °C cure resin matrices, Narmco 5208 and Ferro CE-9000 have been examined in a [0/90] ply orientation over the temperature range from –55 °C to 150 °C. The results indicate that hygrothermal degradation of Kevlar-reinforced composites can be extensive. Hygrothermal degradation which is linear with increasing temperature and moisture content is observed until the matrix transition temperature (T_g) is depressed to the test temperature by moisture solvation. Above T_g, the rubbery matrix does not provide the lateral support required by the Kevlar filaments, and the composites fail at low loads by massive filament buckling. Below matrix T_g the [0/90] laminates exhibited a filament buckling induced yield which progressed in depth into the specimen until final failure occurred in tension. Failure of Kevlar 49/epoxy in bending is not catastrophic. Rather, failure corresponds to a drop in the maximum load carrying capacity. Once an outer ply failed in tension, the material could still sustain load, but at a lower level than before the ply failure.

At the most severe conditions tested (5.0 weight percent moisture at 150°C), the laminates show a strength loss of 60–70 percent and a stiffness loss of 40 percent compared to the values obtained on dry specimens at 21°C. Moisture at room temperature also reduced strength substantially. The 35–40 percent loss in room temperature dry strength appeared to be due to matrix plasticization effects which allowed the filament buckling mechanism to operate more easily. At low temperatures (–55°C), the addition of moisture resulted in greatly increased strengths. Microscopic examination indicated that the filament buckling mechanism in the low temperature, wet specimens is inhibited from operating except in the unconstrained surface ply.

The substantial hygrothermal strength losses seen in the [0/90] Kevlar 49 fabric/epoxy laminates examined in this study indicate that highly stressed designs should be limited to temperatures of 120°C if they are exposed to atmospheric moisture prior to heating. Stiffness critical designs would be less sensitive to moisture, and are instead limited by the thermal sensitivity of the Kevlar 49 filaments.

ACKNOWLEDGMENT

J. E. Perea and K. D. Boultinghouse conducted the experimental aspects of this study. F. P. Gerstle, E. D. Reedy, and A. K. Miller provided technical review of this manuscript in numerous technical discussions during the course of this study.

REFERENCES

1. Hertz, J., "Moisture Effects on the High-Temperature Strength of Fiber-Reinforced Resin Composites," *Proceedings of the 4th National SAMPE Technical Conference.* Azusa, CA: Society for the Advancement of Material and Process Engineering, 1–7 (1972).
2. Young, H. L. and Greever, W. L., "High Temperature Strength Degradation of Composites During Aging in the Ambient Atmosphere, in *Composite Materials in Engineering Design,* ed., B.R. Norton, Metals Park, OH: American Society for Metals, 695–715 (1972).
3. *Environmental Effects on Advanced Composite Materials.* ASTM STP 602, American Society for Testing and Materials (1976).
4. *Advanced Composite Materials—Environmental Effects.* ASTM STP 658, American Society for Testing and Materials (1978).
5. Browning, C. W., "The Mechanisms of Elevated Temperature Property Losses in High Performance Structural Epoxy Resin Matrix Materials after Exposures to High Humidity Environments," *Polym. Engr. and Sci., 18*(1), 16–24 (January 1978).
6. Morgan, R. J. and O'Neal, J. E., "The Durability of Epoxies," *Polym-Plast. Techn. and Engr., 10*(1), 49–116 (1978).
7. Browning, C. E., Husman, G. E., and Whitney, J. M., "Moisture Effects in Epoxy Resin Matrix Composites," *Composite Materials: Testing and Design,* ASTM STP 617, American Society for Testing and Materials, 481–496 (1977).
8. Penn L. and Larsen, F., "Physiochemical Properties of Kevlar 49 Fiber," *J. Appl. Poly. Sci., 23*(1), 59–74 (January 1979).
9. Allred, R. E. and Lindrose, A. M., "The Room Temperature Moisture Kinetics of Kevlar 49 Fabric/Epoxy Laminates" *Composite Materials: Testing and Design,* ASTM STP 674, American Society for Testing and Materials, 313–323 (1979).

10. Augl, J. M., "Moisture Sorption and Diffusion in Kevlar 49 Aramid Fiber," Naval Surface Weapons Center Report NSWC/TR-79-51, Silver Spring, MD (March 1979).**
11. Smith, W. S., "Environmental Effects on Aramid Composites," E. I. du Pont de Nemours, Textile Fibers Dept. Report, Wilmington, DE (1980).
12. Abbott, N. J., Donovan, J. G., Schoppee, M. M., and Skelton, J., "Some Mechanical Properties of Kevlar and Other Heat Resistant, Nonflammable Fibers, Yarns, and Fabrics," Fabric Research Laboratories Report AFML-TR-74-65, Part III, Dedham, MA (March 1975).
13. Trujillo, R. E. and Engler, B. P., "Chemical Characterization of Composite Prepreg Resins, Part I," Sandia Laboratories Report SAND 78-1504, Albuquerque, NM (December 1978).**
14. Trujillo, R. E. and Assink, R. A., Sandia Laboratories, unpublished data.
15. Allred, R. E. and Hall, N. H., "Volume Fraction Determination of Kevlar 49/Epoxy Composites," *Polym. Engr. and Sci., 19*(13), 907–909 (October 1979).
16. Miller, A. K., Sandia Laboratories, unpublished data.
17. Whitney, J. M. and Husman, G. E., "Use of the Flexure Test for Determining Environmental Behavior of Fibrous Composites," *Expl. Mech., 18*(5), 185–190 (May 1978).
18. Zweben, C., Smith, W. S., and Wardle, M. W., "Test Methods for Fiber Tensile Strength, Composite Flexural Modulus and Properties of Fabric-Reinforced Laminates," *Composite Materials: Testing and Design,* ASTM STP 674, American Society for Testing and Materials, 228–262 (1979).
19. Zweben, C., "The Flexural Strength of Aramid Fiber Composites," *J. Comp. Mat'ls., 12,* 422–430 (October 1978).
20. Zweben, C. and Wardle, M. W., "Flexural Fatigue of Marine Laminates Reinforced with Woven Rovings of E-Glass and of Kevlar 49 Aramid," *Proceedings of 34th Annual Conference Reinforced Plastics/Composites Institute,* Society of the Plastics Industry, New Orleans, LA, Section 4C, 1–6 (1979).
21. Greenwood, J. H. and Rose, P. G., "Compressive Behavior of Kevlar 49 Fibres and Composites," *J. Matl. Sci., 9,* 1809–1814 (1974).
22. Kulkarni, S. V., Rice, J. S., and Rosen, B. W., "An Investigation of the Compressive Strength of Kevlar 49/Epoxy Composites," *Composites,* 217–225 (September 1975).
23. Jones, R. M., "Apparent Flexural Modulus and Strength of Multimodulus Materials," *J. Compt. Mat'ls., 10,* 342–351 (October 1976).
24. "Kevlar 49 Data Manual," E. I. du Pont de Nemours, Wilmington, DE.
25. "Flexural Properties of Plastics," ASTM Designation D 790-71, American Society for Testing and Materials, Philadelphia, PA (1978).
26. Morgan, R. J., O'Neal, J. E., and Miller, D. B., "The Structure, Modes of Deformation and Failure, and Mechanical Properties of Diaminodiphenyl Sulphone-cured Tetraglycidyl 4,4′ Diaminodiphenyl Methane Epoxy," *J. Mat'l. Sci., 14*(1), 109–124 (January 1979).
27. Allred, R. E., Hall, N. H., and Miller, A. K., "Moisture and Post-Cure Effects on the Matrix Glass Transition Temperature of Kevlar/Epoxy Laminates," to be submitted to *Polym. Engr. and Sci.*
28. Ciba-Geigy Corp., Ardlsey, NY, product data.

**Available from: National Technical Information Service, U.S. Department of Commerce, 5285 Port Royal Road, Springfield, VA 22161.

5

The Effect of Temperature and Moisture Content on the Flexural Response of Kevlar/Epoxy Laminates: Part II. [±45,0/90] Filament Orientation

R. E. ALLRED

INTRODUCTION

PART I OF THIS STUDY ADDRESSED TEMPERATURE AND MOISTURE EFFECTS on the flexural behavior of [0/90] Kevlar 49 fabric/epoxy laminates. Results indicated that [0/90] Kevlar 49/epoxy laminates are temperature dependent over the range tested (–55°C to 150°C), and that moisture magnifies the temperature sensitivity. The purpose of this study is to expand on the observations of Part I by examining the temperature and moisture dependence of quasi-isotropic Kevlar/epoxy laminates. Quasi-isotropic filament orientations are of general interest for applications that see biaxial tensile or flexural loading. Quasi-isotropic laminates are generally more sensitive to the properties of the matrix, because of shear effects, and may display phenomena not seen in the [0/90] laminates. Studies conducted on the quasi-isotropic laminates which were not addressed with the [0/90] laminates in Part I include the effects of voids, freeze-thaw thermocycling, and long-term moisture exposure on laminate properties. Load-unload response is also examined.

EXPERIMENT

The same material systems discussed in detail in Part I were selected for the quasi-isotropic laminates. Flat plates 12 in. (305 mm) square by 0.10 in. (2.5 mm) thick were prepared by autoclave molding 12 plies of Kevlar 49 181 style fabric, preimpregnated with either the Narmco 5208 or Ferro CE-9000 epoxy resin. A [±45,0/90]$_6$ ply orientation balanced on the mid-plane was used to achieve quasi-isotropic properties. After being molded to a final cure temperature of 175°C, the laminates were subjected to a 4-hour 190°C post cure in air. The quasi-isotropic laminates contained a nominal filament

43

volume fraction of 54 percent and a void content of less than 1 percent as determined by chemical dissolution [1].

The three-point bend geometry used to test the [0/90] laminates in Part I was also used for the [±45,0/90] specimens. Because a discussion of the three point bend test with respect to testing Kevlar-reinforced composites and details of the experimental procedures used are contained in Part I, the reader is referred to that paper prior to examining this work. A span-to-depth ratio of 27.5 was used to test specimens 5.0 in. × 1.0 in. × 0.1 in. (127 mm × 25.4 mm × 2.5 mm). Moisture was introduced by immersion of the specimens in 90 °C distilled water. The flexural specimens were allowed to equilibrate four minutes with the thermally prestabilized test chamber prior to the application of load. Approximately four additional minutes were required to load the samples to maximum load. Midspan deflections were monitored with a linear variable differential transformer.

RESULTS AND DISCUSSION

Load-Deflection Response

The load-deflection behavior of quasi-isotropic Kevlar fabric/CE-9000 laminates tested at the extremes of this study are plotted in Figure 1. The load-deflection curves plotted in Figure 1 are from individual specimens which exhibited a response near the mean of the specimens tested at each condition. The flexural tests were terminated at the end points of the curves as shown in Figure 1. Failure in this case is characterized by an inability to sustain increasing load. Catastrophic failure in the form of specimen separation did not take place; rather, load slowly dropped off in the case of the 21 ° and 150 °C specimens. At –55 °C, load dropped rapidly from the maximum and arrested at approximately 70 percent of maximum load. The –55 °C specimens then reloaded to a lower maximum before exhibiting another rapid drop in load, etc.

As in the [0/90] laminates, the quasi-isotropic laminate load-deflection curves are nonlinear and the maximum load depends upon test temperature. The effect of moisture on load carrying capacity is also temperature dependent (Figure 1). At 150 °C, moisture reduces the maximum load, whereas little effect is seen on maximum load at 21 °C. The addition of moisture at low test temperatures (–55 °C) results in increased load carrying capacity. Similar behavior is seen with the 5208 epoxy matrix laminates at all conditions examined.

To investigate the nonlinear behavior of Kevlar/epoxy in bending, two specimens were loaded into the nonlinear region and unloaded prior to failure. Test conditions were 21 °C, dry. The load-deflection behavior for unloading is shown in Figure 2. The specimen labeled A in Figure 2 was loaded to failure. Specimen B was loaded just below the maximum of A and specimen C was loaded to 75 percent of B. As load was removed, both specimens exhibited a drop in load with little change in deflection. After the initial drop in load, the unloading curves become nearly linear in deflection

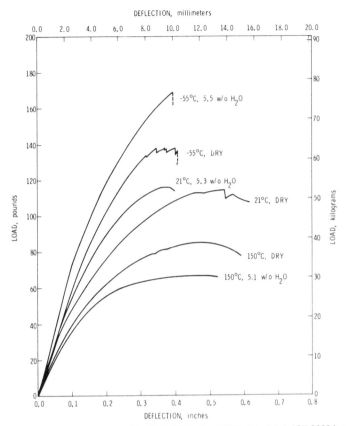

Figure 1. *Load-deflection response of [±45,0/90] Kevlar 49 181 style fabric/CE-9000 laminates as a function of temperature and moisture content.*

until near zero load. A substantial residual deflection is seen in the specimens after load is removed completely (Figure 2). The set in specimen C (75 percent of maximum load) represents 18 percent of the nonlinear (post proportional limit) portion of the loading curve, or 12 percent of total deflection. Specimen B (near maximum load) displayed a residual set of 25 percent of the nonlinear portion of the loading curve, or 22 percent of total deflection. Neither specimen B or C exhibited any surface damage.

Strength Behavior

The apparent flexural strength (from simple beam theory) and the proportional limit (yield) of dry and moisture saturated [±45,0/90] Kevlar fabric/epoxy laminates are plotted versus temperature in Figure 3. Dry strengths of laminates with the CE-9000 and 5208 epoxide matrices are

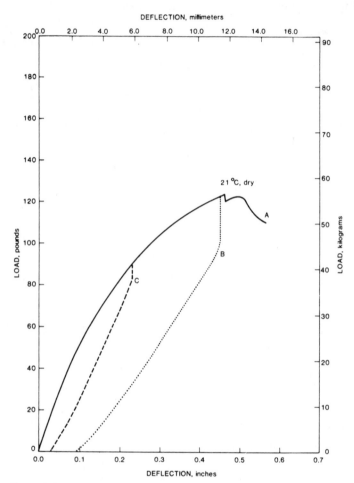

Figure 2. *Unloading behavior of [±45,0/90] Kevlar 49 fabric/CE-9000 laminate.*

statistically equivalent throughout the temperature range examined in this study, and results from both specimen types were averaged to give the dry values plotted in Figure 3. Eight specimens were tested at each condition to determine statistical scatter.

Like the [0/90] ply orientation, the [±45,0/90] laminates also display a temperature sensitivity; however, significant differences exist between the two filament orientations. The quasi-isotropic laminates show a greater ultimate strength retention at 150°C (23 percent loss*), and also show a strength in-

---

*For purposes of discussion, property changes resulting from various environments will be compared to the room temperature, dry condition.

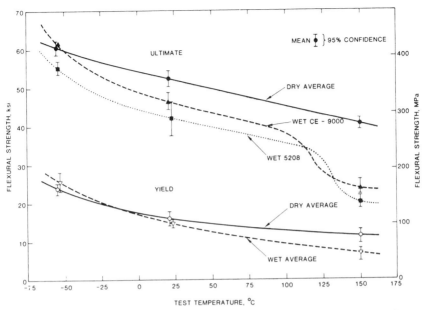

Figure 3. *Flexural strength versus test temperature for [±45,0/90] Kevlar 49 fabric/epoxy laminates dry and with near saturation moisture content.*

crease at low temperature (13 percent gain at –55°C) not seen with the [0/90] laminates. The strength trends shown by the [±45,0/90] laminates probably reflect the temperature dependence of the epoxy matrix. The strength behavior with temperature shown in Figure 3 is similar to neat epoxy resins [2] and indicates the matrix sensitivity of the quasi-isotropic filament orientation. Yield strengths (proportional limits) of the dry quasi-isotropic laminates (Figure 3) are more temperature sensitive than the ultimate strengths. At 150°C, yield strength shows a 31 percent loss, while a 36 percent increase in yield strength is seen at –55°C.

The temperature dependence of dry quasi-isotropic laminates is magnified by moisture. As seen in Figure 3, the 5208 matrix laminates are more affected by moisture than the CE-9000 matrix laminates. At 150°C with near saturation moisture content, the CE-9000 matrix laminates have lost 55 percent of their 21°C dry strength. For the same conditions, the 5208 matrix laminates have lost 62 percent of their room temperature dry ultimate strength.

High moisture concentrations also affect the 21°C properties. With near saturation moisture content, the CE-9000 matrix laminates show a 13 percent reduction in ultimate strength and the 5208 matrix laminates a 21 percent loss. An increase in strength is seen at low temperature (–55°C) with high moisture content. As seen in Figure 3, the CE-9000 matrix laminates show a 16 percent strength increase at –55°C and high moisture content while the 5208 laminates show a 4 percent increase.

Examination of the specimen surfaces with scanning electron microscopy revealed that all of the 5208 matrix laminates tested at –55 °C and 21 °C with saturation moisture content had microcracked. Approximately 30 percent of the 150 °C wet test specimens exhibited microcracking. The extent of the microcracking may be seen in Figure 4. Figure 4a shows an as-fabricated surface and 4b shows the surface after 5500 hours of immersion in 90 °C water. The viewed surfaces were taken outside the span length of the three-point bend specimens. Similar microcracking was observed in untested specimens. Examination of additional specimens revealed that the 5208 matrix laminates microcracked between 1300 and 5500 hours of immersion. Near saturation moisture content was attained after approximately 500 hours immersion. Microcracking due to moisture and thermal spiking in tetra-glycidylmethelene dianiline cured with diamino diphenyl sulfone has been previously observed [3,4]. Microcracking has also been reported in graphite/5208 composites immersed in 82 °C water [5]. No change was observed in the surface appearance of the CE-9000 matrix laminates after 5500 hours immersion. The long-term immersion at 90 °C is an unusually harsh environment which would not be seen in most applications of these materials. It may be observed, however, that the CE-9000 formulation is much more resistant to microcracking than the 5208 formulation.

The effect of surface microcracking on the properties of the saturated 5208 matrix laminates at 21 ° and –55 °C presented in Figure 3 is unknown, but is believed to be minor. At 150 °C, the microcracked specimens displayed the same behavior as uncracked specimens with similar moisture content, and the [0/90] laminates tested in Part I were not microcracked and showed a 40 percent loss in ultimate strength. As such, the 21 percent loss shown by the

 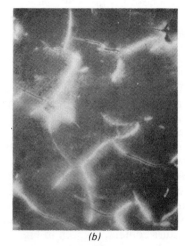

(a)   (b)

Figure 4. Scanning electron micrographs of Kevlar 49 fabric/5208 epoxy laminate surfaces, 333X. (a) as fabricated, and (b) 5500 hour 90°C water immersion.

[±45,0/90] 5208 matrix specimens, with high moisture at 21°C is probably not a surface effect.

The temperature dependence of yield strength is also magnified by high moisture content. Yield strength with both the 5208 and CE-9000 matrices appeared equivalent over the -55°C to 150°C temperature range with near saturation moisture content. The average of both laminate types is plotted in Figure 3. A 63 percent loss in yield strength is seen at 150°C with moisture while a 47 percent gain is seen at -55°C with moisture. At 21°C, a 16 percent loss in yield strength is observed.

Comparison of the data in Figures 1 and 3 reveals that load carrying capacity is not as sensitive as strength to environmental conditions. As with the [0/90] laminates discussed in Part I, moisture induced through-the-thickness swelling strains, which enter the flexural strength equation as the reciprocal squared, cause a strength reduction which is greater than the reduction of maximum load for the [±45,0/90] laminates.

To examine the effect of intermediate moisture concentrations, specimens with both matrix resins were tested with moisture contents varying between dry and saturation. Figure 5 presents a plot of apparent flexural strength versus moisture content for tests conducted at 21 and 150°C. Each data point in Figure 5 is for an individual specimen. The curves shown in Figure 5 are visual best fits to the data. Little effect of moisture on the quasi-isotropic laminates is seen at 21°C until high moisture contents are attained. As

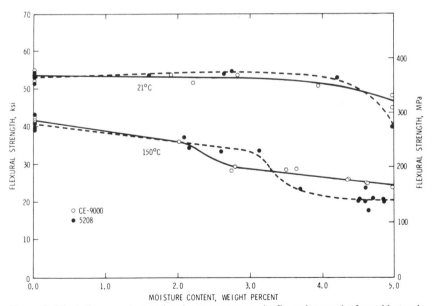

Figure 5. *Effect of temperature and moisture content on the flexural strength of quasi-isotropic Kevlar 49 fabric/epoxy laminates.*

previously discussed, the 5208 matrix specimen at 5 weight percent moisture may have been adversely affected by surface microcracks, and may be conservative.

At 150°C, the results are similar to those obtained for [0/90] laminates. The CE-9000 matrix laminates go through T_g at about 2.0 percent moisture and their strength above T_g is higher than that of the 5208 matrix laminates, which go through T_g at about 3.0 percent moisture. Numerous data points are plotted for both systems with high moisture contents. Those data correspond to various long-term immersions in 90°C water up to 7000 hours. A near-saturation moisture content was reached in the specimens after 500 hours immersion. No further strength degradation was evident in either matrix system with longer immersion times at the 150°C test temperature.

Void Effects

Voids are generally present to some degree in composites and are known to affect mechanical properties and moisture absorption behavior. A review of the effects of voids on the properties of high performance composites is given in Reference [6]. The effect of voids on the apparent ultimate flexural strength of Kevlar fabric/CE 9000 laminates as a function of temperature and moisture is shown in Figure 6. No effect of voids can be seen at 21°C. The majority of composite systems are affected by the volume percent of voids at room temperature [6]. Those systems generally are reinforced with brittle

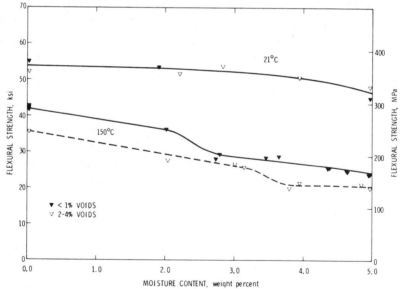

Figure 6. The effect of voids on the flexural strength of quasi-isotropic Kevlar 49 fabric/CE-9000 laminates as a function of temperature and moisture content.

filaments such as graphite, boron, or glass. That the ultimate flexural strength of the Kevlar 40-reinforced laminates shown in Figure 6 is not affected by voids may be due to the toughness of the Kevlar filaments or to the distinctive failure modes which they exhibit.

At 150°C, the 2–4 percent void content specimens show an initial strength loss of approximately 20 percent in the dry condition. As moisture is added, the strength values of the high void specimens parallel the results of low void specimens with the same offset as in the dry condition. The higher void specimens show what appears to be a higher T_g or inflection in the strength curve. The low void specimens display an inflection in strength near 2 weight percent moisture, while the higher void content specimens do not show the inflection until above 3 percent moisture at 150°C (Figure 6). It is believed that this effect is an artifact produced by plotting the data as a function of total moisture content. The specimens with voids have free water in the pores that has not diffused into the matrix. Thus, at a given weight percent moisture, the low void specimens have more moisture in the epoxide matrix and a lower T_g.

Freeze-Thaw Cycling

To examine the possible detrimental effect of freeze-thaw cycling on specimens containing voids and moisture, specimens with a high void content (10–15 percent) were exposed in a temperature-humidity chamber. The cycle consisted of 2 hour holds at the temperature extremes of –30 and 50°C with a 2-hour transition time between the temperature extremes. Humidity was controlled at a nominal 50 percent at 10°C. After four months exposure (360 cycles), ultimate flexural strength at 21°C had dropped 23 percent in both the CE-9000 and 5208 matrix laminates. An additional nine months exposure (1170 total cycles) further reduced the ultimate strength of both laminate types to 63 percent of the initial strength at 21°C. Elastic modulus did not appear to be affected by the freeze-thaw cycling.

The results given in Figures 5 and 6 indicate that moisture without thermocycling would have little effect on laminate strength at 21°C. The substantial strength losses seen in the thermally cycled specimens imply that a freeze-thaw mechanism is degrading the laminate. Moisture content of the cycled specimens varied from 4 to 7 weight percent, which is more than would be expected for exposure to the humidity levels in the environmental chamber [7]. Moisture in excess of the equilibrium level for the composite would exist as free water in voids. The expansion of free moisture upon freezing could create internal stresses in the laminates and initiate cracks or delaminations. No additional swelling over the normal amount due to moisture absorption was observed in the thermocycled specimens. It was expected that delaminations would result in increased swelling. Since no additional swelling was observed, it may be hypothesized that damage to the composite from freeze-thaw cycling produces crack initiation sites rather than large delaminations.

Desorbed Strengths

To examine the reversibility of moisture effects in Kevlar/epoxy laminates, specimens with both CE-9000 and 5208 matrices were immersed for 7000 hours in 90°C water and then dried under vacuum prior to testing. Strengths of the dried CE-9000 specimens tested at 21°C did not recover to initial dry values, but rather displayed yield and ultimate values typical of the 21°C saturated specimens plotted in Figure 3. At 150°C, the dried CE-9000 specimens did display strength values typical of the 150°C, dry material shown in Figure 3.

Ultimate strengths of the saturated and dried 5208 matrix material showed only a small amount of recovery. At 21°C, the strength of the dried 5208 matrix specimens averaged 46 ksi (317 MPa), the same value as determined for the dried CE-9000 matrix specimens. An ultimate strength of 32 ksi (220 MPa) was determined for the dried 5208 matrix material at 150°C. That value is considerably lower than the 41 ksi (283 MPa) strength determined for dry as-fabricated laminates (Figure 3) and for the dried CE-9000 matrix specimens at 150°C. Yield strengths of the dried 5208 matrix specimens were equivalent to the dry yield data presented in Figure 3.

The lower strength values displayed by the saturated and then dried laminates indicate that some permanent chemical or physical changes occurred in the composite during the long-term moisture exposure. The resin matrix or filament-matrix interface are likely to be the affected constituents. Fourier transform infrared spectroscopic studies of similar epoxy resins exposed to moisture and elevated temperature have shown that bond rupture and further curing occur as competing processes under those conditions, and that bond cleavage is the dominant mechanism [8]. Moisture has also been shown to promote crazing in tetraglycidyl methylene dianiline cured with diamino diphenyl sulfone [4], which is close to the composition of 5208. Further studies would be required to determine the mechanism responsible for the reduced strengths displayed by the Kevlar/epoxy laminates examined in this work.

Stiffness Behavior

The effect of temperature and moisture on the elastic modulus of [±45,0/90] Kevlar fabric/epoxy laminates is given in Figure 7. Moduli were determined from the slope of the flexural stress-strain curve prior to yield. In the dry condition, the 5208 and CE-9000 matrix laminates are equivalent in stiffness and are averaged in Figure 7. The temperature dependence of stiffness is similar to the strength dependence: a 20 percent loss in stiffness is seen at 150°C and an 11 percent gain is seen at –55°C.

Moisture also magnifies the temperature sensitivity of modulus in quasi-isotropic Kevlar/epoxy laminates (Figure 7). Specimens with each resin matrix appear similar at 21°C and 150°C and near saturation moisture content. At 21°C, a 10 percent drop in laminate stiffness is seen with near satura-

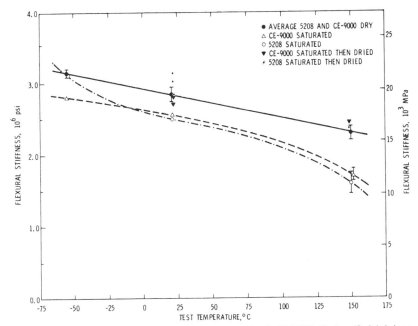

Figure 7. *Elastic modulus versus test temperature for [±45,0/90] Kevlar 49 fabric/epoxy laminates as a function of environmental history.*

tion moisture content. As temperature is increased to 150°C, a 43 percent loss in stiffness is seen compared to the 21°C, dry condition. Modulus is thus not as sensitive as strength to the combined effects of temperature and moisture. Since stiffness is a filament dominated composite property, stiffness should follow trends exhibited by the filaments themselves. If the wet filament tensile modulus data determined in Reference [9] are extrapolated to 150°C, a loss of 34 percent of the 21°C, dry modulus is predicted, which is in the range of the 43 percent loss observed for the laminates in this study. No effect of moisture at 21°C on filament modulus was seen in Reference [9].

At low temperature (-55°C) with saturation moisture content, the stiffness of the CE-9000 matrix specimens is slightly below (2 percent) the 21°C dry value. The 5208 matrix material shows an increase in stiffness to a value near that of the -55°C, dry condition.

The effect of long term moisture exposure followed by drying is also shown in Figure 7. Virtually no change was evident in the stiffness behavior with the CE-9000 matrix when tested at 21°C. A slight increase was seen with the CE-9000 tested at 150°C. The saturated and dried 5208 matrix specimens exhibited an increased stiffness at 21°C and a similar value at 150°C compared to as-fabricated dry laminates (Figure 7). An increase in stiffness may indicate that further curing took place while the material was in the moisture swollen state.

Failure Modes

The macroscopic failure appearance for all test conditions, with the exception of 150 °C with moisture, is shown in Figure 8. Failure initiated as edge delaminations on the tensile side of the specimen and propagated at a 45° angle towards the specimen center where failure occurred by tearing of the surface ply. At 21 °C and –55 °C, there was no change in the visual appearance of the specimen surface on the compression side with both the CE-9000 and 5208 matrices. In addition to the tensile failure appearance shown in Figure 8, slight ridges appeared around the loading ram in the center of the specimen width at 150 °C in the dry condition. The compression ridges were more pronounced with the CE-9000 than with the 5208 matrix. Tests conducted at 150 °C with near saturation moisture content exhibited no apparent tensile damage but were characterized by large compression ridges on either side of the loading ram.

The failure zones of the quasi-isotropic flexure specimens were sectioned at the mid-point of the specimen width and at the edge, and then polished to a one micron diamond finish for microscopic examination of composite failure modes. The locations of the sections prepared for microscopic examination are indicated by the arrows in Figure 8.

In specimens tested at 21 °C dry, failure occurred by delamination on the tensile side below the second ply (a 0/90 orientation) as shown in Figure 9. The macroscopic appearance of the tensile surface indicated that different failure processes were occurring at the edge and center of the specimen width (Figure 8); however, microscopic examination revealed a similar failure appearance for both sections. The delamination failure mode is characteristic of all test conditions below matrix T_g. Extensive matrix cracking on the tensile

Figure 8. Macroscopic tensile failure surface appearance of [±45,0/90] Kevlar 49 181 style fabric/5208 epoxy laminate tested in bending at 21°C.

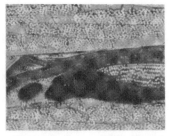

Figure 9. Cross section through tensile surface of [±45,0/90] Kevlar 49 style 181 fabric/CE-9000 flexural specimen showing delamination failure mode, 25X. Test conditions: 21°C, dry.

side between the outer three plies were also evident. No evidence of 0° filament buckling could be found in specimens tested at 21 °C until near saturation moisture levels. No evidence of damage could be seen in sections of the
dry specimens which were unloaded prior to failure (Figure 2).

As moisture was introduced into the 21 °C test specimens, the amount of
delamination observed in the edge sections was higher than in the center section. Delaminations were observed below the first 0/90 ply (second layer) and
second 0/90 ply (fourth layer) in the edge sections and only below the first
0/90 ply in the center sections. Edge delamination may have been promoted
by tensile stresses induced by moisture gradients in the specimens below
saturation moisture content. Matrix cracking on the tensile side of the laminates was prevalent in specimens tested at 21 °C with moisture. As moisture
content approached saturation, previously unseen failures were observed
which included delamination separation around the first 0/90 ply on the compression side and buckling in the 0° filaments of the first and second 0/90
plies. Buckling was limited to the specimen edge sections. No damage on the
compression side of the center sections was seen. Extensive delamination also
occurred on the tensile side above, below, and within the first 0/90 ply. The
onset of filament buckling appears to be responsible for the drop in the 21 °C
flexural strength at high moisture content shown in Figure 5.

Specimens tested at –55 °C displayed no evidence of filament buckling,
even with saturation moisture content. Failures from tests at –55 °C were
characterized by delamination below the first 0/90 ply on the tensile surface
and extensive matrix cracking on both tensile and compression surfaces. The
150 °C dry specimens displayed a slight amount of 0° filament compressive
buckling in the first 0/90 ply in addition to the tensile delaminations
characteristic of the lower temperature tests. A reduced amount of matrix
cracking on the tensile side was seen at 150 °C. As moisture was introduced at
150 °C, a mixed mode delamination and buckling failure occurred (Figure 10).
Buckling of the 0° filaments increased in magnitude with increasing moisture
content until the matrix T_g was depressed to the 150 °C test temperature. At

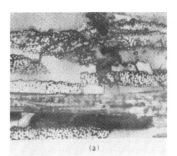 

Figure 10. *Cross sections through [±45,0/90] Kevlar 49 style 181 fabric/CE-9000 flexural
specimen tested at 150°C with 2 w/o moisture content, 25X. (a) tensile zone, and (b) compression zone.*

that point, failure occurred by massive filament buckling as seen in the [0/90] laminates discussed in Part I. Some evidence of delamination on the tensile surface in the center sections was found in the above T_g specimens (>2 percent moisture with CE-9000 matrix, Figure 5) until near saturation moisture concentrations.

It is important to note that for both resin systems the ultimate strengths above T_g for either resin are equivalent for both filament orientations. The above T_g strength with a CE-9000 resin matrix at 150 °C is near 25 ksi (170 MPa). Above T_g strength for the 5208 matrix laminates is near 20 ksi (135 MPa) at 150 °C. That strength is independent of ply orientation for a given matrix above T_g implies that the resin stiffness controls the filament buckling mechanism under those conditions.

The delamination failure seen in the [±45,0/90] laminates is an unexpected result, as is the lack of filament buckling at lower temperatures. Delamination appears to be the result of an edge effect caused by the difference in Poisson's ratios between the ±45 and 0/90 plies as discussed by Pagano and Pipes [10]. That delamination becomes more extensive with moisture present suggests that the interface between the filaments and matrix is being weakened by moisture. Additional experiments are required to quantify the stress-state at delamination. Experiments are also being conducted to explain the cause of nonlinearity in the quasi-isotropic specimen response which takes place with an apparent lack of filament buckling, and to determine the actual mechanism of the buckling itself when it is exhibited as a failure mode.

SUMMARY AND CONCLUSIONS

Results of this investigation show that hygrothermal degradation can be extensive in quasi-isotropic Kevlar 49 fabric/epoxy laminates. At 150 °C with near saturation moisture content, a 55–60 percent loss of 21 °C dry ultimate and yield strengths is observed. The reverse behavior is seen at low temperatures where increased strengths are measured at –55 °C. Moisture also affects strength at 21 °C. Losses of 13–21 percent from dry properties occur with near saturation moisture content. Elastic modulus is not as sensitive as strength to the combined effects of temperature and moisture.

Analysis of failure modes as a function of temperature and moisture revealed an unexpected delamination failure on the tensile side of the specimens. This was characteristic of all test conditions below matrix glass transition temperature (T_g). Delamination generally occurred below the outer ±45 ply or within the second ply (a 0/90 orientation). Very little evidence of 0° filament compressive buckling was found in tests below matrix T_g. This is in contrast to the [0/90] orientation discussed in Part I which exhibited a buckling induced yield below T_g. At 21 °C and saturation moisture content, some buckling was observed in the quasi-isotropic laminates, and was associated with a drop in flexural strength. A slight amount of 0° filament buckling was observed in 150 °C dry specimens. Above T_g, failure is characterized by massive filament buckling. The delamination failure mode appears to be initiated by an edge effect resulting from the quasi-isotropic stacking sequence. Delamination increased with moisture, which suggests that moisture degrades the interfacial region and thereby promotes delamination. Extensive nonlinearity is seen in the load-deflection response of specimens which do not display permanent filament buckling. The cause of the nonlinearity in those specimens is unknown.

Voids reduce elevated temperature strength, but have little effect at room temperature. Freeze-thaw cycling of high void specimens containing moisture reduces composite strength. No additional swelling is observed in thermocycled specimens. Degradation of cycled specimens may be due to the initiation of cracks at voids during the freezing phase of the cycle.

Specimens which were moisture saturated and then dried did not recover initial strength values. Stiffness of the dried specimens did not return to control values. The lower strengths of the dried specimens indicate that permanent chemical changes may be occurring either in the matrix or at the interface during long-term moisture exposure.

These results, in conjunction with the results for [0/90] laminates discussed in Part I of this study, indicate that one must allow for environmental effects when designing with Kevlar/epoxy composites. An understanding of the expected composite moisture content and service temperature is important to the design process.

ACKNOWLEDGMENT

J. E. Perea and K. D. Boultinghouse conducted the experimental aspects of this study. F. P. Gerstle, E. D. Reedy, and A. K. Miller provided technical review of this manuscript and participated in numerous technical discussions during the course of this study.

REFERENCES

1. Allred, R. E. and Hall, N. H., "Volume Fraction Determination of Kevlar 49/Epoxy Composites," *Polym. Engr. and Sci., 19* (13), 907–909 (October 1979).
2. Kaelble, D. H., "Physical and Chemical Properties of Cured Resins," in *Epoxy Resins Chemistry and Technology,* C.A. May and Y. Tanaka, eds. New York: Marcel Dekker, 327–369 (1973).
3. Browning, C. W., "The Mechanisms of Elevated Temperature Property Losses in High Performance Structural Epoxy Resin Matrix Materials after Exposures to High Humidity Environments," *Polym. Engr. and Sci., 18* (1), 16–24 (January 1978).
4. Morgan, R. J. and O'Neal, J. E., "The Durability of Epoxies," *Polym.-Plast. Tech. and Engr., 10* (1), 49–116 (1978).
5. Shirrell, C. D., Leisler, W. H., and Sandon, F. A., "Moisture-Induced Surface Damage in T300/5208 Graphite/Epoxy Laminates," *Nondestructive Evaluation and Flaw Criticality for Composite Materials,* ASTM STP 696, American Society for Testing and Materials, 209–222 (1979).
6. Judd, N. C. W. and Wright, W. W., "Voids and Their Effects on the Mechanical Properties of Composites—An Appraisal," *SAMPE J., 14* (1), 10–14 (January/February 1978).
7. Allred, R. E. and Lindrose, A. M., "The Room Temperature Moisture Kinetics of Kevlar 49 Fabric/Epoxy Laminates," *Composite Materials: Testing and Design,* ASTM STP 674, American Society for Testing and Materials, 313–323 (1979).

8. Levy, R. L., Fanter, D. L., and Summer, C. J., "Spectroscopic Evidence for Mechano-chemical Effects of Moisture in Epoxy Resins," *J. Appl. Poly. Sci., 24,* 1643–1664 (1979).
9. Abbott, N. J., Donovan, J. G., Schoppee, M. M., and Skelton, J., "Some Mechanical Properties of Kevlar and Other Heat Resistant, Nonflammable Fibers, Yarns, and Fabrics," Fabric Research Laboratories Report AFML-TR-74-65, Part III, Dedham, MA (March 1975).
10. Pagano, N. J. and Pipes, R. B., "Some Observations on the Interlaminar Strength of Composite Laminates," *Int. J. Mech. Sci., 15,* 679–688 (1973).

6

Effects of Temperature and Moisture on Sheet Molding Compounds

G. S. SPRINGER

INTRODUCTION

OWING TO THEIR FAVORABLE PERFORMANCE CHARACTERISTICS, LIGHT weight composite materials have been gaining wide applications in commercial, space, and military applications. For this reason, in recent years several investigators have measured the properties of glass fiber reinforced sheet molding compounds (SMC). In this report a summary is given of the effects of temperature and moisture on the engineering properties of SMC materials. In presenting the results emphasis is placed on the main features and characteristics of the data. Readers interested in details of the material behavior are referred to the appropriate references quoted in the text, figure captions, and table headings. A brief summary of static properties at room temperature is given in Table 1. Further information regarding room temperature properties may be found in reference [1].

Sheet molding compounds consist of polyester (or, less frequently, vinylester or epoxy) resins reinforced with glass fibers. The fibers may be randomly oriented (designated as SMC-R) or may be continuous (SMC-C and XMC). A material may also contain a combination of chopped and continuous fibers (SMC-C/R and XMC-3). Numbers added after the letters R and C indicate the weight percent of chopped and continuous fibers, respectively. XMC contains 75% glass fibers by weight. Typical formulations and densities of different types of materials are given in Tables 2 and 3.

STATIC PROPERTIES

Tensile Strength and Modulus. The environment has a marked effect on the tensile strength and modulus. Generally, both the ultimate tensile strength and the tensile modulus decrease at elevated temperatures (Figure 1) and dur-

Table 1. Room temperature tensile strength and modulus (S_t and E_t), Poisson's ratio (v), compressive strength and modulus (S_c and E_c), flexural strength and modulus (S_f and E_f), in plane shear strength and modulus (S_{LT} and E_{LT}), and short beam shear strength and modulus (S_S and G_S). L-longidutinal, T-transverse direction. Strength in MPa and modulus in GPa (refs. 2–7).

Material	S_t	E_t	v	S_c	E_c	S_f	E_f	S_{LT}	E_{LT}	S_S	G_S
XMC-3(L)	561	35.7	0.31	480	37	973	34.1	91.2	4.47	55	—
XMC-3(T)	70	12.3	0.116	160	14.5	139	6.8				
SMC-C20/R30(L)	289	21.4	0.3	306	20.4	645	25.7	85.4	4.09	41	—
SMC-C20/R30(T)	84	12.4	0.18	166	12.2	165	5.9				
SMC-R25	82	13.3	0.25	183	11.7	220	4.8	79	4.48	30	5
SMC-R50	164	15.8	0.31	225	15.9	314	14.0	62	5.94	25	7
SMC-R57	160	16.5	—	—	—	—	—	—	—	—	—
SMC-R65	227	14.7	0.26	241	17.9	403	5.7	128	5.38	45	—
EA SMC-R30	30	8.7	0.30	—	—	—	—	—	—	—	—
VE-SMC-R50	165	7.0	—	—	—	—	—	—	—	50	4
VE-SMC-C40/R10(L)	426	—	—	—	—	—	—	—	—	—	—
VE-SMC-C40/R10(T)	57	—	—	—	—	—	—	—	—	—	—
VE-XMC-3(L)	648	—	—	—	—	—	—	—	—	—	—
VE-XMC-3(T)	74	—	—	—	—	—	—	—	—	—	—

Table 2. Material formulations and densities of SMC materials. (PPG-PPG Industries, OFC-Owens Corning Fiberglas) (refs. 2,3).

Material	Ingredient	Type	Weight %	Density kg/m³
XMC-3	Continuous Glass Fibers- ± 7.5°, X-Pattern	PPG XMC Strand Type 1064	50	
	2.54 cm Chopped Glass Fibers	PPG XMC Strand Type 1064	25	
	Resin	PPG Selectron RS-50335 Isophthalic Polyester	21.5	1970
	Monomer	Styrene	2.4	
	Thickener	PPG Selectron RS-5988	0.8	
	Catalyst	TBPB	0.2	
	Mold Release	Zinc Stearate	0.1	
SMC-C20/R30	Continuous Glass Fibers—Aligned	OCF 433AB Roving	20	
	2.54 cm Chopped Glass Fibers	OCF 433AB Roving	30	
	Resin	OCF-E980 Polyester	32.3	
	Filler	Calcium Carbonate	16.1	1810
	Mold Release	Zinc Stearate	0.8	
	Thickener	Magnesium Oxide	0.5	
	Catalyst	TBP	0.3	
	Inhibitor	Benzoquinone	Trace	

(continued)

60

Table 2. (continued).

Material	Ingredient	Type	Weight %	Density kg/m³
SMC-R25	2.54 cm Chopped Glass Fibers	E-Glass (OCF 951 AB)	25	
	Resin	Polyester (OCF E-920-1)	29.4	
	Filler	Calcium Carbonate	41.8	
	Internal Release	Zinc Stearate	1.1	
	Catalyst	Tertiary Butyl Perbenzoate	0.3	1830
	Thickener	Magnesium Hydroxide	1.5	
	Pigment	Mapico Black	0.8	
SMC-R50	2.54 cm Chopped Glass Fibers	OCF 433AB	50	
	Resin	OCF-E980 Polyester	32.3	
	Filler	Calcium Carbonate	16.1	
	Mold Release	Zinc Stearate	0.8	1870
	Thickener	Magnesium Oxide	0.5	
	Catalyst	TBP	0.3	
	Inhibitor	Benzoquinone	Trace	
SMC-R57	Formulated Epoxy Resin	Epoxy Sheet Molding Compound (Gulf 1057)	43	
	1.27 cm Chopped Glass Fibers	E-Glass (OCF-495)	57	1740
SMC-R65	2.54 cm Chopped Glass Fibers	E-Glass (PPG 518)	65	
	Rigid Resin	Polyester (PPG 50271)	16	
	Flexible Resin	Polyester (PPG 50161)	16	1820
	Thickener, etc.			
EA-SMC-R30	2.54 cm Chopped Glass Fibers	E-Glass (OCF 956)	28	
	Resin	Polyester	19.9	
	Filler	Calcium Carbonate	41	1830
	Thickener	Balance	11.1	

ing exposure to different types of fluids (Tables 4–5). The decrease in properties depends on the temperature, the type of fluid, and the length of exposure. Interestingly, under some conditions there is a slight ($\sim$10%) increase in both the tensile strength and the tensile modulus. The increase is probably due to plasticization of the material.

Compressive Strength and Modulus. Both the compressive strength and modulus depend on the material composition, on the fiber orientation, and on the temperature, as shown in Figure 2. As expected, the strength and the modulus are highest along the fiber direction of composites containing continuous fibers (XMC-3 and SMC-C20/R30).

Shear Strength and Modulus. In most cases there is a significant decrease in the shear strength and in the shear modulus at elevated temperature (Table 6) and during exposure to humid air and to different types of liquids

Table 3. VE-SMC-R50 paste formulation.

Component	Parts
A-SIDE	
XD-9013.03	10
TBPB	1
Camelwite	92
Zinc Stearate	3
B-SIDE	
Derakane* 470-45	100
Maglite D	50
Camelwite	100
Pump 10-14/1 by weight A/B	
Max viscosity = 6000 cps	
(90F, RVT, # 4 Spindle, 20 rpm)	

*Registered Trademark of Dow Chemical Company.

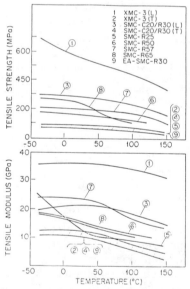

Figure 1. The effect of the amount of chopped fibers on the tensile strength of XMC-3 composites. Total fiber content by weight = 75 percent (ref. 8).

(Tables 7,8). As in the case of tensile properties, shear properties also increase slightly under some conditions. Again, this increase is caused by plasticization on the material.

Flexural Strength and Modulus. The flexural strength and modulus decrease with increasing temperature, as illustrated in Figure 3.

Table 4. Tensile strength retained (percent) after immersion in different fluids for 30 and 180 days. (L-longitudinal, T-transverse direction) (refs. 3,4).

Fluid	XMC-3 30 days	SMC-C20/R30 30 days	SMC-R25 30 days	SMC-R25 180 days	SMC-R50 30 days	SMC-R50 180 days	VE-SMC-R50 30 days	VE-SMC-R50 180 days
Water, 23C	89 (L) 107 (T)	103 (L) 73 (T)	—	—	—	—	—	—
Humid Air, 23C, 50% r.h	—	—	100	90	100	100	100	105
Humid Air, 93C, 50% r.h	—	—	95	95	100	90	105	102
Humid Air, 23C, 100% r.h	—	—	90	80	95	80	105	75
Humid Air, 93C, 100%, r.h	—	—	95	55	95	65	95	55
Salt Water 23C	95 (L) 107 (T)	90 (L) 87 (T)	95	65	105	80	102	80
Salt Water, 93C	—	—	70	45	105	50	85	53
No. 2 Diesel, 23C	—	—	90	90	98	98	103	105
No. 2 Diesel, 93C	—	—	95	90	98	98	101	102
Motor Oil, 23C	95 (L) 110 (T)	100 (L) 108 (T)	95	80	95	95	95	102
Motor Oil 93C	—	—	90	80	90	97	105	102
Antifreeze, 23C	95 (L) 110 (T)	78 (L) 108 (T)	95	80	95	95	95	105
Antifreeze, 93C	—	—	75	30	85	30	105	45
Gasoline, 23C	97 (L) 108 (T)	101 (L) 96 (T)	90	90	100	100	100	95
Gasoline, 93C	—	—	75	70	95	70	105	85
Transmission Fluid, 23C	99 (L) 120 (T)	82 (L) 110 (T)	—	—	—	—	—	—
Break Fluid, 23C	97 (L) 93 (T)	97 (L) 109 (T)	—	—	—	—	—	—

Table 5. Tensile modulus retained (percent) after immersion in different fluids for 30 and 180 days (ref. 4).

Fluid	SMC-R25 30 days	SMC-R25 180 days	SMC-R50 30 days	SMC-R50 180 days	VE-SMC-R50 30 days	VE-SMC-R50 180 days
Humid Air, 23C, 50% r.h	105	110	100	90	98	95
Humid Air, 93C, 50% r.h	120	110	90	80	95	90
Humid Air, 23C, 100% r.h	100	95	90	80	95	90
Humid Air, 93C, 100% r.h	120	110	85	80	90	90

(continued)

63

Table 5. (continued).

Fluid	SMC-R25		SMC-R50		VE-SMC-R50	
	30 days	180 days	30 days	180 days	30 days	180 days
Salt Water, 23C	90	95	90	80	95	90
Salt Water, 93C	110	90	85	65	90	85
No. 2 Diesel, 23C	110	115	90	90	95	95
No. 2 Diesel, 93C	120	95	95	90	90	90
Motor Oil, 23C	95	110	80	90	90	95
Motor Oil, 93C	110	115	90	90	95	95
Antifreeze, 23C	90	110	85	80	90	95
Antifreeze, 93C	85	85	80	50	90	75
Gasoline, 23C	95	90	85	85	95	90
Gasoline, 93C	80	80	88	60	85	75

FATIGUE

The effect of temperature on tension—tension fatigue life is illustrated in Figures 4 and 5. For chopped fiber composites (SMC-R25 and SMC-R65) an increase in temperature from 23C to 93C results in about a two-fold decrease in fatigue strength of the material. The fatigue strengths of materials containing continuous fibers (XMC-3 and SMC-C20/R30) seem

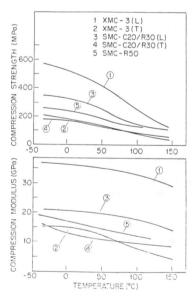

Figure 2. The effect of temperature on the compression strength and compression modulus. (L-longitudinal, T-transverse direction) (ref. 3).

Table 6. Losses in in-plane shear strength (S_{LT}), shear modulus (E_{LT}), and ultimate shear strain (ε_{LT}) when the temperature is raised from 23C to 93C (ref. 3).

| | LOSS (Percent) | | |
	S_{LT}	E_{LT}	ε_{LT}
XMC-3	38	48	13
SMC-C20/R30	40	44	0.7
SMC-R50	—	22	—

Table 7. Short beam shear strength retained (percent) after immersion in different fluids for 30 and 180 days (ref. 4).

| | SMC-R25 | | SMC-R50 | | VE-SMC-R50 | |
Fluid	30 days	180 days	30 days	180 days	30 days	180 days
Humid Air 23C, 50% r.h	105	120	110	110	100	103
Humid Air 93C, 50% r.h	120	110	98	110	105	120
Humid Air 23C, 100% r.h	110	110	90	95	95	95
Humid Air 93C, 100% r.h	102	95	92	85	95	95
Salt Water 23C	110	100	95	95	99	98
Salt Water 93C	95	65	80	35	95	75
No. 2 Diesel 23C	115	120	90	75	95	110
No. 2 Diesel 93C	125	125	103	110	100	105
Motor Oil 23C	100	125	85	95	90	115
Motor Oil 93C	110	120	90	110	95	120
Antifreeze 23C	105	110	90	100	95	105
Antifreeze 93C	75	55	50	15	85	50
Gasoline 23C	95	95	80	95	98	99
Gasoline 93C	70	75	85	85	75	75

Table 8. Short beam shear modulus retained (percent) after immersion in different fluids for 30 and 180 days (ref. 4).

| | SMC-R25 | | SMC-R50 | | VE-SMC-R50 | |
Fluid	30 days	180 days	30 days	180 days	30 days	180 days
Humid Air 23C, 50% r.h	110	125	105	100	105	115
Humid Air 93C, 50% r.h	125	115	95	95	110	120
Humid Air 23C, 100% r.h	100	115	85	85	90	95
Humid Air 93C, 100% r.h	115	120	95	85	105	105
Salt Water 23C	110	100	85	90	100	95
Salt Water 93C	85	80	65	50	90	95
No. 2 Diesel 23C	125	120	95	95	105	115
No. 2 Diesel 93C	110	115	95	95	110	110
Motor Oil 23C	105	125	75	90	90	110
Motor Oil 93C	105	125	85	95	90	115
Antifreeze 23C	105	115	75	90	90	110
Antifreeze 93C	80	75	54	15	85	55
Gasoline 23C	95	90	80	95	105	110
Gasoline 93C	60	85	70	70	80	75

65

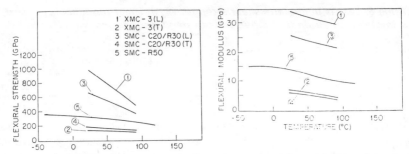

Figure 3. The effect of temperature on the flexural strength and flexural modulus. (L-longitude, T-transverse direction) (ref. 3).

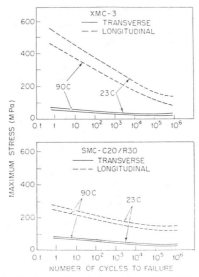

Figure 4. Tension-tension fatigue results. R = 0.05 (ref. 3).

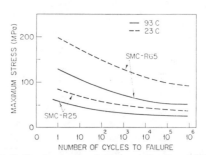

Figure 5. Tension-tension fatigue results. R = 0.05 (ref. 2).

66

to be affected less by changes in temperature than by the orientation of the fibers.

CREEP

The results of static creep tests are presented in Figures 6–12. The curves are average values. There is considerable scatter in the actual data. An arrow at the end of a curve indicates that the specimen did not fail at the end of the test, while a cross indicates specimen failure.

As expected, the strain increases with load, temperature, relative humidity, and time. The increase in strain with time is not uniform. Step "jumps" occur in strain at random times. Because of these unpredictable

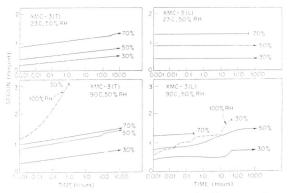

Figure 6. Creep of XMC-3 at 70, 50, and 30 percent of static ultimate tensile strength (L-longitudinal, T-transverse direction) (ref. 3).

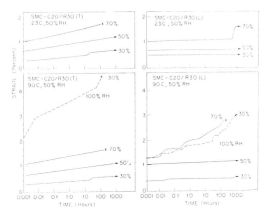

Figure 7. Creep of SMC-C20/R30 at 70, 50, and 30 percent of static ultimate tensile strength. (L-longitudinal, T-transverse direction) (ref. 3).

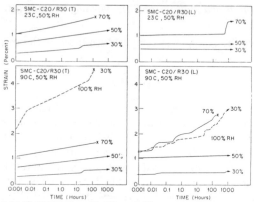

Figure 8. Creep of SMC-C20/R30 at 70, 50, and 30 percent of static ultimate tensile strength. (L-longitudinal, T-transverse direction) (ref. 3).

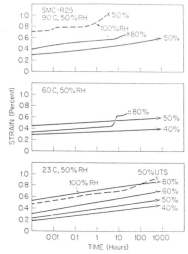

Figure 9. Creep of SMC-R25 under different loads (percent of static ultimate tensile strength) (ref. 2).

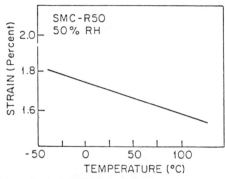

Figure 10. Strain (elongation) of SMC-R50 at failure as a function of temperature (ref. 5).

68

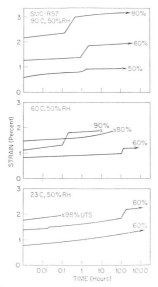

Figure 11. Creep of SMC-R57 under different loads (percent of static ultimate tensile strength) (ref. 2).

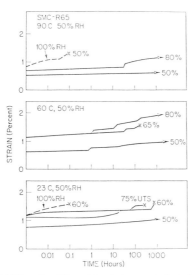

Figure 12. Creep of SMC-R65 under different loads (percent of static ultimate tensile strength) (ref. 2).

jumps, the strain cannot be described by simple viscoelastic models.

Heimbuch and Sanders [2] investigated the stress rupture of SMC-R25, SMC-R57 and SMC-R65 composites in air at 23, 60, 90C and at 50% and 100% relative humidities. Owing to the large scatter in the data, the effect of the environment on stress rupture cannot be ascertained from the results of these tests.

ADHESIVE BONDED SINGLE LAP JOINTS

The results presented in this section were obtained with single lap joints bonded with a two part urethane adhesive, characterized in detail in reference [9].

Moisture Absorption Characteristics. Typical moisture absorption data obtained with XMC-3 to SMC-R50 joints are given in Figure 13. Data for SMC-R50 to SMC-R50 joints exhibit similar trends. At 23C both XMC-3 to SMC-R50 and SMC-R50 to SMC-R50 joints seem to approach asymptotically the same maximum moisture content (Mm) when immersed in the same fluid. During a two month test period Mm is reached only in air. In water and in 5% NaCl-water mixture the maximum moisture contents are not attained. The Mm values can be estimated by extrapolating the data, giving 0.18, 1.5 and 2.0 percent for air, salt water, and water, respectively.

At 93C (immersion in water) the maximum moisture level is not approached asymptotically. Here the weight increases for about the first 100 hours and then decreases at a rapid rate. This indicates that the material deteriorates during exposure. At 93C both bonded and unbonded test specimens behave similarly, suggesting that degradation is mostly in the composite and not in the adhesive.

Joints loaded up to 30 percent of their strength did not show appreciable change in their moisture absorption characteristics.

Lap Shear Strength. Lap shear strengths of adhesive bonded single lap joints are given in Table 9. Changes in baseline strength and modulus during

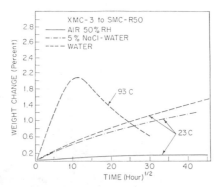

Figure 13. *Moisture absorption of adhesive bonded XMC-3 to SMC-R50 single lap joints.*

Table 9. Baseline ("as received") lap shear strengths of adhesive bonded single lap joints.

	Strength (MPa)	
	23C	93C
XMC to SMC-R50	6.55	3.89
SMC-R50 to SMC-R50	6.11	2.12
SMC-R25 to SMC-R25	3.83	—

environmental exposure are illustrated in Table 10. Neither the strengths nor the moduli change significantly when the joints are exposed to room temperature fluids. In some cases the strength improves slightly (10–15%) during environmental conditioning. The beneficial effects of fluid and temperature are likely due to plasticization. The strength of joints immersed in hot (93C) water and in salt water for 30 days decrease by a factor of two. Loading (up to 30 percent of the baseline strength) during exposure does not seem to affect the strength.

The joints may fail by delamination of the composite or by separation of the adherent. In these tests, most failures occurred by delamination of the adherent. Separation of the adhesive was predominant only at higher (93C) temperatures.

Fatigue. Wang et al [9] conducted fatigue life tests on SMC-R25 to SMC-R25 and SMC-R50 to SMC-R50 single lap joints. During the tests the stress levels were 30, 50, 70 and 90 percent of the static shear strength. Prior to the fatigue tests the specimens were soaked for 30 days at room temperature in the following liquids: 50% by weight salt water, motor oil, transmission fluid, and gasoline. The ranges of data are shown in Figure 14. The data are not shown separately for specimens immersed in the different fluids because the fluids did not have a significant effect on the fatigue life.

Table 10. Changes in lap shear strength (S/S_B) and modulus (E/E_B) of adhesive bonded single lap joints after 30 days of environmental exposure at 23C (B-baseline value) (ref. 9).

Fluid	SMC-R25 to SMC-R25		SMC-R50 to SMC-R50	
	S/S_B	E/E_B	S/S_B	E/E_B
Air	1.00	1.00	1.00	1.00
Motor Oil	0.89	1.05	0.85	0.92
Transmission Fluid	0.92	1.01	0.91	0.91
Gasoline	0.95	0.80	1.20	0.77
Salt Water	0.97	0.72	0.98	0.78
Brake Fluid	0.95	0.87	0.96	0.85
Antifreeze	0.96	1.15	0.81	0.81

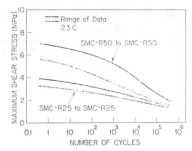

Figure 14. Maximum shear stress of adhesive bonded single lap joints (SMC-R50 to SMC-R50 and SMC-R25 to SMC-R25) during tension-tension fatigue (ref. 11).

The residual strengths and moduli were also measured for specimens surviving for one million cycles [9]. Cyclic stressing at 30 percent of ultimate strength does not degrade appreciably either the strength or the modulus; in general, both the strength and the modulus retained at least 80 percent of their initial value.

Creep. Creep deformations of adhesive bonded single lap joints under static and cyclic loadings are shown in Figures 15 and 16. These figures illustrate the effects of material, fluid, temperature, and applied load on creep behavior. The type of material used in forming the joints has smaller effect on creep than does the type of fluid, the temperature, and the applied load. The creep is lowest in air, and is higher in water, in salt water, and in hydrocarbons. The creep also increases with temperature and with applied load. For example, at 23C none of the XMC-3 to SMC-R50 or SMC-R50 to SMC-R50 joints failed during static creep. In air at 93C only one of the

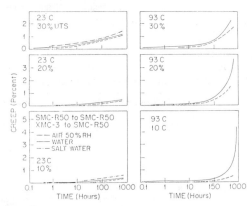

Figure 15. Creep of adhesive bonded single lap joints (SMC-R50 to SMC-R50 and XMC-3 to SMC-R50) immersed in air, water, and 5 percent NaCl-water mixture under different loads (percent of static ultimate tensile strength).

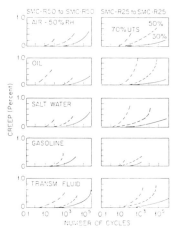

Figure 16. *Creep of adhesive bonded single lap joints (SMC-R50 to SMC-R50 and SMC-R25 to SMC-R25) during tension-tension fatigue under different loads (-30 percent UTS, -50 percent UTS, -70 percent UTS) (ref. 9).*

joints failed, this occurring at 30 percent load level. During water immersion (at 93C) all but 3 coupons failed before the end of the 715 hours test. During cyclic creep, only joints with 30 percent load survived for one million cycles. At higher loads none of the joints survived for one million cycles.

VIBRATION DAMPING

The vibration damping properties may be characterized by two parameters (the loss factor and the storage modulus) obtained by exciting the material with forced sinusoidal oscillations and by measuring the input stress and output strain [3]. The loss factor is the tangent of the phase angle between the stress and the strain, and is equal to the ratio between the energy dissipated and the energy stored in the material. The storage modulus is the in-phase component of the ratio of input stress to output strain.

The effects of temperature and soaking in different types of liquids are illustrated in Tables 11 and 12. An increase in temperature (from 23C to 120C) increases the damping and reduces the stiffness. Soaking in liquids has similar effects. Soaking for 1000 hours considerably increased the damping of chopped fiber composites (SMC-R25 and SMC-R60), while their stiffness decreased slightly. The damping characteristics of continuous fiber composites (XMC-3 and SMC-C20/R30) change little in the fiber direction. It is noteworthy that both temperature and moisture-induced changes in the vibration properties appear to be reversible [11].

MOISTURE ABSORPTION

Glass fiber reinforced organic matrix composites absorb moisture when exposed to humid air or to liquids. The weight changes of different types of

Table 11. Loss factor and storage modulus at 23C, and maximum changes in these parameters when the temperature is increased from 23C to 120C (L-longitudinal, T-transverse direction) (refs. 3,10).

Material	Loss Factor at 23C 0.1 Hz 10 Hz	Storage Modulus at 23C (GPa)	Max. Change Loss Factor	Percent Storage Modulus
XMC-3(L)	0.028 0.025	36	+ 129	− 7
XMC-3(T)	0.063 0.053	—	+ 355	− 54
SMC-C20/R30(L)	0.034 0.029	—	+ 80	− 5
SMC-C20/R30(T)	0.051 0.049	—	+ 204	− 44
SMC-R25	0.037 0.035	4	+ 471	− 48
SMC-R65	0.039 0.034	8	+241	− 30
Steel	∿0.001			

Table 12. Maximum changes in dynamic properties during 1000 hours of soak (L-longitudinal, T-transverse direction) (ref. 11).

Fluid	Material	Maximum Change (percent) Loss Factor	Storage Modulus
Distilled Water 22C	XMC-3(L)	+ 53	0
	SMC-C20/R30(L)	+ 38	0
	SMC-R25	+ 193	− 20
	SMC-R65	+ 180	− 7
Distilled Water 50C	XMC-3(L)	+ 88	0
	SMC-C20/R30(L)	+ 112	0
	SMC-R25	+ 210	− 10
	SMC-R65	+ 247	− 12
Salt Water 23C	XMC-3(L)	+ 55	0
	SMC-C20/R30(L)	+ 50	0
	SMC-R25	+ 117	− 4
	SMC-R65	+ 178	− 7
Motor Oil 23C	XMC-3(L)	0	0
	SMC-C20/R30(L)	0	0
	SMC-R25	+ 25	0
	SMC-R65	+ 24	0
Antifreeze	XMC-3(L)	0	0
	SMC-C20/R30(L)	0	0
	SMC-R25	+ 29	0
	SMC-R65	+ 22	0
Gasoline 22C	XMC-3(L)	0	0
	SMC-C20/R30(L)	0	0
	SMC-R25	178	− 15
	SMC-R65	30	0

SMC composites exposed to different types of fluids are presented in Figures 17-19. The weight change (M) is defined as

$$M = \frac{\text{wet weight-dry weight}}{\text{dry weight}} \times 100 \text{ percent}$$

The data show that, in general, when the dry material is submerged in the fluid the weight at first increases then levels off for some length of time. Both the initial rate of weight increase and the value at which the weights level off depend on a) the material, b) the temperature, and c) the environment (relative humidity of air or the type of liquid used). The data also show that in some instances the weight does not remain constant after it reaches a level value but keeps either increasing or decreasing. This suggests that under some conditions the moisture transport is by a non-Fickian process. One reason for the non-Fickian behavior may be that moisture transfer through the resin does not proceed by a process that can be described by Fick's law. Another plausible explanation of the observed non-Fickian absorption process is as follows. Owing to the moist, high temperature environment, microcracks develop on the surface and inside the material. Moisture rapidly enters the material, causing an increase in weight. As the cracks grow larger, material, most likely in the form of resin particles, is actually lost. In fact, such material loss is frequently observed after a few hours of exposure to the moist environment. As long as the moisture gain is greater than the material loss, the weight of the specimen increases. Once the weight of the lost material exceeds the weight of the absorbed moisture, the weight of the specimen decreases. Of course, when the material is lost, the measured weight change no longer corresponds to the moisture content of the material.

The foregoing results were all obtained with unstressed specimens. However, the moisture absorption characteristics of stressed and unstressed SMC materials do not differ appreciably.

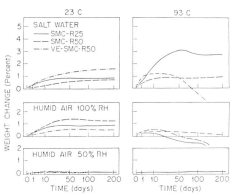

Figure 17. *Weight change during immersion in humid air and in saturated salt water (ref. 4).*

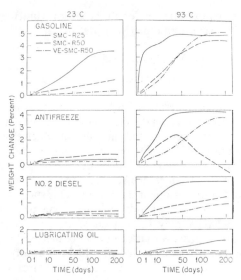

Figure 18. *Weight change during immersion in different types of hydrocarbons (refs. 4, 12).*

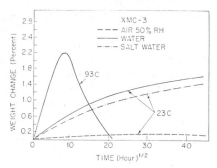

Figure 19. *Weight change of XMC-3 immersed in humid air, water, and in 5 percent NaCl-water mixture (ref. 4).*

THERMAL EXPANSION

The dimensional changes of the material during temperature cycles must be taken into account in the design process. The thermal expansion coefficient values reported by Heimbuch and Sanders [2] and by Riegner and Sanders [3] are reproduced in Table 13. As expected, the thermal expansion coefficients are lowest along the fiber direction of composites containing continuous fibers (XMC-3 and SMC-C20/R30). The coefficient is high in the transverse direction of these materials and also for SMC-R25 composites. More comprehensive data on the variation of the thermal expansion coefficient with temperature are not yet available.

Table 13. Thermal expansion coefficient α at room temperature (L-longitudinal, T-transverse direction) (refs. 2,3).

Material	α(μm/m °C)
XMC-3(L)	8.7
XMC-3(T)	28.6
SMC-C20/R30(L)	11.3
SMC-C20/R30(T)	24.6
SMC-R25	23.2
SMC-R50	14.8
SMC-R65	13.7

ACKNOWLEDGMENTS

This work was supported by the Materials Laboratory, U.S. Air Force Systems Command, Wright Patterson Air Force Base, Dayton, Ohio.

REFERENCES

1. Springer, G. S., "Properties of Organic Matrix Short Fiber Composites", Materials Laboratory, Air Force Systems Command, Report AFWAL-TR-82-4004, Wright Patterson Air Force Base, Dayton, Ohio (1982).

2. Heimbuch, R. A. and Sanders, B. A., "Mechanical Properties of Chopped Fiber Reinforced Plastics", in *Composite Materials in the Automotive Industry*, American Society of Mechanical Engineers, pp. 111–139. (1978).

3. Riegner, D. A. and Sanders, B. A., "A Characterization Study of Automotive Continuous and Random Glass Fiber Composites", Report GMMD 79-023, General Motors Corporation, Manufacturing Development, GM Technical Center, Warren, Michigan, 48090 (1979).

4. Springer, G. S., Sanders, J. A., and Tung, R. W. "Environmental Effects on Glass Fiber Reinforced Polyester and Vinylester Composites", *J. Composite Materials*, *14*, pp. 213–232. (1980).

5. Denton, D. L., "Mechanical Properties Characterization of an SMC-R50 Composite", 34th Annual Technical Conference, Reinforced Plastics/Composites Institute, The Society of the Plastics Industry, 1979, Section 11-F; also SAE Paper 790671 (1979).

6. Enos, J. H., Erratt, R. L., Grancis, E. and Thomas, R. E., "Structural Performance of Vinylester Resin Compression Molded High Strength Composites", 34th Annual Technical Conference, Reinforced Plastics/Composites Institute, The Society of Plastics Industry, 1979, Section 11-E.

7. Adams, D. F. and Walrath, D. E., "Iosipescu Shear Properties of SMC Composite Materials", Department of Mechanical Engineering, The University of Wyoming, Larrabee, Wyoming, 82738 (1981)

8. Ackley, R. H. and Carley, E. P., "XMC-3 Composite Material Structural Molding Compound", 34th Annual Technical Conference, Reinforced Plastics/Composites Institute, The Society of Plastics Industry, 1979, Section 21-D.

9. Wang, T. K., Sanders, B. A. and Lindholm, U.S., "A Loading Rate and Environmental

Effects Study of Adhesive Bonded SMC Joints'', Report GMMD80-044, General Motors Corporation, Manufacturing Development, GM Technical Center, Warren, Michigan, 48090 (1980).

10. Seiffert, V. W., "Review of Recent Activities and Trends in the Field of Automobile Materials", in *Worldwide Applications of Plastics*, Society of Automotive Engineers, SP-482 pp. 1-6 (1981).

11. Gibson, R. F., Yau, A. and Riegner, D. A. "The Influence of Environmental Conditions on the Vibration Characteristics of Chopped-Fiber-Reinforced Composite Materials", Presented at AIAA/ASME/ASCE/AHS 22nd Structure, Structured Dynamics and Materials Conference (1981).

12. Loos, A. C., Springer, G. S., Sanders, B. A. and Tung, R. W., "Moisture Absorption of Polyester-E Glass Composites", *J. Composite Materials*, *14*, pp. 142–154 (1980).

7

The Influence of Environmental Conditions on the Vibration Characteristics of Chopped-Fiber-Reinforced Composite Materials

R. F. GIBSON, A. YAU, E.W. MENDE, AND W. E. OSBORN

INTRODUCTION

IT IS NOW RECOGNIZED THAT THE CHARACTERIZATION OF MODERN FIBER-reinforced plastics must include testing to determine the effects of environmental conditions on mechanical properties. Design engineers need to have such information available if these materials are to be put to optimum use. The effects of elevated temperatures, moisture absorption, and chemical attack are the primary concerns at present.

Environmental testing to date has been centered on static loading of continuous fiber composites for aerospace applications. For example, static flexure tests of graphite/epoxy beams showed that moisture and temperature can induce a change in the failure mode [1]. Finite element analysis has been used to show that moisture abosorption has a significant effect on residual stresses in a composite material [2]. In [3], at least two mechanisms for moisture degradation in epoxy matrix composites were identified: a lowering of the glass transition temperature and a change in residual stress distribution due to matrix swelling. The implication for dynamic behavior is that the dynamic mechanical properties (particularly internal damping) should be sensitive to these effects.

The effects of temperature and moisture on the dynamic stiffness and internal damping of graphite/epoxy composites were reported in [4], and a continuation of this work was reported in [5]. It was concluded in [4] that damping increased for some laminate configurations, but decreased for others in a hot, moist environment. In [5], it was concluded that damping changes in the composites due to hygrothermal effects were about the same as those produced in aluminum alloy calibration specimens.

The effects of environmental conditions on static mechanical properties [6]

79

and weight gain [7] of E-glass/polyester automotive chopped fiber composites (or sheet molding compounds (SMC)) have been investigated, but up to now no work has been done on effects of environmental conditions on dynamic properties. In the preceding phase of the research reported here, the dynamic stiffness and damping of a number of E-glass/polyester chopped fiber automotive composites were measured under the room conditions [8,9]. In a related paper, measured complex moduli of SMC materials were compared with theoretical bounds [10]. The present paper describes the second phase of this work—the measurement of the effects of elevated temperature, moisture absorption, and chemical attack on the dynamic properties of these same chopped fiber composites. The materials are described in Table 1.

EXPERIMENTAL TECHNIQUE

The forced flexural vibration technique used in [8], and described in [11], required the attachment of electrical resistance strain gages to each specimen. For the present environmental tests, non-contacting proximity transducers (Figure 1) were used instead of strain gages to detect the vibrational response of the specimens, so that the difficulties associated with attachment of strain gages in hot, moist environments were eliminated. Because the probes only sense motion of metallic surfaces, adhesive-backed aluminum foil targets were attached to the specimens (Figure 1). The same resonant dwell method was used, but the damping characteristics were found from measurement of the specimen amplitude ratio at resonance. It is shown in [12] and [13] that the loss factor, η, is related to the resonant amplitude ratio of the cantilever specimen by the equation

$$\eta = C_r \frac{a_b}{a_t}$$

where

Table 1. Description of materials tested.

Material	Weight Percentages of Constituents		
	Chopped E-glass Fibers	Continuous E-glass Fibers	Polyester Resin, Fillers, etc.
PPG SMC-R25[1]	25	0	75
PPG SMC-R65	65	0	35
PPG XMC-3	25	50 (± 7.5°, x-pattern)	25
OCF SMC-R25[2]	25	0	75
OCF C20/R30	30	20 (Aligned)	50

[1]Manufactured by PPG Industries, Fiber Glass Division, Pittsburgh, PA 15222.
[2]Manufactured by Owens-Corning Fiberglas Corporation, Toledo, OH 43659.

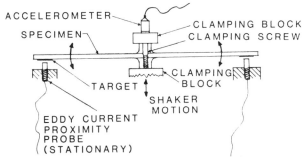

Figure 1. *Specimen-transducer configuration.*

C_r = dimensionless constant depending on resonant mode number
a_b = resonant vibration amplitude at base of specimen
a_t = resonant vibration amplitude at tip of specimen

The storage modulus, E, (dynamic stiffness) was found from the measured resonant frequencey, as in [8]. Using complex modulus notation, the loss factor is the ratio of the loss modulus to the storage modulus.

The data presented in [8] and [10] was obtained by testing specimens under vacuum conditions, so that parasitic damping due to aerodynamic drag did not affect the damping measurements. However, it was also shown in [8] that, for sufficiently small vibration amplitudes (when the ratio of tip amplitude to specimen thickness is less than about 0.2), air damping is negligible for these particular materials. Since the environmental testing reported here could not be carried out in a vacuum, the vibration amplitudes were kept below the threshold for significant air damping. As shown in [8] and [10], dynamic stiffness and damping of these materials are practically independent of vibration frequency. Thus, the environmental testing was done at only one frequency corresponding to the first mode of transverse vibration of each specimen. A detailed description of the new technique, as well as a comparison of the errors generated by using both new and old techniques, is given in [12] and [13].

The environmental tests were divided into three general categories: (1) elevated temperatures, (2) moisture absorption and chemical attack, and (3) moisture absorption at elevated temperatures. The complex moduli of chopped fiber reinforced PPG SMC - R25, PPG SMC-R65, and OCF SMC-R25, along with hybrid chopped/continuous fiber reinforced PPG XMC-3, and OCF C20/R30 (see Table 1) were measured under all of these conditions. All specimens were oven dried, weighed, and tested at room temperature before further testing as described later on.

ELEVATED TEMPERATURE TESTS

The purpose of the elevated temperature tests is to show how the complex moduli change with temperature. Thus, the complex modulus of a given material under a certain uniform specimen temperature is desired. In order to check the actual temperature distribution in the specimen, six holes were drilled in the side of a specimen along the specimen length and six iron-constantan thermocouple sensors were attached in the holes to record the steady-state temperature along the specimens simultaneously with a data acquisition system. Specimen dimensions are given in Table 2 and representative temperature profiles for one material are shown in Figure 2. The recorded data

Table 2. Description of specimens used in elevated temperature tests. *

Material	Average Dimensions, mm (in.)			Density, 10⁻⁷ (kg/m³)(lb/in³)	Frequency (Hz)***
	Width	Thickness	Length**		
PPG SMC-R65	18.9(0.743)	3.25(0.128)	196.8(7.75)	4.98(0.067)	44.5
PPG SMC-R25	18.8(0.740)	3.28(0.129)	196.8(7.75)	5.70(0.070)	39.6
OCF SMC-R25	18.8(0.742)	3.30(0.130)	196.8(7.75)	4.76(0.064)	36.7
OCF C20/R30(L)	18.5(0.73)	3.61(0.142)	196.8(7.75)	4.98(0.067)	59.7
OCF C20/R30(T)	19.9(0.743)	3.86(0.152)	196.8(7.75)	4.76(0.064)	32.3
PPG XMC-3(L)	18.9(0.743)	2.62(0.103)	196.8(7.75)	5.57(0.075)	49.1
PPG XMC-3(T)	18.8(0.742)	2.89(0.114)	196.8(7.75)	5.35(0.072)	23.2

*Measurements made after drying of specimens, before elevated temperature tests.
**Length of cantilever.
***First mode.

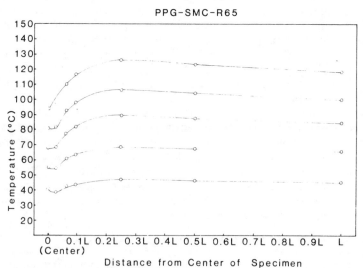

Figure 2. Temperature profiles of PPG-SMC-R65 specimen.

shows that the temperature distribution along the specimen is not uniform, even after equilibrium has been established. The temperature is lowest at the center of the specimen and increases along the length to a maximum value and then decreases gradually to the end. The temperature gradient is greatest at the center. The reason for this distribution is that the specimen is clamped at the center with aluminum clamping blocks, which are in turn mounted on the shaker armature. Thus, the clamp acts as a heat sink and heat is transferred away from the specimen more rapidly at the center than at the ends.

Although plastic insulation strips were inserted between the clamping blocks to decrease heat transfer, the temperature gradient was still relatively large at the center. The range of the temperature difference is greater for a higher center (controller) temperature (T_c). The maximum and minimum temperatures of the PPG SMC-R65 specimen versus controller temperatures are shown in Figure 3. Nearly identical results were found for the other materials tested. The temperature is also slightly different at the two sides of the double cantilever beam because the temperatures of the heaters are not uniform. However, this difference is very small and negligible compared with the temperature gradient due to the heat transferred by the shaker armature.

All of the elevated temperature tests were done under such non-uniform temperature distributions, and the results presented later must be qualified by the use of Figure 3, which shows the temperature range for a given center temperature.

The linear range of the proximity probe is from 0.5 mm (0.02 in.) to 3.1 mm (0.122 in.). The gap between the probe and specimen should be within

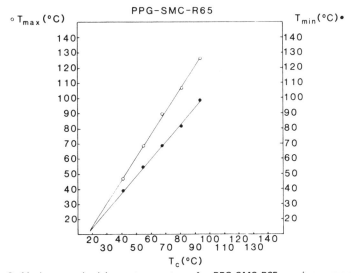

Figure 3. *Maximum and minimum temperatures for PPG-SMC-R65 specimen versus center temperature.*

this range for accurate measurement, but it usually changed with temperature. When temperature increased, sometimes the gap decreased and sometimes it increased. In addition, the gaps of the two sides usually changed differently. This effect depends on the thermal properties of the materials and residual stresses caused by molding and cutting specimens. During the elevated temperature tests, it was necessary to measure the gap (by monitoring the D.C. probe voltage) and to readjust it when it was out of the linear range. This occurred several times during the experiments.

Specimens for the elevated temperature tests (Table 2) were initially dried at 65 °C (150 °) in a forced-air oven until no further weight loss was observed. This usually required approximately two weeks. After oven-drying, specimens were stored in a vacuum desiccator chamber at room temperature until the elevated temperature tests began. Aluminum foil targets were attached to the specimens just before testing began.

During the elevated temperature tests, the temperature set-point was increased from room temperature to 120 °C in increments of 20 °C. After the test was done at room temperature, the set temperature was changed to 40 °C and the test was done at least one hour later, after the specimen reached equilibrium. Then the set temperature was reset to 60 °C and the above procedure was repeated till the last test was done at 120 °C set temperature. Note that the set temperature is higher than the actual center temperature at the higher temperatures. As temperature increased, the loss factor increased and the storage modulus decreased for all materials. Representative data are shown in Figures 4–6. Similar results for all materials can be found in [14]. Note that all properties are shown as functions of the center temperature in Figures 4–6. The temperature range for a given center temperature may be found from Figure 3. Maximum changes in damping (loss factor) and stiff-

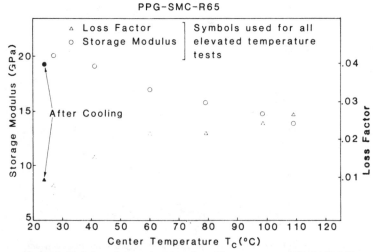

Figure 4. Variation of dynamic properties with temperature: PPG-SMC-R65.

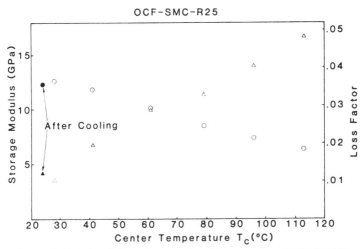

Figure 5. *Variation of dynamic properties with temperature: OCF-SMC-R25.*

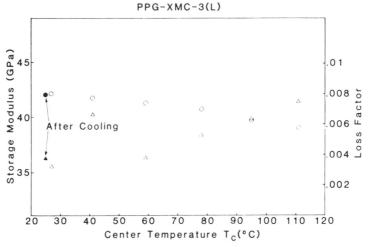

Figure 6. *Variation of dynamic properties with temperature: PPG-XMC-3(L).*

ness (storage modulus) during the tests are tabulated in Table 3. It is apparent that damping is much more sensitive to temperature change than stiffness is, and that the chopped fiber reinforced materials are more sensitive than the hybrid chopped/continuous fiber reinforced composites. Note that the longitudinal (L) specimens of the hybrid composites are much less sensitive than the transverse (T) specimens. These changes are consistent with expected

Table 3. Maximum changes in dynamic properties during elevated temperature tests (2 0° C-12 0° C).

	Maximum Change, Percent	
Material	Loss Factor	Storage Modulus
PPG SMC-R65	+241	−30
PPG SMC-R25	+471	−48
OCF SMC-R25	+375	−49
OCF C20/R30(L)	+ 80	− 5
OCF C20/R30(T)	+204	−44
PPG XMC-3(L)	+124	− 7
PPG XMC-3(T)	+355	−54

behavior of viscoelastic materials below their glass transition temperatures (T_g). That is, the test temperature range is much closer to the T_g of the polyester matrix resin than the T_g of the glass fibers, and therefore the matrix-controlled materials should show greater sensitivity to temperature changes than the fiber-controlled materials. The chopped fiber composite specimens and the transverse (T) hybrid composite specimens are considered to be matrix-controlled, whereas the longitudinal (L) hybrid composite specimens are considered to be fiber-controlled.

In order to check for possible irreversible effects of elevated temperatures, all specimens were allowed to cool to room temperature after the 120 °C test, then tested again (the points labelled "After cooling" in Figures 4–6). These results indicate little or no permanent change in stiffness or damping after exposure to elevated temperatures. The small changes observed may be due to relaxation of residual stresses in the materials.

SOAKING TESTS

The effects of moisture absorption and chemical attack on dynamic properties were found by soaking specimens in several common automotive fluids and removing them from the soaking baths periodically for vibration testing as described previously. In this manner, any changes in dynamic properties could be correlated with exposure time and amount of fluid absorbed. Since absorption of the various fluids occurs primarily by diffusion, the soaking test is an accelerated test in comparison with some actual automotive exposure levels (i.e., humid air, gasoline fumes, etc.). In order to obtain data up to the saturation point, the test duration was set at approximately 1000 hours (41.7 days). Previous moisture absorption data on these same materials indicated that, in most cases, saturation occurred before 1000 hours of soaking time [7].

Test Procedure

Due to the large number of tests involved (i.e., six different soaking baths and five different materials), only one specimen of each material was tested. Only the longitudinal (*L*) configurations of the hybrid composites were tested, since this would be the orientation for most applications. Specimen dimensions were similar to those given in Table 2. The fluids used were a) distilled water at 21°–24°C, b) distilled water at 49°–51°C, c) 5 percent by weight salt (NaCl) in distilled water at 21°–24°C, d) Texaco 10w-30 motor oil at 21°–24°C, e) Amoco unleaded gasoline at 21°–24°C, and f) Dow-Gard anti-freeze solution at 21°–24°C. The test procedure for each specimen was as follows:

1. oven-dry at 65° (150 F) until no further weight loss was observed (approximately two weeks).
2. store in vacuum desiccator chamber at room temperature until soaking tests began.
3. remove from vacuum chamber, attach aluminum foil targets, weight and vibration test at room temperature.
4. place in appropriate soaking bath (completely immersed).
5. remove from soaking bath, weigh, vibration test, and replace in bath at approximately 1, 2, 3, 6, 11, 20, 30, 35, and 41 days soaking time.
6. after approximately 1000 hours in bath, re-dry at 65°C (150°F) in oven for approximately two weeks, store in desiccator until temperature reaches room temperature, then weigh and vibration test (this step carried out for specimens soaked in distilled water and salt water only).

Soaking was done in fiberglass pans at room temperature (21°–24°C). The one exception was the elevated temperature distilled water bath, which was maintained at 49°–51°C by a constant temperature circulator. For safety reasons, the gasoline bath was placed in a fume hood with forced air exhaust.

Results

The variations of loss factor, storage modulus, and percent weight gain with soaking time for all materials in distilled water are shown in Figures 7, 8, and 9 respectively. Similar results for the remaining soaking baths may be found in [14]. Maximum changes in damping and stiffness during all of the tests are tabulated in Table 4. Only the statistically significant changes are shown in Table 4. As shown in [12] and [13], the "worst case" ratios of standard deviations to mean values for loss factor and storage modulus measurements were

$$\frac{S_\eta}{\overline{\eta}} = 0.0685 \text{ and } \frac{S_E}{\overline{E}} = 0.0071$$

Assuming that the data follows normal (Gaussian) distributions, there is a 99.7 percent probability that any measured percent changes in properties

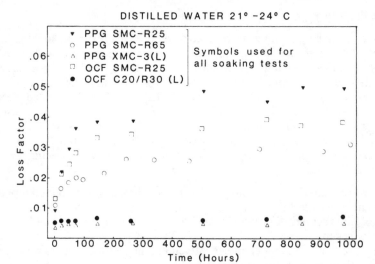

Figure 7. Variation of loss factor with soaking time: all materials in distilled water 21°-24°C.

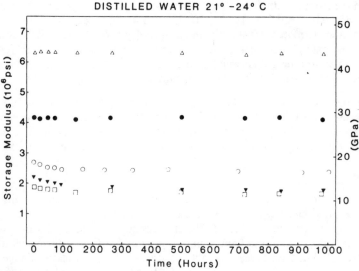

Figure 8. Variation of storage modulus with soaking time: all materials in distilled water 21°-24°C.

within the so-called "three-sigma" limits are due to scatter in the data. The three-sigma limits are therefore

$$\pm 3\left(\frac{S_\eta}{\bar{\eta}}\right) = \pm 0.20 \text{ and } \pm 3\left(\frac{S_E}{\bar{E}}\right) = \pm 0.021$$

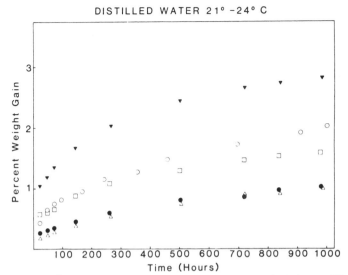

Figure 9. *Percent weight gain versus soaking time: all materials in distilled water 21°-24°C.*

Therefore, any changes in loss factor less than ± 20 percent and any changes in storage modulus less than ± 2.1 percent were considered insignificant and designated as "nil" in Table 4.

It is apparent that the damping in the chopped fiber composites (i.e., SMC-R25 and SMC-R65) increases markedly (+ 527 percent maximum) as a result of exposure to moisture, but that the corresponding effect is much less pronounced (+ 112 percent maximum) for the hybrid composites. A more detailed discussion of the correlations between damping and moisture content is presented in the next section. As expected, elevated temperature water (49°–51°) causes more rapid diffusion of moisture, and, consequently, more rapid increases in damping than the room temperature water. In addition, the maximum loss factors are higher at the elevated temperature, as expected.

As shown in Table 4, none of the soaking fluids had a significant effect on the stiffness of the hybrid composites. In general, small reductions in stiffness of the chopped fiber composites were associated with large increases in damping, and the materials showing the greatest increases in damping show the greatest reduction in stiffness. Thus, damping is much more sensitive to moisture absorption than stiffness is.

The effects of motor oil and anti-freeze on damping of chopped fiber composites were just barely significant, and the corresponding effect on hybrid composites was insignificant. Exposure to motor oil and anti-freeze had no effect on stiffness of any material. Gasoline had a significant effect on two of the chopped fiber composites, but no effect on the hybrids. As shown later, gasoline also shows an excellent correlation between loss factor and percent weight gain.

Table 4. Maximum changes in dynamic properties during 1000 hour soaking tests.

Bath	Material	Maximum Change, Percent	
		Loss Factor	Storage Modulus
Distilled	PPG SMC-R65	+180	− 7
Water	PPG SMC-R25	+448	−20
21°–24°C	OCF SMC-R25	+193	−11
	PPG XMC-3(L)	+ 53	nil*
	OCF C20/R30(L)	+ 38	nil
Distilled	PPG SMC-R65	+247	−12
Water	PPG SMC-R25	+527	−20
49°–51°C	OCF SMC-R25	+210	−10
	PPG XMC-3(L)	+ 88	nil
	OCF C20/R30(L)	+112	nil
Salt	PPG SMC-R65	+178	− 7
Water	PPG SMC-R25	+400	−16
21°–24°C	OCF SMC-R25	+117	− 4
	PPG XMC-3(L)	+ 50	nil
	OCF C20/R30(L)	+ 55	nil
Motor	PPG SMC-R65	+ 24	nil
Oil	PPG SMC-R25	+ 29	nil
21°–24°C	OCF SMC-R25	+ 25	nil
	PPG XMC-3(L)	nil*	nil
	OCF C20/R30(L)	nil	nil
Gasoline	PPG SMC-R65	+ 30	nil
21°–24°C	PPG SMC-R25	+147	− 3
	OCF SMC-R25	+178	−15
	PPG XMC-3(L)	nil	nil
	OCF C20/R30(L)	nil	nil
Anti-	PPG SMC-R65	+ 22	nil
Freeze	PPG SMC-R25	+ 24	nil
21°–24°C	OCF SMC-R25	+ 29	nil
	PPG XMC-3(L)	nil	nil
	OCF C20/R30(L)	nil	nil

*When the maximum change in the properties is not significantly greater than the scatter in the data, the change is designated as "nil."

The overall ranking of the soaking fluids, in decreasing order of effects on dynamic behavior of the composites is as follows:

Greatest effect—Distilled water   (49°–51°C)
           Distilled water   (21°–24°C)
           Salt water   (21°–24°C)
           Gasoline   (21°–24°C)
Least effect—Motor oil, anti-freeze   (21°–24°C)

It is important to note that, in all cases where stiffness was reduced, damping was magnified. A simultaneous loss of stiffness and damping is obviously undesirable, and this did not happen during the tests. Another important result concerns the reversibility of these effects. In order to check reversibility, the specimens that were soaked in fluids having greatest effects (distilled water and salt water) were re-dried after approximately 1000 hours soaking time and tested again (step 6 in procedure). It was found that the moisture-induced changes in damping were reversible in all cases, and this implies that no permanent structural changes occurred. In all but two cases there were no permanent changes in stiffness. Slight increases in stiffness for OCF-SMC-R25 in distilled water and salt water without corresponding reductions in damping appear to be anomalous, since all previous data indicates that changes in damping and stiffness have opposite signs, and that damping is much more sensitive than stiffness is. The fact that no permanent changes in damping occurred in any case is considered to be highly significant, in view of the high sensitivity of damping.

Regression Analysis of Loss Factor Versus Percent Weight Gain

In the previous section, it was concluded that the materials showing the greatest moisture absorption showed the greatest changes in damping. This is confirmed by plotting loss factor versus percent weight gain, as in Figure 10. Figure 10 shows that there is a correlation between loss factor and amount of moisture absorbed for the chopped fiber SMC materials, but not for the hybrids. Based on this observation, a linear regression analysis of each set of

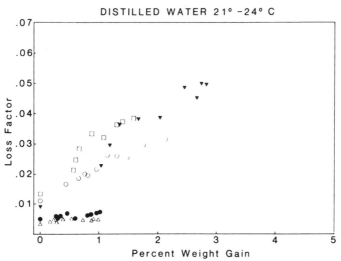

Figure 10. *Loss factor as a function of percent weight gain: all materials in distilled water 21°-24°C.*

loss factor-weight gain data for the SMC materials was carried out, and the results are given in Table 5. The most significant finding here is the magnitude of the correlation coefficients (a correlation coefficient of zero means no correlation, 1.0 means perfect correlation). For example, all SMC materials show excellent correlation for distilled water and salt water, and two of the materials show excellent correlation for gasoline. All materials show poor correlation for motor oil and anti-freeze. Thus, there is a definite correlation between increased moisture content and corresponding magnification of damping in these materials.

The actual mechanism for this moisture-induced damping effect is unknown, but several hypotheses are suggested by findings in previous publications. In the introduction, two mechanisms for moisture degradation were discussed [2]. A lowering of the glass transition temperature, T_g, by increased moisture content should produce changes which are consistent with present findings. That is, as the temperature of a viscoelastic material is increased through the glass transition range, damping peaks and stiffness drops off drastically. When the operating temperature is below T_g, a moisture-

Table 5. Linear regression parameters for loss factor versus percent weight gain.*

$$\eta = a + bw$$

Bath	Material	Intercept, a	Slope, b	Correlation Coefficient
Distilled Water 21°-24°C	PPG SMC-R65	0.0126	0.0093	0.98
	PPG SMC-R25	0.0117	0.0140	0.97
	OCF SMC-R25	0.0150	0.0159	0.96
Distilled Water 49°-51°C	PPG SMC-R65	0.0163	0.0051	0.85
	PPG SMC-R25	0.0223	0.0097	0.84
	OCF SMC-R25	0.0249	0.0062	0.78
Salt Water 21°-24°C	PPG SMC-R65	0.0097	0.0098	0.98
	PPG SMC-R25	0.0149	0.0147	0.92
	OCF SMC-R25	0.0141	0.0233	0.89
Motor Oil 21°-24°C	PPG SMC-R65	0.0085	0.0038	0.71
	PPG SMC-R25	0.0081	0.0032	0.69
	OCF SMC-R25	0.0133	0.0022	0.37
Gasoline 21°-24°C	PPG SMC-R65	0.0092	0.0056	0.65
	PPG SMC-R25	0.0076	0.0150	0.99
	OCF SMC-R25	0.0121	0.0150	0.99
Anti-Freeze 21°-24°C	PPG SMC-R65	0.0072	0.0025	0.69
	PPG SMC-R25	0.0089	0.0056	0.55
	OCF SMC-R25	0.0129	0.0043	0.64

*η = Loss factor.
w = Percent weight gain.

induced reduction of T_g would shift stiffness and damping curves to lower temperatures. This would cause the damping to increase and the stiffness to decrease at the operating temperature. Reference [2] does not indicate whether the reduction in T_g is reversible or not. The present findings indicate that the changes in damping and stiffness are reversible. The second mechanism suggested in [2], which is a change in the residual stress distribution due to matrix swelling, could cause increased friction at matrix-fiber interfaces, thus increasing damping. It would seem that this mechanism would also be reversible, in agreement with present findings. In conclusion, additional work is needed before the actual mechanisms for the moisture-induced changes in dynamic behavior can be clearly identified.

CONCLUSIONS

1. At elevated temperatures ($20°$–$120°C$), dynamic stiffness decreased and damping increased for all materials, but damping is much more sensitive to temperature changes than stiffness is.
2. The temperature-induced changes in dynamic properties are essentially reversible.
3. Damping in the chopped fiber composites (SMC-R25 and SMC-R65) increases markedly with prolonged exposure to moisture, while stiffness drops off slightly.
4. Damping and stiffness of the hybrid chopped/continuous fiber composites show relatively little change with exposure to moisture.
5. Linear regression analysis shows that a definite correlation exists between increased moisture content and corresponding magnification of damping in SMC-R25 and SMC-R65 materials.
6. Of the fluids tested, water causes the most significant changes in dynamic properties of the composites, while motor oil and anti-freeze have the least effects.
7. The moisture-induced changes in dynamic properties are reversible, and this indicates that no permanent damage occurs.
8. The exact mechanism (or mechanisms) of moisture-induced changes in damping and stiffness of the materials is unknown, but a lowering of the glass transition temperature and matrix swelling are possible causes.
9. Damping is much more sensitive to environmental effects than stiffness is, and damping measurements should be very useful in further studies of these effects.

ACKNOWLEDGMENTS

The authors gratefully acknowledge the financial support provided by the Manufacturing Development Group at General Motors Technical Center. We would also like to thank PPG Industries and Owens-Corning Fiberglass Corporation for supplying materials and technical data. We are indebted to Darrel Brown of the Mechanical Engineering Department at the University of

Idaho for his work on the experimental apparatus and the material specimens.

REFERENCES

1. Whitney, J. M. and Husman, G. E., "Use of the Flexure Test for Determining Environmental Behavior of Fibrous Composites," *Experimental Mechanics, 18* (5), 185–190 (May 1978).
2. Adams, D. F. and Miller, A. K., "The Influence of Material Variability on the Predicted Environmental Behavior of Composite Materials," *Journal of Engineering Materials and Technology, ASME, 100* (1), 77–83 (Jan. 1978).
3. Browning, C. E., Husman, G. E., and Whitney, J. M., "Moisture Effects in Epoxy Matrix Composites," *Composite Materials: Testing and Design (Fourth Conference),* ASTM STP 617, 481–496 (1977).
4. Maymon, G., Briley, R. P., and Rehfield, L. W., "Influence of Moisture Absorption and Elevated Temperature on the Dynamic Behavior of Resin Matrix Composites: Preliminary Results," *Advanced Composite Materials—Environmental Effects,* ASTM STP 658, 221–233 (1978).
5. Rehfield, L. W. and Briley, R. P., "A Comparison of Environmental Effects on Dynamic Behavior of Graphite/Epoxy Composites with Aluminum Alloys," ASME Paper 78-WA/Aero-10 (1978).
6. Heimbuch, R. A. and Sanders, B. A., "Mechanical Properties of Automotive Chopped Fiber Reinforced Plastics," General Motors Manufacturing Development Report No. MD 78-032 (1978).
7. Loos, A. C., Springer, G. S., Sanders, B. A., and Tung, R. W., "Moisture Absorption of Polyester-E Glass Composites, *Journal of Composite Materials, 14,* 142–154 (April 1980).
8. Gibson, R. F., Yau, A., and Riegner, D. A., "Vibration Characteristics of Automotive Composite Materials," presented at ASTM Symposium on Short Fiber Reinforced Composites, Minneapolis, Minnesota, April 1980, in print in ASTM STP 772 (1982).
9. Gibson, R. F. and Yau, A., "Final Report—Phase 1: Dynamic Stiffness and Internal Damping of Automotive Chopped Fiber Reinforced Plastic Materials," submitted to General Motors Manufacturing Development, University of Idaho (July 31, 1979).
10. Gibson, R. F. and Yau, A., "Complex Moduli of Chopped Fiber and Continuous Fiber Composites: Comparison of Measurements with Estimated Bounds," *Journal of Composite Materials, 14,* 155–167 (April 1980).
11. Gibson, R. F. and Plunkett, R., "A Forced Vibration Technique for Measurement of Material Damping," *Experimental Mechanics, 11* (8), 297–302 (Aug. 1977).
12. Yau, A., "Experimental Techniques for Measuring Dynamic Mechanical Behavior of Composite Materials," M.S. Thesis, Mechanical Engineering Department, University of Idaho (July 1980).
13. Gibson, R. F., Yau, A., and Riegner, D. A., "An Improved Forced Vibration Technique for Measurement of Material Damping," *Experimental Techniques, 6* (2), 10–14 (April 1982).
14. Gibson, R. F., Yau, A., Mende, E. W., and Osborn, W. E., "The Influence of Environmental Conditions on the Dynamic Mechanical Behavior of Automotive Fiber Reinforced Plastics," Final Report submitted to General Motors Manufacturing Development, University of Idaho (August 1980).

8

Thermal and Photo-Degradation Behaviors of Glass-Fiber Reinforced Rigid Polyurethane Foam

K. MORIMOTO, T. SUZUKI, AND R. YOSOMIYA

1. INTRODUCTION

LOW-EXPANSION POLYURETHANE FOAM UNIFORMLY REINFORCED WITH glass fiber (FRU) has outstanding characteristics as a lightweight structural material. However, there is no report detailing its thermal and photo-degradation behaviors which are essential to its practical application.

These behaviors have been studied rather extensively for rigid polyurethane foam (PUF), which is the matrix of FRU; for example, in papers [1] and [2] for thermal degradation, and papers [3], [4] and [5] for photo-degradation. Photo-degradation, in particular, has been an issue of critical concern in the practical application of many of the PUF moldings derived from aromatic polyisocyanates.

This paper reports the results of a study conducted on the effects of thermal and photo-degradation upon the bonding strength between matrix and fiber in FRU and upon its flexural characteristics and other properties.

2. EXPERIMENTAL

2.1 Materials

FRU and PUF samples used in this experiment are of the same types as in our previous papers [6], [7] and [8].

They were molded to $1000 \times 1000 \times 3$ mm size. FRU-L were reinforced with glass-fiber continuous strand mat, while FRU-M and FRU-S were reinforced with 13mm and 3mm length glass-fiber chopped strands, respectively. The properties and the reinforcements of them are summarized in Table 1.

Specimens for determination of interlaminar shear strength were cut from molded boards of $200 \times 200 \times 15$ mm size. The glass fiber content of short beam shear specimens cut from these plates was approximately 9% by weight.

2.2 Heat Treatment

Each sample was heated in a thermostatic chamber held at 50 °C, 75 °C, 100 °C and 125 °C (in air) for predetermined periods.

2.3 WOM Exposure

Each test piece (50mm wide and 3mm thick) was held by a frame with 150mm clamping length and allowed to stand in a weatherometer (WOM) under the following operating conditions for definite periods. The apparatus used for the test was WEL-6X-HC Xenon Long-life Weatherometer (Suga Testing Instrument Co., Ltd.).

Operating conditions:
Temperature   40 °C
Amount of irradiation 0.17×10^7 Joule/cm^2 · hr
Shower   18 minutes (120-minute interval)

Table 1. Properties and reinforcements of test samples.

| | Properties of Samples (at 25 °C) | | | Reinforcements | | |
| | | Flexural Properties | | Glass Fiber Strand | | |
Sample No.	Apparent density (g/cm³) obs.	Modulus (kg/mm²) obs.	Strength (kg/mm²) obs.	Length (mm)	Vol. Cont. (%) calc.	Wt. Cont. (%) calc.
L-1	1.35	603	16.0	Continuous	8.5	15.7
L-2	1.26	540	13.6	↑	↑	16.9
L-3	1.05	410	9.75	↑	↑	20.0
L-4	0.84	301	7.1	↑	↑	25.3
M-1	1.30	450	13.0	13	9.5	17.7
M-2	1.22	410	11.3	↑	↑	19.5
M-3	1.05	336	8.4	↑	↑	23.3
M-4	0.95	298	7.1	↑	↑	24.7
S-1	1.25	403	8.9	3	8.0	15.6
S-2	1.16	320	7.15	↑	↑	17.2
S-3	0.94	177	4.9	↑	↑	21.8
S-4	0.74	106	2.2	↑	↑	26.0
PUF-1	1.05	265	7.2	—	0	0
PUF-2	0.85	163	4.6	—	↑	↑
PUF-3	0.72	128	3.2	—	↑	↑
PUF-4	0.53	70	1.8	—	↑	↑

FRU-L (L-1–L-4), FRU-S (S-1–S-4)

2.4 Bending Test

The bending test was carried out according to the Three-point Bending Test specified in JIS K-6911 by using a Shimadzu AUTOGRAPH DSS-2000. The size of test pieces was 75 × 25 × 3mm, at a cross head speed of 5 mm/min., and bending span length 50 mm. The test pieces subjected to the WOM exposure test were set to the apparatus so that tensile load will be applied to the deteriorated side during bending.

2.5 Determination of Interlaminar Shear Strength

Similar to the experiment described in our previous paper [6], the interlaminar shear strength was determined using a test specimen 25mm wide and 15mm thick according to 3 point-bending test by short beam method.

2.6 Measurement of IR Spectrum

A −302 IR spectrophotometer (Japan Spectroscopic Co., Ltd.), a conventional diffraction grating instrument, was used for IR analysis. The specimens were prepared by the method of Yosomiya et. al [9] and measured by the KBr disc method.

3. RESULTS AND DISCUSSION

3.1 Degradation of Flexural Characteristics by Heat Treatment and Its Activation Energy

The relationship of the flexural modulus, E_b, and flexural strength, σ_b, of heat-treated FRU and PUF samples to various heating time was obtained at different heating temperature levels. Typical examples are shown in Figure 1 and Figure 2. The same tendency was observed for degradation in both E_b and σ_b; little change was noticed under heating conditions up to 50 °C and up to 1000 hr, and some of the samples showed a slight increase in E_b and σ_b at the initial stage of treatment when heated at 50 °C and 75 °C. This is probably caused by the progress of crosslinking of the polyurethane resin. For the samples treated at 100 °C or higher, on the other hand, it is clear that degradation started to proceed from the beginning.

Figure 3 shows the relationship between heating temperature and retention of σ_b after 500 hours of treatment. As can be seen from this figure, σ_b declined more sharply in the order of:

$$FRU\text{-}S > FRU\text{-}M > FRU\text{-}L \doteq PUF$$

The shorter the length of glass fiber incorporated, the sharper the drop in σ_b and this drop is greater than the decline in σ_b of the matrix PUF. This fact suggests that, for FRU reinforced with short fiber, the decrease in the bonding strength between fiber and matrix has a significant effect on the decline in

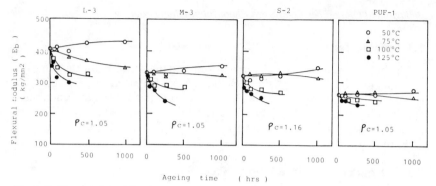

Figure 1. Thermal-degradation behavior of flexural modulus under various temperature for FRU and PUF.

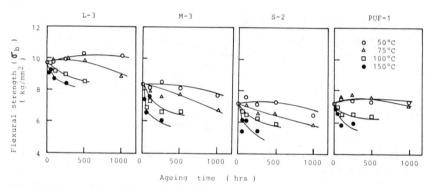

Figure 2. Thermal-degradation behavior of flexural strength under various temperature for FRU and PUF.

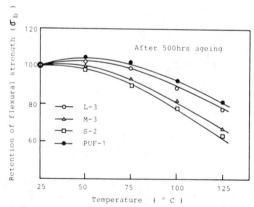

Figure 3. Retention of flexural strength under various temperature treatment after 500 hours for FRU and PUF.

98

σ_b of the reinforced materials. When reinforced with long fiber, on the contrary, the decrease in bonding strength is probably very low.

Degradation of FRU and PUF would be ascribable to oxidative degradation of matrix due to increases in ambient temperature. It is well known that oxidation of a polymer generally causes chain scission or forms new bonds, changing its physical properties. In connection with this heat-induced degradation of polymers, a method to assess their service lives has been proposed on the basis of kinetic theory [10], [11]. It is known that the following equation holds for a n'th order reaction

$$\frac{dC}{dt} = -kC^n \tag{1}$$

where C is concentration of reactant, k is reaction constant and t is time. Assuming that equation (1) can also apply to a physical quantity P like σ_b, similar to concentration C, and putting "1" as the reaction order n, one may well interpret the results of thermal degradation by this relationship. In addition, reaction constant k is expressed by

$$k = A \; exp(-\Delta E_a / RT) \tag{2}$$

where A is constant, ΔE_a is activation energy of the reaction, R is gas constant and T is absolute temperature.

Putting equation (2) into equation (1) and substituting concentration C equation (1) with a physical quantity P, we obtain equation (3) as follows:

$$\ell nt = \ell n \left[\frac{1}{A} \ell n \frac{P_o}{P}\right] + \frac{\Delta E_a}{RT} \tag{3}$$

where P_o is the value of P when $t = 0$. Hence, the time t_e required for P_o to lower to a given level of P_1 can be represented by

$$\ell n \; t_e = A' + \frac{\Delta E_a}{RT} \tag{4}$$

As an example, putting σ_b in Figure 2 as physical quantity P in equation (3), finding deterioration time t_e, which corresponds to σ_{b1} after 75 °C × 1000 hr. treatment, for different heat treating temperatures, and plotting $\ell n \; (t_e)$ against 1/T, we obtained a linear relationship as shown in Figure 4. The activation energy for deterioration ΔE_{ao} was found from the gradient of this curve. Table 2 lists the values of ΔE_{ao} for all the samples obtained in the same way. Similarly the values of activation energy (ΔE_{ab}) for deterioration of flexural modulus E_b were determined and shown in Table 2. No significant difference was observed in the values of ΔE_{ab} and ΔE_{ao} between FRU and PUF: about 12 ∼ 21 Kcal/mole for ΔE_{ab} and about 12 ∼ 17 Kcal/mole for ΔE_{ao}.

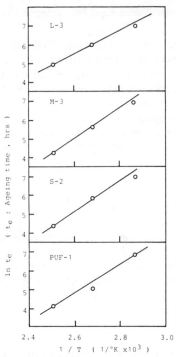

Figure 4. Temperature dependence of the ageing time, t_e at the same extent of degradation.

Table 2. Activation energies calculated from flexural strength and flexural modulus of thermal degradation.

Sample No.		ΔE_{ao} (Kcal/mole)	ΔE_{ab} (Kcal/mole)
FRU-L	L-1	14.5	16.2
	L-2	15.6	15.8
	L-3	12.1	15.4
	L-4	13.1	13.7
FRU-M	M-1	16.6	13.9
	M-2	16.2	17.8
	M-3	17.0	21.2
	M-4	14.1	17.8
FRU-S	S-1	15.4	20.2
	S-2	16.2	14.5
	S-3	16.8	14.6
	S-4	13.3	12.5
PUF-1		15.2	10.1
PUF-2		15.2	10.9
PUF-3		12.5	11.9
PUF-4		13.9	15.0

100

The general tendency is that samples with larger apparent density show greater activation energy for degradation.

3.2 Changes in Interlaminar Shear Strength by Heat Treatment

Measurements of interlaminar shear strength, τ_b, of heat-treated FRU-M and FRU-S are shown in Figure 5. It can be seen from the figure that degradation started in the initial stage of treatment when the samples were heated at 90 °C or higher, and that it progressed more rapidly at higher temperatures.

As described in our previous paper (ref.[6]), τ_b can be expressed by the following formulas:

$$\sigma_o = \tau_b / \alpha \tag{5}$$

$$\alpha = [2G_m \, (1-K)/E_f \cdot \ln \, (r_o/r_f)]^{1/2} \tag{6}$$

where σ_o is the stress developed in the fiber while interfacial delamination is in progress, G_m is the shear modulus of matrix, K is a parameter representing slippage of surface, E_f is elastic modulus of the fiber, $2r_o$ is distance between fibers and r_f is the radius of fiber. Equation (7) can be derived from formulas (5) and (6)

$$\tau_b = \frac{1}{2} \, \sigma_o \, [2G_m \, (1-K)/E_f \cdot \ln \, (r_o/r_f)]^{1/2} \tag{7}$$

where $E_f \, \ln(r_o/r_f)$ is a constant independent of heat treatment. Assuming that σ_o also shows no change by heat treatment, we obtain equation (8) from equation (7) by introducing a proportionality factor γ.

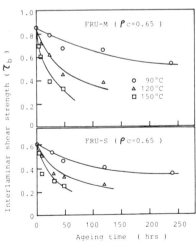

Figure 5. *Thermal-degradation behavior of interlaminar shear strength under various temperature for FRU.*

$$\tau_b = \gamma \sqrt{(1-K)} \sqrt{G_m} \qquad (8)$$

Separately, based on the relationship $G_m = E_m/2\,(1+\nu)$ described in our previous paper [8], we can obtain the change in $\sqrt{G_m}$ of PUF heat-treated at 120°C from the thermal degradation data of the same samples shown in Figure 1. The values of $\sqrt{G_m}$ thus obtained were compared with the values of τ_b in Table 3 and Figure 6. The two curves are both linear within the heating time range tested, but are not proportional to each other. This is probably because the value of K increases with increasing heating time.

3.3 Changes in Flexural Properties after WOM Exposure

Figure 7 shows the retentions of flexural modulus E_b and flexural strength σ_b of the samples subjected to the WOM exposure test. The slight increase in these values observed in the initial stage of the exposure test would be ascribable, as in the case with the heat treatment, to the progress of crosslinking of polyurethane that might proceed at an elevated temperature of 40°C. After 1500 hours of exposure, however, apparent drops in E_b and σ_b are observed for all the samples to a greater or lesser extent. This tendency agrees well with the result of outdoor exposure test conducted by Watanabe et. al [12] on glass-fiber reinforced unsaturated polyesters (FRP).

In addition, it can also be seen from Figure 7 and E_b tends to drop more rapidly than σ_b does and that the speed of decrease in both E_b and σ_b is larger with samples of higher apparent density when compared among those reinforced with the same fiber material, including non-reinforced samples. One possible reason for this is that as the degree of expansion of matrix increases and the porosity of sample becomes greater, more light is scattered on its surface layer, with less light reaching its inside. This means that, in a highly expanded sample, overall deterioration remains comparatively low because photo-degradation is confined mainly to its surface layer (though somewhat extensive in this region). In fact, observation of cut surfaces of samples subjected to the WOM exposure test revealed that, in those of higher apparent

Table 3. Calculations of $\sqrt{G_m}$ from E_m and observation values of τ_b.

Aging Time (hrs.)		0	40	80	120
E_m (kg/mm^2)		60	52	46	44
$G_m = \dfrac{E_m}{2\,(1+\nu)}$ *		22.2	19.3	17.3	16.3
$\sqrt{G_m}$		4.71	4.39	4.13	4.04
τ_b	FRU-M	0.85	0.52	0.40	0.33
(kg/mm^2)	FRU-S	0.61	0.37	0.29	0.24

*$\nu = 0.35$ (Poisson's ratio of PUF).

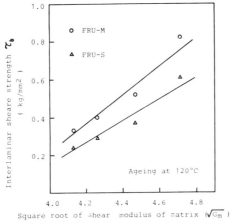

Figure 6. *Relation between interlaminar shear strength of FRU and square root of shear modulus of matrix.*

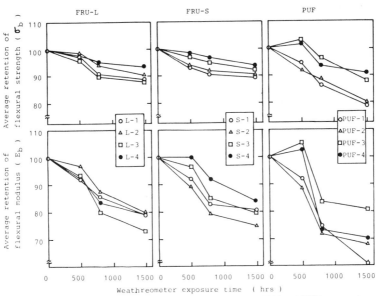

Figure 7. *Retentions of flexural strength and flexural modulus of FRU and PUF vs. weatherometer exposure time.*

density, yellowing spreads over the entire depth, though not so marked in the surface layer. In samples of lower apparent density, on the other hand, although marked yellowing is observed in the surface layer, no discoloration is noticed inside. Such a tendency was most remarkable in nonreinforced PUF, and the difference in deterioration due to difference in apparent density was smaller in the case of FRU-L and FRU-S. This indicates that the presence of glass fiber reinforcement increases the resistance of matrix with higher apparent density to WOM exposure.

3.4 Changes in IR Spectrum Due to Degradation by WOM Exposure

Figure 8 compares the IR spectrum of the deteriorated surface of sample PUF-4 after exposure test with that before exposure. Shollenberger et. al [4] and Beachell et. al [13] have already reported the changes in IR spectrum of polyurethane caused by photo-degradation.

It is apparent from Figure 8 that the amount of urethane linkage has been diminished as a result of WOM exposure, because the peaks characteristic to this bond, such as those at 3350cm^{-1}, 1720cm^{-1}, 1540cm^{-1}, 1500cm^{-1} and 1015cm^{-1}, have declined. Little change is observed for the strong peaks at 1080cm^{-1} and 1220cm^{-1} characteristic to ether linkage. On the other hand, the significant reduction in intensity of 815cm^{-1} peak, which is assigned to substituted aromatic compounds, suggests loss, or conversion to other types of substitution, of the aromatic polyisocyanate groups contained. This indicates that the polyether segment in polyurethane was little affected by WOM exposure.

CONCLUSION

Thermal and photo-degradation characteristics of rigid polyurethane foams (PUF) and glass fiber reinforced rigid polyurethane foams (FRU) have been studied. The results obtained may be summarized as follows:

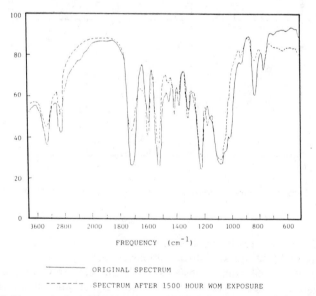

Figure 8. Changes in the infrared spectrum of PUF-4 on weatherometer exposure in air.

(1) Reduction in flexural strength and flexural modulus by heat treatment was more marked with the samples containing glass fiber of shorter length.

(2) No significant difference was observed in activation energy of thermal degradation between FRU and PUF: about $12 \sim 21$ Kcal/mole for flexural modulus, and about $12 \sim 17$ Kcal/mole for flexural strength. With a few exceptions, samples of higher apparent density tended to show slightly higher activation energy.

(3) A similar tendency was observed also for the decline in interlaminar shear strength by heat treatment. Probably degradation of this type stems from both the reduction in shear modulus of matrix and that in bonding strength between fiber and matrix.

(4) Reduction in flexural properties by WOM exposure was nearly the same for FRU and PUF, but was slightly larger with samples of higher apparent density.

(5) IR analysis proved a decrease of urethane linkage after WOM exposure.

ACKNOWLEDGMENTS

This work has been supported by Nishiarai Chemical Laboratory in Nisshin Spinning Co., Ltd.

The authors wish to express their thanks to Professor Akio Nakajima of Kyoto University for his continued advice and helpful discussions.

The authors also acknowledge the great help and advice from Mr. K. Kimura, Executive Director of Chemical Division of the Company, during the study and thanks are given to him for approving publication of the paper.

REFERENCES

1. Remakrishanan, M., "Pyrolysis and Thermal Degradation of Rigid-Urethane Foams," University of Utah, Salt Lake City, 1975.

2. J. R. Ward and L. J. Decker, "Determination of the Thermal Decomposition Kinetics of Polyurethane Foam by Guggenheim's Method," Ind. Eng. Chem. Prod. Res. Dev., Vol. 21, pp. 460–461, 1982.

3. H. C. Beachell and I. L. Chang, "Photodegradation of Urethane Model Systems," *Journal of Polymer Sciences:* part A-1, Vol. 10, pp. 503–520, 1972.

4. C. S. Shollenberger and F. D. Stewart, "Thermoplastic Urethane Structure and Ultraviolet Stability," *J. Elastoplastics*, Vol. 4, pp. 294–331, October, 1972.

5. N. S. Allen and J. F. McKellar, "Photochemical Reactions in an MDI-Based Elastomeric Polyurethane," *Journal of Applied Polymer Science*, Vol. 20, pp. 1441–1447, 1976.

6. K. Morimoto, T. Suzuki and R. Yosomiya, "Flexural Properties of Glass Fiber Reinforced Rigid Polyurethane Foam," Ind. Eng. Chem. Prod. Res. Dev., in press.

7. K. Morimoto, T. Suzuki and R. Yosomiya, "Stress Relaxation of Glass Fiber Reinforced Rigid Polyurethane Foam," *Polym. Eng. Sci.*, in press.

8. K. Morimoto, T. Suzuki and R. Yosomiya, "Adhesion between Glass Fiber and Matrix of Glass Fiber Reinforced Rigid Polyurethane Foam under Tension," Polym. Plas. Tech. Eng., submit.

9. K. Noma and R. Yosomiya "A discussion on the surface of resin by IR analysis," Resin Finishing, Japan, Vol. 12, pp. 35-40, 1963.

10. H. Aoki and A. Yoshida, "The Kinetic Evaluation of Plastic Films at their Deterioration," Zairyo, Japan, Vol. 21, pp. 309-314, 1972.

11. J. Shimada, "The Mechanism of Oxidative Degradation of ABS Resin. Part I. The Mechanism of Thermooxidative Degradation," *Journal of Applied Polymer Science*, Vol. 12, pp. 655-669, 1968.

12. T. Watanabe, "Weathering Test of Fiberglass Reinforced Plastic under Flexural Load," Kobunshi Ronbunshu, Japan, Vol. 39, pp. 1-6, 1982.

13. H. C. Beachell and C. P. Ngoc Son, "Color Formation in Polyurethanes," *Journal of Applied Polymer Science*, Vol. 7, pp. 2217-2237, 1963.

9

Buffer Strips in Composites at Elevated Temperature

C. A. BIGELOW

INTRODUCTION

FIBROUS COMPOSITE MATERIALS LIKE GRAPHITE/POLYIMIDE ARE LIGHT, stiff, and strong. They have great potential for reducing weight in aircraft structures. However, fibrous composites are usually notch sensitive and, therefore, lose much of their strength when damaged [1]. Buffer strips have been shown to greatly improve the damage tolerance of graphite/epoxy laminates loaded in tension [2]. These narrow, parallel buffer strips are made into the laminate itself by interrupting and replacing piles of the laminate with another material or layup. These buffer strips can arrest fracture and increase the residual strength of the damaged laminate. However, the buffer strip concept has been evaluated only at room temperature [2]. The purpose of the present investigation was to evaluate this concept at elevated temperature.

Center-notched graphite/polyimide buffer strip panels were tested at room and elevated (177 ± 3 °C) temperatures. One layup was used: $[45/0/-45/90]_{2S}$. All panels had the same buffer strip width and spacing. The buffer strips were made by replacing narrow strips of the 0° graphite plies with strips of 0° S-glass/polyimide on either a one-for-one or a two-for-one basis. The latter panels had twice as many plies of buffer material as the former. During tests, panels were radiographed and crack-opening displacements were recorded to indicate fracture onset, fracture arrest, and the extent of damage in the buffer strip after arrest.

The residual strengths and failure strains of the buffer strip panels were compared to the ultimate strengths and strains of the plain (no buffer strips) panels at room and elevated temperatures. Fracture arrest and the effect of the number of buffer strip plies were examined at room and elevated

107

temperatures. The shear-lag analysis developed in Reference [2] for buffer strip panels was also used to examine the data.

EXPERIMENTAL PROCEDURES

Materials and Specimens

The specimen configuration is shown in Figure 1. The specimens were made with Celion 6000*/PMR-15 graphite/polyimide** in a 16-ply, quasi-isotropic layup, $[45/0/-45/90]_{2S}$. The buffer strips were made from S-glass/PMR-15 tape and were made by replacing narrow strips of the 0° graphite plies with 0° S-glass on a one-for-one or a two-for-one basis. (The cross-section shown in Figure 1(c) illustrates a two-for-one replacement.) Only the 0° graphite plies were interrupted by the buffer material; the ±45° and 90° graphite plies were continuous throughout the panels. All of the panels were made with 5-mm-wide buffer strips spaced 20 mm apart, as shown in Figure 1. Half of the panels were made with buffer strip replacement on a one-for-one basis, and half were made with buffer strip replacement on a two-for-one basis. One side of each panel was made flat. The length of the test section of all panels was greater than twice the panel width. Six panels of each configuration were tested.

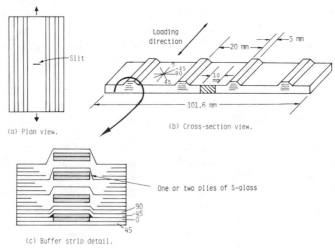

(a) Plan view.

(b) Cross-section view.

(c) Buffer strip detail.

Figure 1. *Buffer strip configuration.*

---

*Celion 6000: Registered trademark of Celanese Corporation.
**The use of trade names in this article does not constitute endorsement, either expressed or implied, by the National Aeronautics and Space Administration.

Slits 10 mm long and 0.020 (±0.002) mm wide were cut through each panel to represent damage (see Figure 1). The slits were located at the center of each panel and, since the slit length was less than the buffer strip spacing, no buffer strips were cut.

Also, small coupon-type specimens, notched and unnotched, were made from the plain 16-ply laminate to measure its ultimate tensile strength, moduli, and fracture toughness. These were tested at room and elevated temperatures.

Test Procedures and Equipment

The panels were loaded to failure in uniaxial tension at a rate of about 1333 N/s. They were tested in a servo-controlled, closed-loop testing machine with load as the feedback signal. For the elevated temperature tests, an oven was mounted on the test stand and closed around the test section of the specimen. Approximately 178 mm of the panel length was enclosed in the oven. The specimens were heated to 177 ± 3 °C for at least an hour before testing to insure thermal equilibrium during each test. Half of the panels were tested at this temperature.

Loads, strains, and the opening displacement of the slit (commonly called crack-opening displacement or COD) were recorded on a digital data acquisition system during each test. Strains were measured at a location remote from the slit using strain gages bonded to the specimen. The ring gage[†], shown in Figure 2, measured the COD. This ring gage was approximately 31 mm in

(a) Close-up view of COD gage.   (b) Gage mounted in buffer strip panel.

Figure 2. *Ring gage used to measure crack-opening displacements.*

[†]Designed and manufactured by John A. Shepic, 142 South Ingalls, Lakewood, CA 80226.

diameter and could easily operate in the confined space of the oven. It was designed to withstand temperatures higher than 260°C. During the room-temperature tests, when audible and visual evidence of crack extension developed, the load was held constant while the region containing the slit and the two center buffer strips was radiographed in situ. An X-ray opaque dye-penetrant, zinc dioxide, was used to enhance the image of the damaged area. Radiographs were not taken during the elevated-temperature tests because the oven surrounded the specimen. Also, since all specimens were loaded until catastrophic failure occurred, no radiographs were made after the tests.

RESULTS AND DISCUSSION

Typical Fracture Arrest Results

Figure 3 shows two radiographs of a buffer strip panel tested at room temperature with two-for-one replacement. The vertical dark strips in the pictures are the S-glass buffer strips. The very dark figure in the center of the picture is an end view of the ring gage used to measure COD. The first radiograph was taken at a load of 78 kN, 65 percent of the ultimate load. A small amount of damage can be seen as some delamination in the ±45° plies and some transverse cracking in the 90° plies. The second radiograph was taken shortly before failure at a load of 113 kN, 94 percent of the ultimate load. In this radiograph, the delamination has expanded considerably, reaching well into the adjacent buffer strips, with much more transverse cracking in the 90° plies. The damage and crack growth seen in the radiographs in Figure 3 are not as well defined as that reported in Reference [2]. Radiographs reported [2] for graphite/epoxy laminates showed only a small amount of delamination at the slit tips, with a well-defined crack extending into the buffer strip. In contrast, radiographs here show large fan-shaped areas of delamination and matrix cracking at the slit tips. Here, the arrested fractures were not well-defined cracks but included large areas of damage at the slit ends. In the room temperature tests, jumps in COD indicated damage growth, such as

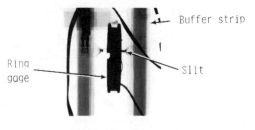 

(a) P = 78 kN, 65 percent   (b) P = 113 kN, 94 percent
    of ultimate load.   of ultimate load.

Figure 3. Radiographs of a panel with buffer strips (two-for-one replacement).

delamination and matrix cracking, at the slit ends, rather than clearly defined crack extension. This type of damage growth is typical of graphite/polyimide laminates.

Effects of Temperature

Some typical test results for buffer strip panels tested at room and elevated temperatures are shown in Figure 4. The remote strain is plotted against slit length calculated from COD for two panels. The slit length was calculated from the crack-opening displacement using compliance methods. For comparison, the failing strains of panels without buffer strips are also shown. The failing strains for panels without buffer strips were found using coupon data from unnotched and notched specimens made from the basic laminate. The unnotched tensile data are shown in Table 1 and the notched data are shown in Table 2. The two curves of failing strains for panels without buffer strips were fitted to these tabulated data.

For the room-temperature tests (open symbols), Figure 4 shows an increase in the calculated slit length for strain levels below the curves for panels without buffer strips. Since the slit length was calculated from COD measurements, any sudden increase in COD results in a corresponding increase in the calculated slit length. However, as mentioned in the previous section, abrupt jumps in COD were caused by damage growth such as delamination and matrix cracking, not by crack extension. This was confirmed by the

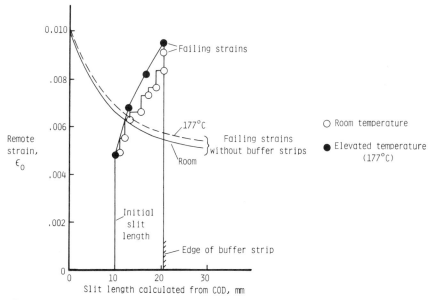

Figure 4. *Typical fracture arrest results for panels with buffer strips (two-for-one replacement).*

Table 1. Unnotched tensile properties.

	Longitudinal Elastic Modulus (GPa)	Ultimate Strength (MPa)	ε_c
Elevated	48.2	493	0.0109
Temperature	48.0	495	0.0105
(177°C)	(48.1)[a]	(494)	(0.0107)
Room	46.1	445	0.0099
Temperature	43.4	426	0.0103
	(44.8)	(435)	(0.0101)

[a]Parenthetical numbers indicate averages of data shown.

Table 2. Fracture properties of panels without buffer strips.

	Slit Length (mm)	Width (mm)	Residual Strength (MPa)	ε_c
Elevated			283	0.0064
Temperature	10.2	50.8	304	0.0072
(177°C)			(294)[a]	(0.0068)
			242	0.0056
	25.4	101.6	246	0.0055
			(244)	(0.0056)
Room			306	0.0066
Temperature	10.2	50.8	303	0.0066
			(305)	(0.0066)
	25.4	101.6	251	0.0052

[a]Parenthetical numbers indicate averages of data shown.

radiographs of the damaged area. The damage grew until it reached the buffer strip and then stopped. The load was increased and the panel failed at a strain of about 0.0092, an increase of over 60 percent above the failing strain of a panel without buffer strips. This improvement in failure strains due to the use of buffer strips is comparable to results reported [2] for graphite/epoxy panels. An improvement in the damage tolerance of the buffer strip panels was also seen in the elevated-temperature tests (solid symbols in Figure 4). In the elevated-temperature tests, the damage grew stably until it reached the buffer strip and then the panel failed at a strain of about 0.0095, an increase of over 60 percent compared to the failing strain of a plain panel. All of the panels tested showed behavior similar to that shown in Figure 4. The failing strains and strengths for the buffer strip panels are listed in Table 3.

Figure 5 shows two curves of COD versus load at room and elevated temperatures for a buffer strip panel with two-for-one replacement.

Table 3. Fracture properties of buffer strip panels.

	$\dfrac{h_b}{h_o}$	Residual Strength (MPa)		ε_c	
	1	369		—	
	1	404	(385)[a]	0.0101	(0.0098)
Elevated	1	381		0.0095	
Temperature	2	361		0.0095	
(177°C)	2	348	(355)	0.0092	(0.0094)
	2	356		0.0096	
	1	400	(405)	0.0095	(0.0095)
	1	409		—	
Room	2	370		0.0084	
Temperature	2	376	(376)	0.0091	(0.0089)
	2	383		0.0092	

[a]Parenthetical numbers indicate averages of data shown.
h_b/h_o = ratio of total thickness of 0° plies in basic laminate to total thickness of plies of buffer strip material.

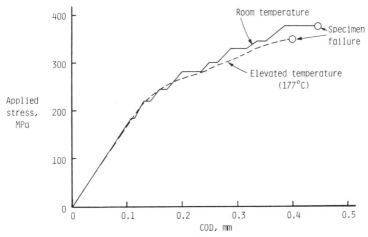

Figure 5. *Typical crack-opening displacements (COD) for a buffer strip panel (two-for-one replacement).*

For the elevated-temperature tests, as also shown in Figure 4, the damage grew stably with increasing load until the crack tip reached the buffer strip, then the panel failed. The COD curve for the elevated-temperature test is relatively smooth with no discontinuities in slope, while the room-temperature curve has several distinct jumps in COD indicating damage growth, delamination and matrix cracking. Note that both curves on the average have nearly the same slopes. These results are typical of all panels tested, including the panels without buffer strips.

Although the buffer strip panels had higher failing strains when tested at 177 °C, they had lower failing stresses than found in the room-temperature tests, as shown in Table 3. Typical remote stress-strain curves from one specimen are shown in Figure 6 for 177 °C and room temperature. As can be seen in the figure, the failure strains for elevated temperature are slightly higher (~1 percent) than for room temperature, while the corresponding failure stress is about 8 percent lower. The elevated temperature lowered the modulus by softening the laminate, resulting in higher strains but lower strengths. The same trend was observed for plain graphite/polyimide panels with slits. However, the unnotched tensile coupons tested at 177 °C showed higher strains (~6 percent) and also higher strengths (~12 percent) than the coupons tested at room temperature, as shown in Table 1. This does not agree with results reported in Reference [1]. Data in Reference [1] are for the same layup as, but a different stacking sequence than, the laminate used in this report. This difference should not significantly affect the tensile data. Data from Reference [1] show slight decreases (<10 percent) in the tensile modulus and strength when the temperature went from room temperature to 177 °C. Strains were not reported in Reference [1]. Although the same layup was not tested, Reference [3] also reported slight decreases in the tensile modulus and tensile strength, with a slight increase in the failing tensile strains, for increasing temperature in a laminate similar to the one used in this study. Based on the data given in References [1] and [3] the strength at 177 °C (Table 1) is slightly higher than expected although the modulus at 177 °C is consistent with data in Reference [1]. The room temperature modulus and strength shown in Table 1 were lower than results reported in Reference [1].

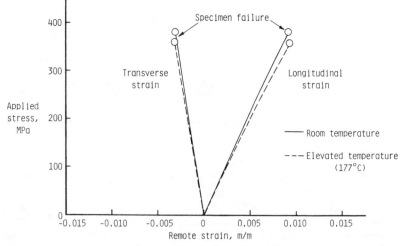

Figure 6. *Typical stress-strain curves for a buffer strip panel (two-for-one replacement).*

The fractures in most of the panels tested at room temperature were self-similar; that is, they followed a path colinear with the slit. After the panels failed, the S-glass was delaminated from the graphite/polyimide for most of the specimen length, like the results reported in Reference [2]. At elevated temperature, the fractures were similar, but with more delamination and splitting in the S-glass plies.

To determine the effect of elevated temperature on the behavior of the buffer strip material, some unidirectional 6-ply S-glass/polyimide unnotched tensile coupons were tested at room temperature and at 177 °C. The elevated temperature had virtually no effect on the ultimate strength or modulus of the S-glass, although the higher temperature did increase the strain to failure. This is consistent with the material behavior in the buffer strip panel.

Effect of Number of Buffer Strip Plies

Figure 7 illustrates the effect of the number of buffer strip plies on the panel residual strength. For comparison, the estimated residual strength is shown for a panel without buffer strips; this panel had a crack equal in length to the buffer strip spacing of the other panels. Also for comparison, results from Reference [2] are shown for graphite/epoxy buffer strip panels having the same layup and configuration as the present buffer strip panels. Each bar in the figure represents the average of two or three tests.

For both types of buffer strips, the panel residual strengths exceeded those of the panels without buffer strips by at least 40 percent, as shown in Figure 7. The residual strengths of the panels with two-for-one buffer strip replacement were slightly lower than those of the one-for-one replacement panels.

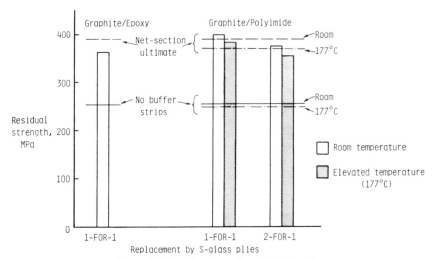

Figure 7. *Strengths of panels with buffer strips.*

This differs from the results reported in Reference [2]. However, the difference is rather small (~5 percent) and the residual strengths of the two-for-one panels were probably limited by the net-section ultimate strength of the panel. The strengths of the graphite/polyimide buffer strip panels at room temperature were higher than the strengths of the graphite/epoxy buffer strip panels. This was expected since plain graphite/polyimide panels show higher strengths than plain graphite/epoxy panels reported in Reference [2].

ANALYSIS

The shear-lag analysis developed in Reference [2] for graphite/epoxy buffer strip panels was used to analyze the data from the graphite/polyimide buffer strip panels. The model can account for the differences in buffer strip materials, the number of plies of buffer material, the matrix damage at the crack-tips, the constraint of cross-plies (i.e., differences in layup), and the width and spacing of the buffer strips. Only the important results of the analysis are presented here.

In Figure 8, the reciprocal of the strain concentration factor, $1/K_T$, is plotted against the buffer strip spacing ($W_a = 20.32$ mm) multiplied by a stiffness ratio, $h_o E_o / (h_b E_b)$. Here, h_o and h_b are the thicknesses of the 0° plies of graphite and buffer material, respectively, while E_o and E_b are the Young's moduli of the 0° graphite and buffer material, respectively. The values used for E_o at room and elevated temperature were 128 GPa and 130 GPa, respectively [1]. Log scales were used in Figure 8 for convenience. For comparison, test values of the ratio of the remote failing strain of the panel to the ultimate tensile strain of the buffer strip material, $\varepsilon_c / \varepsilon_{tub}$, were also plotted in Figure 8. Each symbol represents the average of several tests. The value of ε_{tub} used for the 0° S-glass was 0.0281 [2]. Figure 8 shows that the

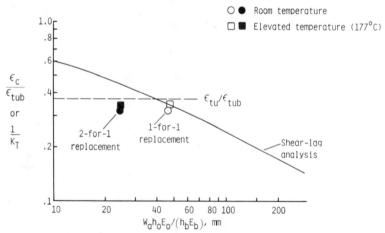

Figure 8. Correlation of experimental and analytical results.

analytic values of $1/K_T$ correlated very well with the test values of $\varepsilon_c/\varepsilon_{tub}$ for the panels with one-for-one buffer strip replacement (open symbols) at both room temperature and 177 °C. The test values of $\varepsilon_c/\varepsilon_{tub}$ for the panels with two-for-one buffer strip replacement did not correlate as well with the analytic values of $1/K_T$. As seen in Figure 8, the failing strains of the panels with two-for-one replacement were limited by the ultimate unnotched tensile strain of the laminate. The value of the ratio $\varepsilon_{tu}/\varepsilon_{tub}$ is shown as a dashed line in the figure where the value of ε_{tu} was taken from Table 1.

CONCLUSIONS

Graphite/polyimide panels with a center slit and S-glass buffer strips parallel to the loading direction were tested to measure the effect of elevated temperature (177 ± 3 °C) on their residual tensile strength. Panels were made with a $[45/0/-45/90]_{2S}$ layup. The buffer strips were made by replacing narrow strips of the 0° graphite plies with strips of 0° S-glass on either a one-for-one or a two-for-one basis. Slits were cut at the panel center between buffer strips to represent a crack. This study led to the following conclusions:

1. The buffer strips were effective at a temperature of 177 °C and increased panel strength by at least 40 percent over plain panels.
2. Buffer strip panels tested at 177 °C had slightly lower residual strengths but higher failure strains than similar panels tested at room temperature.
3. Unlike the panels tested at room temperature, the buffer strip panels tested at 177 °C exhibited stable crack growth until the crack grew into the buffer strip.
4. Panels with two-for-one buffer strip replacement had slightly lower residual strengths (by 5 percent) than panels with one-for-one replacement.
5. A shear-lag type analysis developed for room temperature tests of graphite/epoxy panels correctly predicts remote failing strains at room temperature and 177 °C for graphite/polyimide buffer strip panels with a one-for-one buffer strip replacement.
6. Panels with a two-for-one buffer strip replacement failed at strains less than those predicted by shear-lag analysis. The effectiveness of the buffer strips was probably limited by the net section ultimate strength of the panel.

NOMENCLATURE

h_o, h_b total thickness of 0° plies in basic laminate and in buffer strip, respectively, m

E_o, E_b Young's modulus of 0° Gr/Pi and 0° buffer material, respectively, Pa

K_T effective strain concentration factor at failure

W_a buffer strip spacing, m

ε_c remote panel strain at failure

ε_o remote panel strain
ε_{tu} ultimate tensile strain of basic laminate
ε_{tub} ultimate tensile strain of buffer material

REFERENCES

1. Garber, D. P., Morris, D. H., and Everett, R. A., Jr., "Elastic Properties and Fracture Behavior of Graphite/Polyimide Composites at Extreme Temperatures," *Composites for Extreme Environments*, N. R. Adsit, ed., ASTM STP 768, American Society for Testing and Materials, 73-91 (1982).
2. Poe, C. C., Jr., and Kennedy, J. M., "An Assessment of Buffer Strips for Improving Damage Tolerance of Composite Laminates," *Journal of Composite Materials Supplement, 14,* 57-70 (1980).
3. Cushman, J. B., and McCleskey, S. F., "Design Allowables Test Program, Celion 3000/PMR-15 and Celion 6000/PMR-15 Graphite/Polyimide Composites," Boeing Aerospace Co., NASA Contract NAS1-15644, NASA CR-165840 (1982).

10

Predicting the Time-Temperature Dependent Axial Failure of B/A1 Composites

J. A. DiCarlo

INTRODUCTION

THE ABILITY TO PREDICT THE FAILURE MODES OF STRUCTURAL MATERIALS under various environmental conditions has both a practical and fundamental significance. On the practical side, the predictive theory and associated equations will allow the design engineer to estimate structural failure for conditions not covered in available test data. On the fundamental side, verification of the predictive theory by comparison with experimental data will confirm the mechanistic models employed to derive the theory. Such confirmation not only may allow future theoretical modifications which yield a more accurate predictive ability but also may lead to the development of practical techniques for improving structural performance.

The objective of this paper is to demonstrate that predictive equations can be developed that adequately describe the effects of time, temperature, and stress on the axial failure modes of B/A1 composites. For many metal matrix composite systems the reinforcing fibers deform elastically so that time-temperature effects arise mainly from the mechanical properties of the matrix. However, for B/A1 composites, both the matrix and fiber make major contributions to the time and temperature dependence of composite failure. This is due to the fact that in contrast to other ceramic fibers such as silicon carbide and alumina, boron fibers display a low temperature creep which has a significant effect on fiber fracture. For this reason the approach taken in this paper will be to first investigate boron fiber creep and fracture data in order to establish a model and equations that accurately describe the axial failure modes of as-produced fibers. Once this is accomplished, composite data will be examined to determine how these equations might be modified to describe fiber creep and fracture within B/6061 A1 composites.

119

As part of this examination, a general metal matrix composite fracture theory will be developed based on the primary fiber and matrix mechanisms which contribute to time and temperature-dependent axial composite failure.

PROCEDURE

Background

The low temperature deformation of a boron fiber has been observed to be characteristically anelastic [1,2,3]. That is, in a creep test upon application of a constant tensile stress σ at time $t = 0$, the total strain ε in the fiber increases with time t and temperature T according to

$$\varepsilon(t, T, \sigma) = \varepsilon_e(T, \sigma) + \varepsilon_a(t, T, \sigma) \qquad (1)$$

Here, $\varepsilon_e = \sigma/E(T)$ is the time-independent elastic strain which depends on T only through the elastic Young's modulus E. The anelastic creep strain ε_a is zero at $t = 0$ but increases with time, temperature, and stress. No evidence of plastic strain or strains other than anelastic has been found in boron fiber deformation for temperatures up to 800 °C. Three properties which characterize ε_a and distinguish it from plastic strain are:

I. Linearity: ε_a is directly proportional to σ.
II. Equilibrium: After passage of sufficient time, ε_a reaches or relaxes to a unique equilibrium value.
III. Recoverability: Upon removal of σ, the developed ε_a completely disappears at a rate which is time and temperature dependent.

Because of property I it is convenient to introduce a stress-independent anelastic strain function A defined by

$$A(t, T) = (\varepsilon/\varepsilon_e) = 1 + (\varepsilon_a/\varepsilon_e). \qquad (2)$$

$A = 1$ for pure elastic behavior, and $A > 1$ for anelastic behavior. With this definition for A, Equation (1) can be written as

$$\varepsilon(t, T, \sigma) = [\sigma/E(T)]A(t, T) \qquad (3)$$

which is simply Hooke's law multiplied by the time-temperature dependent factor A. Clearly, to understand and predict the effects of anelastic deformation on the failure modes of boron fibers and B/Al composites, one must have accurate knowledge of the boron A function. In this paper the results of various deformation experiments will be presented in which Equations (2) and (3) were employed to calculate the A functions for as-produced boron fibers and for fibers within B/Al composites.

Deformation tests

The primary experiments employed to determine the A function for as-produced boron fibers were low-stress flexural stress relaxation (FSR) and flexural internal friction (FIF) tests performed at various temperatures up to 800 °C. Details of the test apparatus and applicable deformation theories are described elsewhere [1]. For the FSR tests, anelastic creep strains were allowed to develop for one hour, whereas for the FIF tests, creep occurred only in a time span of one vibration period which was typically of the order of one millisecond. This large difference in test time coupled with anelasticity theory permitted accurate extrapolation of the A function for time-temperature conditions not covered by the tests.

As predicted by anelasticity property I, the A functions measured by the low-stress flexural tests were found to be independent of the applied stress. However, when the FSR tests were conducted at stress levels above 50 ksi (0.3 GN/m²), an unexpected stress dependence for the A function was observed. To study this effect in greater detail, two types of high stress tensile experiments were performed: room temperature elongation experiments [4] at stress levels of about 400 ksi (2.8 GN/m²), and stress rupture experiments on etched boron fibers (1) at temperatures from 200 to 1000 °C and at stress levels between 200 and 800 ksi (1.4 and 5.6 GN/m²). Since the latter experiments have a direct bearing on a predictive fracture model for boron fibers, the details of the deformation theory involved in its interpretation will be discussed here.

After slightly etching 203 μm boron on tungsten fibers, Smith [5] observed that essentially all cases of fiber fracture could be explained by crack initiation within the region of the tungsten boride core. This result suggests a "composite fiber" fracture model in which an etched boron on tungsten fiber fractures whenever the total axial strain of the boron anelastic sheath becomes equal to the core fraction strain. By assuming a brittle elastic core with a fracture strain independent of time and temperature, one can then use Equation (3) to express this model in the following form:

$$\bar{\varepsilon}_u = [\bar{\sigma}_u(t,T)/E(T)]A(t,T) = CONSTANT \qquad (4)$$

Here $\bar{\varepsilon}_u$ is the average (ultimate) fracture strain of the etched fiber, and $\bar{\sigma}_u$ is the average (ultimate) fracture stress required to obtain $\bar{\varepsilon}_u$. In the stress rupture tests, $\bar{\sigma}_u(t,T)$ were measured for various times and temperatures. These data plus room temperature $\bar{\varepsilon}_u$ data were then inserted into Equation (4) to determine the plus room temperature ε_u data were then inserted into Equation (4) to determine the A function at the $\bar{\sigma}_u$ stress level.

To determine the A function for fibers within B/Al composites, it was necessary to perform a composite test in which essentially all deformation is fiber controlled. One test that fulfills this requirement is the internal friction or damping test conducted on B/Al composites which have been annealed near 400 °C. The experimental and theoretical details of this test are described elsewhere [6]. The annealing treatment essentially eliminates dislocation damping within the aluminum matrix, leaving the fibers as the only source of composite damping [7]. Application of the rule-of-mixtures to axial temperature-dependent damping data for unidirectional B/Al composites allowed accurate calculations of fiber damping as a function of temperature. As previously described for the FIF data [1], these fiber damping results were then used as deformation data to calculate the A function for fibers within B/Al composites. The stress levels of the composite damping test varied between 0.01 and 1 ksi (0.07 and 7 MN/m²). In this range no stress effects on fiber damping were observed, indicating a true stress independence for the A function up to at least 1 ksi.

Specimens

The primary specimens employed for the single-fiber tests were 203 µm (8 mil) and 142 µm (5.6 mil) boron on tungsten fibers supplied by Avco Specialty Materials Division. During the chemical vapor deposition of the boron sheath, the original 13 µm tungsten substrate became completely borided to form a 17 µm diameter core region.

Because fiber coatings may have an effect on the fiber anelasticity, 142 µm boron fibers coated with a 1.5 µm thick silicon carbide layer were also examined. These fibers which also contained the 17µm tungsten-boride core were supplied by Composite Technology, Inc. under the tradename "Borsic."

The unidirectional B/Al composite specimens employed for the damping experiments contained normally 50 volume percent boron or Borsic fiber in a 6061 Al matrix. The specimens reinforced by 203 µm boron fibers were fabricated by TRW whereas those reinforced by the Borsic fibers were fabricated by Avco. Typical fabrication techniques involving diffusion bonding near 500°C were employed for both composite types.

RESULTS AND DISCUSSION

As-Produced Fibers

Fiber-Creep—In previous work [1] it was determined that anelastic creep in boron fibers is a thermally-activated process. As such, the time and temperature conditions required to produce a certain anelastic strain ε_a are not independent variables. That is, fixing the test temperature fixes the deformation time at which the given strain will be reached. The results of the low-stress FSR and FIF test indicated that for boron fibers, the relationship between the time and temperature variables was best expressed in terms of a q parameter given by

$$q = (\ln t + 33.7) \, T(10^{-3}).   \tag{5}$$

Here t is deformation time in seconds and T is test temperature in degrees

Kelvin. Thus, ε_a and the anelastic function A depend only on the one variable q rather than the two variables t and T. This result greatly simplified the experimental procedures required to determine the A function. For example, since q is weakly dependent on time *(lnt)* and directly proportional to temperature, deformation tests were typically conducted by holding the deformation time constant and measuring the development of anelastic creep strain as a function of test temperature. The deformation strains and Equations (2) and (3) were then used to calculate $A(q)$ at the q value corresponding to the particular time-temperature test conditions.

The $A(q)$ results from the low-stress flexural tests on as-produced fibers are shown as curve A_L in Figure 1. The subscript L refers to the fact that the measurements were made at low stresses below 50 ksi (0.3 GN/m²) where $A(q)$ was observed to be stress independent. Actual data points are not indicated because they have negligible error and were measured almost continuously ($\Delta q \cong 0.3\ K$). To put these results in perspective, a 60 sec test at room temperature corresponds to a q value of 28K. Since $A_L \cong 1$ for q values less than $15K$, it follows that at low stresses and short times, boron fibers deform essentially elastically at room temperature and below. However, at longer times or higher temperatures, these fibers will display anelastic creep, even at very low stress.

As previously discussed, raising the stress level above 50 ksi (0.3 GN/m²) produced an unexpected increase in the A function. To study this effect, tensile elongation and fracture tests were conducted on single fibers. In Figure 2 the A function results from the tensile and flexural tests are plotted as a function of stress for q values of 11, 20, and 29K. For tests of one minute duration these values roughly correspond to test temperatures of 20°, 250°, and 500°C. Although the stress effect data are limited, the Figure 2 curves were drawn assuming a discontinuous behavior for the boron fiber A function. That is, as stress increases, the A function remains constant at the A_L level until at some transition stress σ^* where it rather abruptly increases to a constant

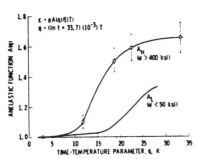

Figure 1. The anelastic strain functions for as-produced boron fibers.

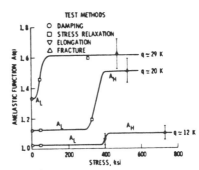

Figure 2. The effects of stress on the anelastic strain function for as-produced boron fibers.

A_H level as measured by the tensile fracture tests on etched fibers. The subscript H refers to the "high" stress level (>400 ksi (2.8 GN/m²)) of these measurements (1). The q dependence of the A_H function is shown in Figure 1. The large error bars for the A_H data points are due primarily to a coefficient of variability of ∼5 percent in the fracture stress data.

Although it is not obvious that the Figure 2 stress effect data support the assumption of an abrupt increase in A from one level to another, there does exist some indirect evidence for such behavior. For example, in previous work it was shown that boron fiber anelasticity could be explained by grain boundary type sliding of small substructural boron units [2,4]. Based on this microstructural model, the maximum A function to be expected for boron fibers is given approximately by the A_H curve of Figure 1. The fact that A_L is less than A_H suggests that at low stress all boron units cannot participate in creep, due possibly to the existence of some unknown internal "locking" mechanism. The stress-induced increase from A_L to A_H indicates that high stress can unlock the immobilized units, giving rise to maximun anelastic creep. If the locking mechanisms are all of one type, one might expect that the unlocking should occur over a narrow stress range. Experimental support for this may be found in the boron fiber torsional damping data or Firle [8] who observed that as shear stress is increased, fiber damping remains constant until some higher shear stress level at which it increases rather abruptly over a small stress range. Thus, the assumption is made that A_L and A_H are constant over certain stress regimes and that the transition from A_L to A_H occurs abruptly at σ^*. As indiated by the Figure 2 curves, σ^* decreases with increasing q or temperature, suggesting that the unlocking mechanism is thermally activated.

Summarizing the practical aspects of the above results, one can now predict creep of as-produced boron fibers by employing the following equation:

$$\varepsilon(t, T, \sigma) = [\sigma/E(T)]A_o(q) \qquad (6)$$

where $A_o = A_L$ for $\sigma < \sigma^*$ and $A_o = A_H$ for $\sigma > \sigma^*$. The a parameter is given by Equation (5) and $\sigma^*(q)$ can be estimated from the Figure 2 results. Accurate data for $E(T)$ were measured during the single fiber damping tests (1,4). These data are shown in Figure 3 in terms of the ratio $E/E(20°C)$ where $E (20°) = 60.5 \times 10^6$ psi (418 GN/m²).

It should be noted that Equation (6) describes the total deformation strain which includes both the elastic and anelastic strain components. One can calculate the anelastic creep strain ε_a simply by replacing A_o by (A_o-1) in Equation (6). However, in any practical fiber creep test, it would be very difficult to observe ε_a directly since like the elastic strain it develops linearly with stress and also begins to recover immediately upon stress removal. For this reason, total deformation strain is considered to be a more practical parameter for understanding and describing boron fiber creep. One final design point is that although considerations of stress effects on A_o may be im-

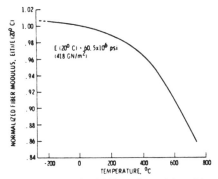

Figure 3. *The normalized axial Young's modulus of boron fibers.*

portant in some situations, one could in many circumstances neglect the stress effects and design for the upper limit boron fiber creep by simply employing A_H in Equation (6).

Fiber Fracture—Smith has observed that the two primary flaw sites responsible for crack propagation in commercially-produced boron fibers are located within the tungsten boride core and on the fiber surface [5]. By slightly etching the as-produced fibers, he was able to remove the surface flaws and thus observe only core flaw-initiated fiber fracture. As described earlier, one can utilize this fact to develop a "composite fiber" fracture model for calculating the high stress A_H from fracture stress data on slightly etched as-produced fibers. The basic question for this model is Equation (4) in which it is assumed fiber fracture occurs at the core fracture strain which is time and temperature independent. Since fracture stress data were employed to determine A_H, it follows that Equation (4) can be transposed to predict $\bar{\sigma}_u$ as a fraction of time and temperature. That is, the average fracture stress of an etched fiber (core controlled fracture) can be calculated from the equation

$$\bar{\sigma}_u(q) = \bar{\sigma}_u(q_o) \frac{E(T)}{E(T_o)} \frac{A(q_o)}{A_H(q)} \qquad (7)$$

Here $\sigma_u(q_o)$ is the average fracture stress measured at some reference q_o condition, such as, a short-time tensile test ($t_o \cong 60$ sec) at room temperature ($T_o = 293\ K$). Although $A_H(q)$ was determined from short-time fracture test data, it should be realized that theoretically Equation (7) can be generalized through the q parameter to predict core-initiated fracture during other time-dependent tests such as impact and long-time stress rupture. At the present time, however, no data exist to confirm the validity of Equation (7) for other than the short-time tensile test.

The derivation of Equation (7) was simplified by the fact that the etched fiber could be treated as a two component composite in which the outer sheath component fractures whenever the inner core component reaches its fracture strain. This composite model may not be valid, however, when the source of fiber fracture are flaws on the fiber surface. In most as-produced commercial boron fibers, Smith [5] has observed only surface flaw and core-initiated fractures. He found that the two flaw types can be practically distinguished by the fact that core-initiated fractures generally produce strength values greater than 600 ksi (4.1 GN/m²) whereas surface flaw-initiated fractures produce strength values less than 500 ksi (3.4 GN/m²). Since commercial boron fiber spools are generally quoted at average strengths of 500 ksi, it follows that surface flaws do exist in these fibers. Thus, the question arises whether Equation (7) can be utilized to predict fracture stress of unetched as-produced fibers.

This question was examined empirically by plotting in Figure 4 the short-time temperature-dependent fracture stress data of Veltri and Galasso [9] for unetched as-produced boron fibers. To better compare these data with the theoretical predictions of Equation (7), the fracture stress values were normalized by dividing $\bar{\sigma}_u$ by the room temperature value, $\bar{\sigma}_u(q_o) = 500$ ksi (3.4 GN/m²). The range of the theoretical estimates based on Equation (7) and the errors in A_H (cf. Figure 1) are shown by the dashed lines. Comparing these with the experimental data, one finds that although surface flaws were most probably controlling fiber fracture, it does appear that Equation (7) predicts quite well the fracture stress of both etched and unetched as-produced fibers. From this result it follows that Equation (7) can be employed as the general equation for predicting boron fiber fracture stress regardless of the flaw type responsible for fracture initiation. However, it should be realized that anelastic creep effects on fiber fracture are contained in the A_H function used in Equation (7). Thus, in light of the stress effect curves of Figure 2, one should replace A_H by A_L if fiber flaws should initiate fracture at stress levels below the transition stress σ^*.

B/Al Composites

Fiber creep—The anelastic A function for boron fibers within B/6061 Al composites was determined from composite damping data [6]. The results are

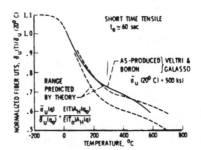

Figure 4. The normalized ultimate tensile strength of as-produced boron fibers.

shown in Figure 5 as curve A_L^{II}. The subscript L again refers to the fact that low fiber stresses (<1 ksi (7 MN/m²)) were used for these measurements. The superscript II is used to distinguish the results for fibers within B/Al composites from the as-produced fiber results which are now labeled with the superscript I.

Comparing the A_L^{II} curve with the as-produced A_L^{I} curve, which is also shown in Figure 5, one observes that the anelastic creep of the as-produced boron fibers is measurably reduced within B/Al composites. Obviously a microstructural change must have occurred in the fiber at some time during composite fabrication. Since the most adverse environmental conditions existed during high temperature diffusion bonding of the composite specimen, the microstructural change most probably occurred during this stage when the boron fibers reacted with the aluminum matrix to form the interfacial bond required for mechanical load transfer. In support of this surface reaction effect on the A function, it was found [7] that Borsic fibers in both the as-produced and composite conditions possess low stress A functions equivalent to the A_L^{II} result in Figure 5. Thus, on the basic level, the boron fiber microstructure responsible for its bulk anelastic deformation character can be measurably affected by surface reactions either with matrices or with fiber coatings. In terms of the boron unit sliding model, it would appear that at the high temperatures of the surface reactions, diffusional processes occur within the fiber which increase the number of immobilized boron units.

Although A_L^{II} can now be inserted into Equation (6) to accurately predict low-stress boron and Borsic creep in B/Al composites, the upper stress limit σ^* at which it can no longer be validly employed remains undetermined. As with the as-produced single fibers, high stress deformation experiments are required from which A^{II} versus stress data can be determined. These experiments could be performed either on single fibers extracted from B/Al composites or on unidirectional B/Al composites in which the fiber contributions can be easily and accurately measured. One obvious method of maximizing fiber effects and minimizing matrix effects in composites would be to

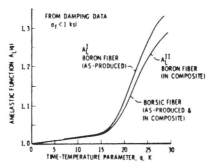

Figure 5. *The effects of composite fabrication and SiC coating on the low-stress anelastic strain function for as-produced boron fibers.*

study specimens with fiber volume fractions of 50 percent or greater. It should be mentioned that high stress creep data for 50 fiber volume percent B/Al composites do exist in the literature [10]. Attempts to employ these data for extracting high stress A^{II} curves were not very fruitful, primarily due to the existence of unknown matrix stress relaxation effects on total composite creep [11]. However, although composite creep data were not useful, it was found that a theoretical study of composite fracture stress versus temperature data could not only shed light on high stress effects on fiber creep but also yield equations for predicting time-temperature effects on B/Al fracture. The theory and results of this study will now be discussed.

Composite fracture—Having established that the A_H function is the appropriate function for predicting high stress fracture of single as-produced fibers, the first question to be answered is whether this function can be used also to predict the high stress creep and fracture of boron fibers within B/Al composites. To examine this question, a literature search was conducted for short-time fracture stress versus temperature data for B/6061 Al composites. In order to minimize matrix loading effects, the search was confined only to composites with nominally 50 percent fiber volume fraction. Summary plots of the literature data which fit this requirement are shown in Figure 6. Although the tensile strengths vary in magnitude from one source to another, one can notice definite trends in the temperature-dependent behavior. For example, below 200 °C the data from all sources indicate essentially no dependence on temperature. Above 200 °C the strength data fall off reaching about 80 percent of the room temperature value near 300 °C.

Regarding fracture of fibers within B/Al composites, one might as a first approximation neglect matrix contributions and assume that the composite stress should drop off at least as rapidly as the fracture stress for single as-produced boron fibers. Examining the experimental results of Figure 4, it is seen that single as-produced fibers fall off to 80 percent of $\bar{\sigma}_u(q)$ near 150 °C, a much faster dropoff than the composite. Thus, from a practical point of

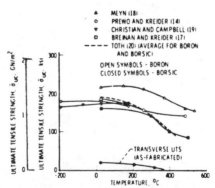

Figure 6. *The ultimate tensile strength of B/6061 A1 composites.*

view, it appears that the Veltri and Galasso data for as-produced fibers cannot be used to understand and predict the temperature dependence of composite strength. From an analytical point of view, one must conclude that the insertion of the A_H in the fiber fracture theory of Equation (7) will not explain the Figure 6 composite data. Since the A_L results of Figure 5 indicate that boron fibers will creep within B/Al composites, the problem of predicting the effects of time and temperature on B/Al axial fracture thus becomes one of not only accounting for matrix plasticity but also of determining the appropriate A function for high stress creep and fracture of the fibers after composite fabrication.

To solve this problem, a theoretical analysis was made of the major factors which affect the temperature-dependent behavior of the fracture stress of metal matrix composites in general and B/6061 Al composites in particular. The axial fracture model chosen is that developed by Rosen [12] in which the fracture modes of fibers within a unidirectional composite are controlled by the fracture of fiber bundles. That is, the composite or fiber bundle completely fails when enough fiber breaks occur so that the load carried by the remaining intact fibers exceeds their strength capability. Due to a distribution of fiber strengths, the weak fibers fracture first leaving the stronger fibers to carry the load. Common practice is to describe the distribution in fiber strengths according to a Weibull distribution [13]. In this case, the average fiber strength $\bar{\sigma}_{uf}$ and the average fiber bundle strength $\bar{\sigma}_{bf}$ are given by

$$\bar{\sigma}_{uf} = \sigma_o \left[\frac{L_s}{d} \right]^{-1/\omega} \Gamma\left(\frac{\omega + 1}{\omega} \right) \tag{8}$$

and

$$\bar{\sigma}_{bf} = \sigma_o \left[\frac{e\omega L_b}{d} \right]^{-1/\omega} \tag{9}$$

Here σ_o is the Weibull distribution scale factor, ω is the Weibull shape factor which describes the scatter in strength values, Γ is the tabulated gamma function, e is the natural base constant, d is the fiber diameter, and L_s and L_b are the test gage lengths of the single fiber and the fiber bundle, respectively. Typically $\omega > 1$, so that $\bar{\sigma}_{bf}$ decreases as the bundle length increases.

When a fiber breaks in a metal matrix composite, the matrix by virtue of its plastic character localizes the loss of load carrying ability of the broken fiber. That is, at an axial distance $\delta/2$ on either side of the break, the stress in the broken fiber returns to the average stress of all the intact fibers. Because of the existence of this "ineffective" length δ, Rosen considers the composite to be made of a series of independent fiber bundles each of length δ. As such, total fracture of the composite occurs whenever any one of these short length bundles fail. If one assumes perfectly plastic behavior for the matrix (no workhardening effects), the length δ of the bundles can be calculated from

$$\delta = \sigma d/2\tau_m \tag{10}$$

where $\sigma = \bar{\sigma}_{bf}$ is the stress in the fibers at composite fracture and τ_m is the shear strength of the matrix. Since τ_m decreases with temperature, it follows that δ will increase, giving rise by Equation (9) to a reduction in $\bar{\sigma}_{bf}$ with temperature.

To express the total stress level at which a metal matrix of composite will fracture, one can employ the above concepts to write the following rule-of-mixtures equation for the average (ultimate) tensile strength of the composite

$$\bar{\sigma}_{uc} = v_f \bar{\sigma}_{bf}(\delta) + v_m \sigma_{ym}(T) \tag{11}$$

Here v_f and v_m are the volume fraction of the fiber and matrix, respectively, and σ_{ym} is the tensile yield strength of the matrix. Equation (11) neglects stress concentration effects of broken fibers on nearby fibers. It also neglects residual stress effects on the fibers due to cooling the composite from fabrication temperature. For soft matrices such as aluminum these effects are small and also tend to oppose each other as temperature is varied. For calculating the temperature dependence of $\bar{\sigma}_{uc}$ it is convenient to normalize each stress term in Equation (11) by dividing by its value for a short-time tensile test at room temperature. The following R parameters are thus defined:

$$R_{uc} = \bar{\sigma}_{uc}(q)/\bar{\sigma}_{uc}(q_o) \tag{12}$$

$$R_{bf} = \bar{\sigma}_{bf}(q)/\bar{\sigma}_{bf}(q_o) \tag{13}$$

$$R_{ym} = \sigma_{ym}(T)/\sigma_{ym}(T_o) \tag{14}$$

In anticipation of applying these equations to B/Al composite behavior, the temperature variable T was replaced, where appropriate, by the more general q variable. Reference conditions (q_o and T_o) are taken as a short-time ($t_o \cong 60$ sec) tensile test at room temperature ($T_o \cong 293\,°K$). Inserting Equations (12), (13), and (14) into Equations (11) one obtains

$$R_{uc} = (R_{bf} + \beta R_{ym})/(1 + \beta) \tag{15}$$

where the constant $\beta = v_m \sigma_{ym}(T_o)/v_f \bar{\sigma}_{bf}(q_o)$. Thus, to predict composite fracture stress $\bar{\sigma}_{uc}(q)$, one simply needs a theoretical formula for R_{bf} plus experimental information for $\bar{\sigma}_{uc}(q_o)$ and $R_{ym}(T)$.

To derive R_{bf} one can utilize Equation (9) to write

$$R_{bf} = \frac{\sigma_o(q)}{\sigma_o(q_o)} \left[\frac{\delta(T)}{\delta(T_o)} \right]^{-1/\omega} \tag{16}$$

Here it is assumed that the flaw distribution as measured by ω does not change during the test. Also, to include fiber creep effects, a q dependence has been assigned to σ_o since by Equation (8), σ_o is the only term responsible for a change in fiber strength with temperature. Assuming matrix shear strength τ_m is directly proportional to matrix yield strength σ_{ym}, one finds from Equation (10) that

$$\delta(T)/\delta(T_o) = R_{bf}/R_{ym} \tag{17}$$

By inserting this result into Equation (16) and manipulating it follows that

$$R_{bf} = \left[\frac{\sigma_o(q)}{\sigma_o q_o)}\right]^{\omega/(1+\omega)} \left[R_{ym}\right]^{1/(1+\omega)} \tag{18}$$

Thus, under the assumptions stated above, one should now be able to employ Equations (15) and (18) to predict the time-temperature dependent fracture stress of metal matrix composites. For those composites reinforced by boron fibers, both terms on the right hand side of Equation (18) must be considered. However, for those composites reinforced by elastic fibers, $\sigma_o(q) = \sigma_o(q_o)$ so that only the second matrix-related term need be considered.

To determine the high stress A function for boron fibers in B/Al composites, Equations (15) and (18) were applied to the experimental strength data of Figure 6. It was assumed that the temperature dependence of R_{ym} for the 6061 aluminum matrix followed that measured by Prewo and Kreider [14] for the transverse tensile strength of as-fabricated B/6061 composites. Their results which are plotted in Figure 6 show a drop in transverse strength from 20 ksi (138 MN/m²) at room temperature to an extrapolated value of 3 ksi (21 MN/m²) at 370°C [21]. Assuming a room temperature matrix shear strength of 14 ksi (97 MN/m²) [15] and a 5.6 mil (142 μm) diameter fiber whose strength falls off by 70 percent at 370°C, it follows from Equation (10) that the ineffective length δ increases from 0.1 to 0.5 inch (2.5 to 12 mm) between 20° and 370°C. To compare the Figure 6 strength data with theory each set of data was normalized by dividing by their room temperature values and the best fit R_{uc} for all sets was calculated. The best fit result is shown by the dashed curve of Figure 7.

Regarding values for the parameters in the theoretical equations, σ_{bf} and σ_{ym} at room temperature were assumed to be 400 and 8 ksi (2.8 and 0.06 GN/m²) [14], so that $\beta = 0.02$ for 50 percent fiber content in as-fabricated 6061 Al. In examining the literature for information concerning the Weibull distribution of boron fibers removed from B/6061 Al composites, ω parameters were found which range between 8 and 12 with $\omega = 10$ as an average value [14,15,16]. Since the boron fibers definitely show anelastic creep within the composites, the assumption was made that fiber fracture could be described by Equation (7) with an appropriate A function. Thus, for Equation (18), one can write

$$R_{bf} = \left[\frac{E(T) A(q_o)}{E(T_o A(q)}\right]^{\omega/(1+\omega)} \left[R_{ym}\right]^{1/(1+\omega)} \tag{19}$$

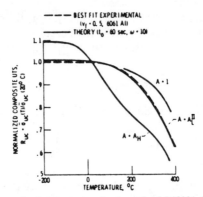

Figure 7. The normalized ultimate tensile strength of B/6061 A1 composites.

With $\omega = 10$, R_{bf} was inserted into Equation (15) and R_{uc} was calculated for three different A functions: $A = 1$ which assumes strictly elastic fiber behavior (no creep), $A = A_H$ which assumes the as-produced high stress condition for the fibers, and A_L^{II} which assumes that the low stress creep behavior does not change for fiber stresses up to $\sim$400 ksi (2.8 GN/m²).

The theoretical results for the three A functions are shown in Figure 7. For elastic fiber behavior, curve $A = 1$ clearly shows that composite strength should fall off simply due to matrix effects in which temperature decreases τ_m and increases the ineffective length δ. If $\omega = \infty$, which implies constant strength fibers, Equation (19) indicates that the matrix would have no effect on fiber bundle strength. For the assumption of as-produced behavior, the A_H curve obviously drops off much too rapidly to explain the as-produced fiber data of Figure 4 with the composite data of Figure 6. Finally, it appears that the A_L^{II} curve gives the best fit to the fracture data. Thus it may be reasonable to assume that fiber creep within B/Al composites is governed by only the A_L^{II} function from zero to at least 400 ksi (2.8 GN/m²). The shift from A_L to A_H observed for the as-produced fiber (cf. Figure 2) apparently does not occur for the B/Al fibers. From a basic point of view one might interpret this as an additional effect of the boron-aluminum interfacial reaction in that besides decreasing the number of mobile boron units, it may also affect the internal locking mechanism in such a way as to increase σ^* to values greater than 400 ksi.

Summarizing the design aspects of the above discussion, it appears that for any fiber stress less than 400 ksi (2.8 GN/m²), the A_L^{II} curve of Figure 2 is the proper A function for describing boron and Borsic creep in B/6061 A1 composites. Thus, equation (6) with $A_o = A_L^{II}$ is the predictive equation for fiber creep. Regarding composite fracture, Equations (15) and (18) with $A = A_L^{II}$ and $\omega = 10$ give good estimates of B/6061 A1 axial fracture strength as a function of temperature. However, because of the thermally-activated nature of boron creep, these equations can be generalized through the q parameter to also include test conditions in which time is the principle variable. For example, in an axial stress rupture test, one can assume $\beta R_{ym} = 0$ (due to rapid stress relaxation in the matrix) and write for the stress rupture strength

$$\bar{\sigma}_{uc}(q) = \bar{\sigma}_{uc}(q_1) \left[\frac{A(q_1)}{A(q)} \right]^{\omega/(1+\omega)} \tag{20}$$

Here $q \geqslant q_1$ and $\sigma_{uc}(q_1)$ is the short-time composite strength at temperature T_1 which is held constant during the test. Since q develops with the log of time, σ_{uc} should show only a weak time dependence. Indeed, limited stress rupture data for B/Al composites confirm such a dependence [17].

The apparent confirmation of the mechanistic theory used to calculate composite fracture stress suggests possible methods of improving the creep and associated stress rupture characteristics of B/Al composites. For example, the fact that fiber creep is reduced by high temperature surface reactions with an SiC coating and a 6061 aluminum matrix suggests that perhaps other fiber secondary treatment processes exist which can either eliminate or drastically reduce anelastic creep in the as-produced boron fibers. These treatment processes, however, should have minimum adverse effects on the as-produced fiber flaw character or else the benefits gained by reducing creep could be lost by a degradation in fiber strength. Regarding the matrix shear strength effect on composite fracture, it follows from Equation (18) that this effect can be minimized by narrowing flaws within the interfacial surface phase that control fiber fracture in metal matrix composites, the largest ω values should result from those bonding conditions which produce the most uniform distribution of interfacial flaws. Thus, a good mechanical interfacial bond may not only produce a stronger composite at room temperature but also at higher temperatures where matrix shear strength falls off.

SUMMARY OF RESULTS AND CONCLUSIONS

Due to linear stress dependence of anelastic strain, it is possible to write simple analytical equations for describing boron fiber creep and fracture as a function of time, temperature, and stress. The primary time-temperature dependent variable in these equations is an anelastic strain function A. From analysis of single fiber deformation and fracture data, A functions for commercial boron fibers were determined. Thus, the creep strain and stress rupture strength of as-produced boron fibers can now be predicted.

Analysis of Borsic fiber and B/Al composite damping data indicates that the creep of SiC coated boron fibers and of boron fibers in B/Al composites is measurably less than the creep of as-produced boron fibers. The reduced A function characteristic of fiber creep in B/Al composites was determined. This function together with a general metal matrix composite fracture theory should now allow fairly good estimates of the effects of time and temperature on the axial failure modes of B/Al composites. Based on the mechanistic models employed in this composite fracture theory, it appears that a good interfacial fiber-matrix bond will not only maximize mechanical load transfer but also may reduce detrimental fiber and matrix effects on B/Al stress rupture properties.

NOMENCLATURE

$A{:}A_L,A_H$	Anelastic strain function: at low stress, at high stress
d	fiber diameter
E	fiber Young's modulus
L_s,L_b	gage length of single fibers and fiber bundle
q	time-temperature parameter
R_{bf},R_{uc},R_{ym}	ratio of test value to short-time value at room temperature for fiber bundle strength, composite tensile strength, and matrix yield strength
t	deformation time (sec)
T	test temperature (K)
v_f,v_m	volume fraction of fibers and matrix
β	ratio of matrix load to fiber load under short-time composite loading at room temperature
δ	"ineffective length" of fiber bundles in composites
$\varepsilon{:}\varepsilon_e,\varepsilon_a$	total fiber strain: elastic component, anelastic component
$\bar{\varepsilon}_u$	average (ultimate) fiber fracture strain
σ_o	Weibull distribution stress factor
σ^*	transition stress for change from A_L to A_H
$\bar{\sigma}_{uf}$	average (ultimate) fiber tensile strength
$\bar{\sigma}_{bf}$	average fiber bundle strength
$\bar{\sigma}_{uc}$	average (ultimate) composite tensile strength
σ_{ym}	tensile yield strength for matrix
τ_m	shear strength of matrix
ω	Weibull distribution shape factor

Superscripts

$I;II$	for as-produced fibers; for fibers within B/Al composites

REFERENCES

1. DiCarlo, J. A., "Time-Temperature-Stress Dependence of Boron Fiber Deformation," *Composite Materials: Testing and Design (Fourth Conference), ASTM STP 617*, American Society for Testing Materials, Philadelphia 443–465 (1977).
2. Prewo, K. M., "Anelastic Creep of Boron Fibers," *J. Compos. Mater., 8*, 411–414 (1974).
3. DiCarlo, J. A., "Anelastic Deformation of Boron Fibers," *Scripta Met., 10*, 115–119 (1976).
4. DiCarlo, J. A., "Mechanical and Physical Properties of Modern Boron Fibers," *ICCM/2, Second International Conference on Composite Materials*, The Metallurgical Society of AIME, New York, 520–538 (1978).
5. Smith, R. J., "Changes in Boron Fiber Strength due to Surface Removal by NASA TN D-8219," Washington, D.C. (1976).
6. DiCarlo, J. A. and Maisel, J. E., "Measurement of Time-Temperature Dependent Dynamic Mechanical Properties of Boron/Aluminum Composites," *Composite Materials: Testing*

and Design (Fifth Conference), ASTM STP 674, American Society for Testing Materials, Philadelphia, 201–227 (1979).

7. DiCarlo, J. A. and Williams W., "Dynamic Modulus and Damping of Boron Silicon Carbide, and Alumina Fibers," NASA TM-81422, Washington, D.C. (1980).

8. Firle, T. E., "Amplitdue Dependence of Internal Friction and Shear Modulus of Boron Fibers," *J. Appl. Phys., 39,* 2839–2845 (1968).

9. Veltri, R. D. and Galasso, F. S., "High-Temperature Strength of Boron, Silicon Carbide, Stainless Steel, and Tungsten Fibers," *J. Am. Ceram. Soc., 54,* 319–320 (1971).

10. Kreider, K. G. and Prewo, K. M., "Boron-Reinforced Aluminum," in *Composite Materials,* K.G. Dreider, ed., Vol. 4. New York: Academic Press, 399–471 (1974).

11. de Silva, A. R. T., "A Theoretical Analysis of Creep in Fibre Reinforced Composites," *J. Mech. Phys. Solids, 16,* 169–186 (1968).

12. Rosen, B. W., "Tensile Failure of Fibrous Composites," *AIAA, J., 2,* 1985–1991 (1964).

13. Corten, H. T., "Micromechanics and Fracture Behavior of Composites," in *Modern Composite Materials,* L.V. Broutman and R.H. Krock, eds. Reading, Mass: Addison-Wesley, 27–105 (1967).

14. Prewo, K. M. and Kreider, K. G., "High Strength Boron and Borsic Fiber Reinforced Aluminum Composites," *J. Compos. Mater., 6,* 338–357 (1972).

15. Wright, M. A. and Wills, J. L., "The Tensile Failure Modes of Metal Matrix Composite Materials," *J. Mech. Phys. Solids, 22,* 161–175 (1974).

16. Wright, M. A. and Welch, D., "Failure of Centre Notched Specimens of 6061 Aluminum Reinforced with Unidirectional Boron Fibers," *Fibre Sci. Technol., 11,* 447–461 (1978).

17. Breinan, E. M. and Kreider, K. G., "Axial Creep and Stress-Rupture of B-Al Composites," *Metall. Trans., 1,* 93–104 (1970).

18. Meyn, D. A., "Effect of Temperature and Strain Rate on the Tensile Properties of Boron-Aluminum and Boron-Epoxy Composites," *Composite Materials: Testing and Design (Third Conference),* ASTM STP 546, American Society for Testing Materials, Philadelphia, 225–236 (1974).

19. Christian, J. L. and Campbell, M. D., "Mechanical and Physical Properties of Several Advanced Metal-Matrix Composite Materials," *Cyrogenic Engineering Conference, Advances in Cryogenic Engineering.* Vol. 18, New York: Plenum Press, 175–183 (1973).

20. Toth, I. J., "Comparison of the Mechanical Behavior of Filamentary Reinforced Aluminum and Titanian Alloys," *Composite Materials: Testing and Design (Third Conference),* ASTM STP 546, American Society for Testing Materials, Philadelphia, 542–560 (1974).

21. *Metals Handbook,* Vol. 1, Metals Partk, Ohio: American Society of Metals, 946 (1961).

11

Detection of Moisture in Graphite/Epoxy Laminates by X-Ray Diffraction

P. PREDECKI AND C. S. BARRETT

INTRODUCTION

THE MOISTURE CONTENT OF RESIN MATRIX LAMINATES IS NORMALLY CAL-culated from theory using Fickian diffusion equations, if the temperature and relative humidity of the laminate is known. Evidence for the validity of this approach has been obtained from moisture weight gain/loss experiments [1]. Moisture content has also been determined by direct chemical methods but these usually require removal of a sample from the laminate. In this paper we describe briefly an X-ray diffraction method for detecting moisture via the changes in residual strains in crystalline filler particles which are embedded in the matrix during layup.

In earlier work [2] we have shown that particles embedded between the first and second plies of unidirectional laminates act as useful indicators of residual and applied stress in the laminates. The principle elastic strains in the particles are measured by X-ray diffraction and are related to laminate stresses by calibration experiments using known applied loads.

In this work, the observed sensitivity of the residual elastic strains in such filler particles to the moisture history of the laminate was investigated further.

EXPERIMENTAL

Six ply, quasi-isotropic $(0°, \pm 60°)_s$ laminates were laid up using Fiberite T300/934 (Fiberite Corp., Winona, Minn.) graphite epoxy prepreg. Aluminum particles of –325 mesh were introduced between the first and second plies during layup by spreading a thin layer of these particles on the first green ply and shaking of particles which did not adhere. Laminates were

cured at 177 °C (350 °F) using procedures recommended for this prepreg.*
Residual stress samples 4.45 × 5.08 × .089 cm thick were then cut from the
laminates and stored in sealed polyethylene bags prior to use.

Samples were exposed to various humidities in a controlled humidity
chamber at 50 °C (Blue M model VP, 100RAT-1, Blue Island, Ill.). Zero
humidity samples were placed in a container with $CaSO_4$ (Drierite) at 50 °C.
Samples were periodically removed and X-rayed or weighed in ambient
laboratory conditions and replaced in the controlled humidity chamber.

The residual strains in the filler particles were measured using procedures
described elsewhere [2,3]. Samples were held in the standard sample holder of
a Siemens Krystalloflex-4 diffractometer. The counter holder was modified
for diffraction angles, 2θ, up to 165° and fitted with a monochromator.
Aluminum powder identical to that used in the laminates was used as a stress
free standard and was positioned in the sample holder in exactly the same
plane as the filler particles of the sample (which were 0.16 mm below the sample surface). Diffraction peak positions of the 333 plus 511 Al peak (around
162.7 °2θ) were determined by fitting a parabola to five step-scanned points
on the top ¼ of the peak from each sample [4] at room temperature. All 2θ's
were corrected to 22.2 °C, using a correction of 0.010 °2θ per degree F, obtained
from the thermal expansion coefficient of aluminum. Each peak determination required 10 to 15 mins. Peaks were measured in three directions relative
to the sample geometry yielding the residual particle strains, $\varepsilon_{\phi\psi}$, at $\phi = 90$
$\psi = 0$, $\phi = 0$ $\psi = 45$ and $\phi = 90$ $\psi = 45$. The angles ϕ and ψ are defined in
Figure 1. From these principle residual strains, ε_{ip}^r, ($i = 1-3$) in the particles
were calculated assuming the laminate orthotropic axes, 1,2,3, were the prin-
cipal axes. The in-plane strains ε_{1r}^r and ε_{2p}^r were almost identical as expected
for the quasi-isotropic laminate. The standard deviation in determining peak
positions was ± .015 °2θ which corresponds to a standard deviation in the
measured strains of ± 56 × 10^{-6}.

RESULTS AND DISCUSSION

The changes in the diffracted peak position, $2\theta_{90,0}$, measured at $\phi = 90$,
$\psi = 0$ with time are shown in Figure 2. Initially at time = 0, the peak position
is quite high, ∼ 163.17 relative to the unstressed Al powder, indicating a dry
sample, as might be expected for the as-cured laminate stored 2 weeks in a
sealed polyethylene bag. Further drying in Drierite at 50 °C further increased
the peak position slightly. After 170 hrs in 0 percent humidity this sample was
placed in 100 percent humid air at 50 °C. There was an immediate and pronounced decrease in peak position accompanied by a weight gain of the
laminate due to moisture absorption. The weight gain reached a plateau at
∼1.3 percent about 230 hrs after the start of the wetting cycle, whereas the
$2\theta_{90,0}$ value went through a minimum and then equilibrated after ∼ 500 hrs at

---

*We are grateful to R. Mirschell, R. Campbell, and B. Burke of Martin-Marietta Co., Denver for fabricating the
samples to our specifications.

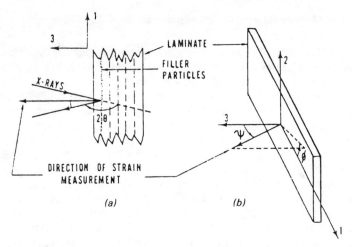

Figure 1. (a) Representation of diffraction conditions, (b) Direction of strain measurement relative to laminate axes.

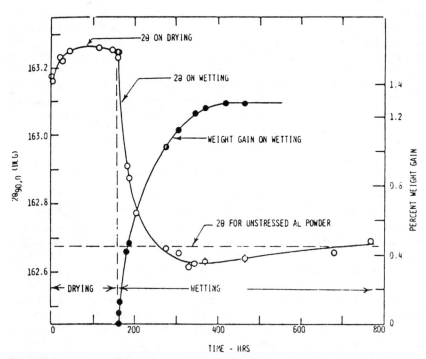

Figure 2. Sensitivity of diffracted beam position (for $\phi = 90$, $\psi = 0$) to environmental moisture at 50°C. Relative humidity was zero during "drying" and 100% during "wetting." The percent weight gain of the laminate during wetting is also shown.

approximately the value for the unstressed Al powder. Samples exposed to more than one such 0 percent RH drying/100 percent RH wetting cycle have shown that the changes in $2\theta_{90,0}$ are reversible and approximately super-posable, provided the samples reach equilibrium in each condition.

The $2\theta_{90,0}$ data for a wetting cycle such as shown in Figure 2, together with $2\theta_{0,45}$ and $2\theta_{90,45}$ data simultaneously taken, were converted to particle strains and the results plotted in Figure 3. It is evident that the in-plane residual strains ε^r_{1p} and ε^r_{2p} which are initially tensile, decrease to zero in 100 to 150 hrs as a result of moisture absorption by the resin. At short times one might expect ε^r_{2p} to decrease more rapidly than ε^r_{1p} since the 0° plies are outer-most and the laminate should experience initially a larger strain in the 2 direc-tion than in the 1 direction [5]. This effect was not detected in the X-ray measurements.

The third strain, ε^r_{3p}, initially compressive, becomes slightly tensile presumably due to swelling of the resin in the 3 direction, before decaying to zero at longer times. This decay is through to be the result of viscoelastic relaxation of the resin in the 3 direction. Using the nomenclature of Yeow et al. [6], the inplane particle strains would be expected to be fiber dominated and therefore time insensitive—apart from the time dependence of moisture uptake. The thickness strain, ε^r_{3p}, is resin dominated and would be expected to show relaxation. This relaxation is also evident in Figure 2 where $2\theta_{90,0}$ peak position continues to decay toward the unstrained state after the weight gain has become constant.

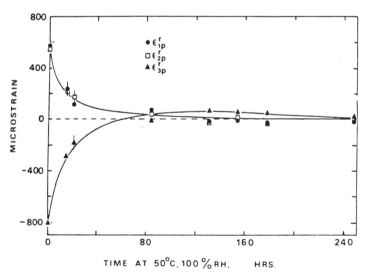

Figure 3. *Changes in principal residual particle strains during the wetting cycle of Figure 2. Error bars in this and subsequent figures are ± one average standard deviation.*

To facilitate use of the method as a probe for laminate moisture content, samples were held at various relative humidities until no further change in diffracted peak position was observed—usually 400–500 hrs. at 50 °C. The resulting plot of equilibrium peak position versus relative humidity (Figure 4) is only linear to a first approximation. The observed scatter was attributable both to sample-to-sample variation and to sample moisture history variation, a scatter that is somewhat more than the statistical error in the measurements. From Figure 4 it is clear, however, that the relative humidity for producing a stress-free state in the Al filler particles at room temperatures is near 100 percent. This is also the percent R.H. for the laminate to be free of residual stresses since the strains in the particles have been shown to be proportional to applied laminate stresses [2]. (Here we are neglecting the contribution to the particle strain due to thermal and moisture expansion mismatch between the Al and the resin.) Calculations by Tsai and Hahn [7] based on single lamina hygrothermal properties for a typical graphite/epoxy laminate, predict the R.H. value for zero residual stress to be around 55 percent. The laminate material considered by Tsai and Hahn had a greater saturation moisture content than found in this study (1.8 percent equilibrium weight gain at 100 percent R.H. versus our 1.3 percent) which may account for the difference.

The effect of annealing the laminates below the curing temperature (177 °C) was explored to see if any residual strain relief is obtained. Since an increase in moisture content reduces the diffraction angle $2\theta_{90,0}$ and therefore the residual strains as shown in Figures 2 and 3, attempts were made to avoid any

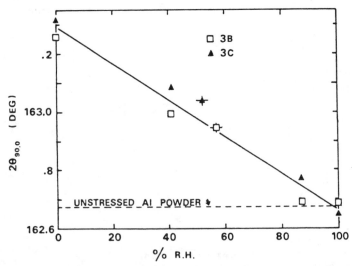

Figure 4. Roughly linear relation between diffracted peak position (for $\phi = 90$, $\psi = 0$) and relative humidity after equilibrium has been reached. Some sample to sample variation is evident.

increase in moisture content during annealing. Two specimens that had been dried in a Drierite desiccator at 50°C (sample 3A for 10 days, sample 3F for 31 days) were annealed for periods of one to two hours at successively higher temperatures; sample 3A in an air oven and sample 3F in a P_2O_5 desiccator. The change in $2\theta_{90,0}$ measured at room temperature as a result of annealing is shown in Figure 5. From Figure 5 one may conclude that the samples started with different moisture contents and that the P_2O_5 was more effective in eliminating moisture during anneals than air annealing until temperatures reached 150°C. There was no evidence of a decrease in diffraction angle and therefore in residual strains as a result of annealing.

Returning to Figure 2, the relation between the X-ray measurements and the moisture weight gain was examined further by plotting the average in-plane particle residual strain; $(\varepsilon_{1p}^r + \varepsilon_{2p}^r)/2$ against the fractional weight gain, $\bar{c}/c_f$, where c_f is the final or plateau moisture concentration in the laminate and c is the average concentration at any time during the wetting cycle. The resulting plot shown in Figure 6 (curve A) is not linear. One might expect that the residual strains would correlate better with the local moisture concentration, $c(x,t)$, at the particular depth $x = x_p = 0.16$ mm below the surface where the particles are located. $c(x_p,t)/c_f$ was obtained assuming Fickian diffusion of moisture in the laminate and using a value of $15.3 \times 10^{-8}mm^2/s$ for the transverse diffusion coefficient calculated from the weight gain data of Figure 2 [3]. The resulting correlation shown as curve B in Figure 6 is similar to and not any better than curve A. The non-linearity may be a consequence of the higher resin content of the layer in which the particles lie. The nonlinearity

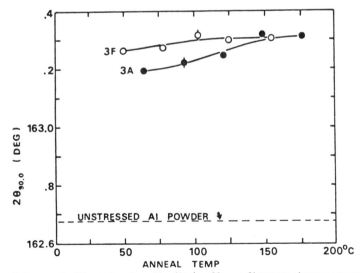

Figure 5. *Increase in diffracted peak position (for $\phi = 90$, $\psi = 0$) measured at room temperature, as a result of annealing laminates at successively higher temperatures; sample 3A always in air, sample 3F always in a P_2O_5 desiccator.*

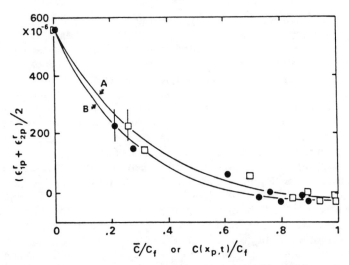

Figure 6. *Decrease in average in-plane residual strain in the Al particles with increasing average moisture content of the laminate, c/c_t, (curve A), and with increasing moisture concentration at the location of the particles, $c(x_p,t)/c_t$, (curve B).*

may also be attributed to the changing gradients in moisture concentration and the accompanying stress gradients. These gradients were avoided in the experiments reported in Figure 4, where approximate linearity was found between percent R.H. and changes in 2θ.

CONCLUSIONS

1. The changes in the diffracted peak positions using $CuK\alpha_1$ radiation and the 333 + 511 planes in Al particles in the laminates are quite sensitive to environmental moisture taken up by the laminates. The reversible change of $0.624°2\theta$ on going from the completely dry to the completely wet condition at equilibrium, compared with an average measurement error of $\pm .015°2\theta$, is sufficiently large to be useful for moisture measurement. The diffracted intensity is sufficient for these measurements to be made at depths up to ~0.38 mm beneath the surface with conventional equipment.
2. The relative humidity which gives zero residual stress at room temperature in the laminates investigated (Fiberite T300/934) is near 100 percent.
3. The correlation between the average in-plane residual strain in the particles and the calculated moisture concentration at the depth below the laminate surface where the particles lie is non-linear.
4. Annealing in either ambient or desiccated air at temperatures below the curing temperature (177 °C) serves mainly to remove additional moisture from the laminates and thereby increases the residual strains.

REFERENCES

1. Shen, C.-H. and Springer, G. S., "Moisture Absorption and Desorption of Composite Materials," *J. Composite Matls., 10,* 2 (1976).
2. Predecki, P. and Barrett, C. S., "Stress Measurement in Graphite/Epoxy Composites by X-ray Diffraction from Fillers," *J. Composite Matls., 13,* 61 (1979).
3. Predecki, P. and Barrett, C. S., "Residual Stresses in Resin Matrix Composites," Presented at the 28th Sagamore Army Materials Research Conference, Lake Placid, New York, July 13–17, 1981 (To be published in conference proceedings).
4. Soc. Automotive Engineers Information Report SAE J 784a.
5. Flaggs, D. L. and Crossman, F. W., "Analysis of Viscoelastic Response of Composite Laminates During Hygrothermal Exposure," *J. Composite Matls., 15,* 21 (1981).
6. Yeow, Y. T., Morris, D. H., and Brinson, H. F., "Time-Temperature Behavior of a Unidirectional Graphite/Epoxy Composite" in *Composite Materials: Testing and Design (5th Conference),* S.W. Tsai, editor. ASTM STP 674, 263–281 (1979).
7. Tsai, S. W. and Hahn, H. T. *Introduction to Composite Materials.* Lancaster, PA: Technomic Publ. Co., 352 (1980).

12

Moisture Detection in Composites Using Nuclear Reaction Analysis

R. J. DeIasi and R. L. Schulte

INTRODUCTION

THE LONG-TERM PERFORMANCE OF POLYMERS AND FIBER-REINFORCED polymer matrix composite materials in aerospace applications depends on the stability of these materials under various environmental conditions. The combined effect of temperature-humidity is of special importance. It is known that epoxy resins and composites absorb moisture during exposure to humidity and that the absorbed water causes degradation of their elevated temperature properties. Thus, it is important to have a thorough knowledge of the rate and mechanism of water transport in polymers and fiber-reinforced composites as a basis for understanding the mechanism of degradation of properties for developing improved materials.

Until recently, studies of water transport in polymers and composites have been based on data obtained from measurements of the rate of moisture absorption under isothermal conditions and the application of analytical diffusion models to explain these data. The diffusion process, however, can be complicated by the presence of fiber reinforcements and the application of repeated thermal cycles and moisture absorption-desorption cycles [1,2].

To obtain the necessary experimental data for a more detailed and direct understanding of water transport mechanisms, a sensitive and reliable method is required for determining the concentration of water as a function of depth of penetration, fiber orientation in composite specimens, relative humidity, time and temperature of exposure, and thermal cycling.

Several experimental approaches have been under development to determine localized moisture content in graphite-reinforced composites. These approaches include chemical analysis of degradation products generated during laser irradiation [3], measurement of cumulative moisture evolution and effusion rate [4], and correlation between bulk moisture content and dielectric constant [5]. In this paper we demonstrate the application of a nuclear reaction technique for directly measuring localized moisture content through the thickness of composite materials.

EXPERIMENTAL APPROACH

Moisture Profiling

Profiling of the localized moisture content is based on conditioning specimens in D_2O instead of H_2O and measuring the resultant deuterium concentrations in the specimens. The measurement of deuterium in the near surface region of materials is accomplished by using the $D(^3He,p)^4He$ nuclear reaction. A finely collimated beam, $12\mu m \times 200\mu m$ of ^3He ions from a Van de Graaff accelerator is incident on the specimen. In the interaction of the ^3He beam with the deuterium in the sample matrix, high energy protons are emitted and these are detected by a silicon surface barrier detector located at an angle of 165 degrees to the incoming beam. The emerging protons form a distinct, isolated group in the charged particle energy spectrum. The proton yield is directly proportional to the deuterium concentration within the reaction volume. Absolute deuterium concentrations can be determined by comparing the proton yield from the sample under study to the yield from a reference sample whose concentration is known after correcting for the different stopping powers of the samples. Two types of deuterated specimens have been successfully used as reference standards. The first is a primary standard prepared by recrystallization of phthalic acid from D_2O or from D_2O-H_2O mixtures. This permits preparation of a range of standards containing as much as 11.1 At. percent deuterium. The second standard is deuterated titanium which yielded 66.6 At. percent deuterium when analyzed relative to a deuterated phthalic acid primary standard. The main advantage of the TiD_2 standard is ease of handling.

For an incident ^3He energy of 1.2 MeV, the probing depth is about 2.5 μm. A beam current of 0.5 na is used to minimize local heating of the specimen. With these parameters, the level of detectability for D_2O is 0.01 At. percent in an epoxy resin.

Specimen Preparation

Two epoxy resin castings, 3501-5 and 5208, and two graphite epoxy composites, 15 ply unidirectional AS/3501-5 ($V_f = 0.64$) and 8 ply AS/3501-5A $(0/\pm45/90)_s$ or $(0/\pm45/0)_s(V_f = 0.61)$, were selected for evaluation. Resin specimens were generally 2.5 cm $\times$ 1.8 cm $\times$ 0.2 cm. The AS/3501-5 specimens were 2.5 cm $\times$ 1.8 cm while the AS/3501-5A specimens were 5 cm $\times$ 2.5 cm. The edges of all specimens were ground and polished. Specimens were dried to constant weight at 82°C and examined for edge microcracks. Moisture conditioning at 77°C was performed in sealed vessels containing D_2O or saturated solutions of NaCl in D_2O. These produced relative humidities (RH) of 100 percent and 75 percent above the liquid. Following moisture conditioning, specimens were cut under liquid nitrogen using a diamond cutoff wheel to expose an inner surface for deuterium analysis.

RESULTS AND DISCUSSION

The first part of this investigation was to ascertain whether the absorption of D_2O in an epoxy resin casting or composite would affect the properties of the epoxy in a manner similar to that due to H_2O absorption. To evalute this effect, we compared the equilibrium moisture content, the glass transition temperature depression, and the moisture diffusivity for 5208 and 3501-5 epoxy resin castings conditioned in D_2O and H_2O.

Specimens of each type of epoxy resin were conditioned at 77°C 100 percent RH (D_2O and H_2O). The equilibrium moisture content, determined from the specimen weight gain, was 6.8 ± 0.3 percent D_2O and 6.0 ± 0.5 percent H_2O for 5208, and 6.8 ± 0.1 percent D_2O and 6.2 ± 0.3 percent H_2O for 3501-5. Correcting for the D/H isotope effect, 6.8 percent D_2O corresponds to 6.1 percent H_2O, indicating that the bulk moisture content is the same for D_2O and H_2O conditioned specimens.

The glass transition temperature, T_g, of 5208 resin specimens conditioned to saturation at 100 percent RH in both D_2O and H_2O was measured by means of thermomechanical analysis. These measurements were performed to assess whether the absorbed D_2O and H_2O are bonded at the same sites within the epoxy resin structure. Representative expansion vs temperature curves are shown in Figure 1. Both the D_2O and H_2O conditioned specimens show an initial deviation from linearity at approximately 120–125°C, which has been referred to as the resin T_g. The average measured values of T_g for specimens conditioned in H_2O and D_2O were 118 ± 4°C and 126 ± 4°C,

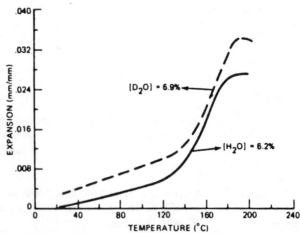

Figure 1. Thermal expansion of 5208 epoxy resin casting saturated in D_2O and H_2O at 100% relative humidity.

respectively. This result indicates that D_2O and H_2O are both hydrogen bonded to the electronegative functional groups on the epoxy resin and contribute equally to a disruption of secondary crosslinking and subsequent reduction in T_g.

Specimens of 3501-5 epoxy resin conditioned at 77 °C, 75 percent, and 100 percent RH (D_2O) were used for D_2O diffusivity evaluation. The equilibrium moisture content for two specimens at each relative humidity was 5.1 percent and 6.9 percent D_2O, corresponding to 4.6 percent and 6.2 percent H_2O, which are in good agreement with the equilibrium H_2O content previously reported [1]. The average D_2O diffusivity, computed from the slope of the initial portion of a plot of weight gain versus square root of time and corrected for edge effects [6], was 1.2×10^{-6} mm$^2/s$. The H_2O diffusivity in 3501-5 castings at the same temperature, as reported in Reference [1], was 1.56×10^{-6} mm$^2/s$.

The excellent agreement in the equilibrium moisture content at both 75 percent and 100 percent RH, the moisture diffusivity, and the glass transition temperature depression indicates equivalence between D_2O and H_2O in epoxy resin castings. Therefore, determination of the D_2O concentration profile through the thickness of epoxy resin or composite will be representative of the H_2O profile on these materials.

A summary of the equilibrium moisture content in two epoxy resins, 3501-5 and 5208, and two graphite epoxy composites, AS/3501-5 and AS/3501-5A, measured by nuclear analysis and gravimetric techniques is contained in Table 1. Moisture content by the gravimetric method represents weight gained by the bulk specimen. On the other hand, with the nuclear method the equilibrium moisture content was determined from a localized measurement of D_2O at the surface of a saturated specimen. We also list under the nuclear method the corresponding H_2O content calculated from the measured D_2O. The bulk D_2O concentration for the two resin castings and those determined by the nuclear technique agree within the errors of the measurements. Because of the lower moisture diffusivity in composites and the longer time needed to reach

Table 1. Equilibrium moisture content measured by gravimetric and nuclear analysis methods.

Specimen	Relative Humidity %	Nuclear Method (% D_2O)	(% H_2O)	Gravimetric Method % H_2O or % D_2O
3501-5	100	6.8 ± 0.4	(6.1 ± 0.3)	6.8 ± 0.1 (D_2O)
5208	100	7.0 ± 0.4	(6.3 ± 0.4)	6.8 ± 0.3 (D_2O)
AS/3501-5	75	1.2 ± 0.1	(1.1 ± 0.1)	1.0 ± 0.1 (H_2O)
AS/3501-5A (0/±45/90)$_S$	100	1.9 ± 0.2	(1.7 ± 0.2)	1.8 ± 0.1 (H_2O)
(0/±45/0)$_S$	100	1.8 ± 0.4	(1.6 ± 0.3)	1.8 ± 0.1 (H_2O)

equilibrium, bulk D_2O content at equilibrium was not measured for the composite specimens. However, bulk equilibrium H_2O concentrations obtained for similar specimens during other investigations are listed.

Deuterium analysis of composite specimen surfaces was always made on the transverse face. A representative reaction zone for a 12 μm × 200μm collimated beam is illustrated in Figure 2. Specimens of unidirectional AS/3501-5 exhibit point-to-point reproducibility across the surface as manifested by the low standard deviation in the nuclear analysis data, and the absolute concentration level is consistent with gravimetric results.

The exterior surface of two AS/3501-5A graphite/epoxy laminates was also investigated to determine the effect of fiber orientation on the measured moisture content. As seen in Table 1, the average value of the moisture level at a saturated exterior surface, as measured by the nuclear method, is in good agreement with that reported from the bulk gravimetric method. The spread in the nuclear data for these composites is larger than that observed for the resins and the unidirectional composite. No direct correlation of D_2O concentrations with fiber orientation were observed for these specimens. It is possible that the spread in these data may be due to the nonuniform fiber content in the region of the ply interfaces arising from the differences in ply to ply fiber orientation. This problem could be overcome by utilizing a wider beam.

The 15 ply unidirectional AS/3501-5 graphite/epoxy was selected for moisture profiling (through the thickness). Specimens were conditioned at 77 °C, 75 percent RH (D_2O), and sectioned under liquid nitrogen to expose the interior. A point-by-point measurement of the D_2O concentration profile was made from the outer edge to the center of the specimen with a resolution of 12 μm in the direction of the moisture gradient. Each data point is the

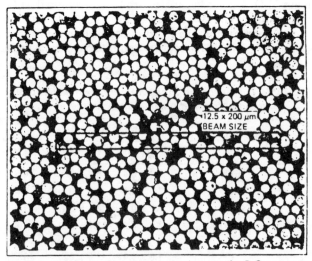

Figure 2. Collimated beam size indicating reaction zone for D_2O measurements.

result of a combined measurement of a multiply-segmented accumulation of counting rate data. The segmentation method monitors the system to ensure that no time dependent changes in moisture content occur during the measurement. The results of these measurements are illustrated in Figures 3 and 4 for two specimens conditioned for 2.5 and 8.7 days, respectively. The concentration of D_2O at the surface was taken from the exterior saturated surface analysis contained in Table 1.

Figures 3 and 4 also show the theoretical prediction of the moisture profile based on the finite difference numerical integration solution of classical diffusion equations. No attempt has been made to acquire a best fit to the data. For these calculations a diffusivity of 1.75×10^{-7} mm²/s was used, which was computed from the bulk weight gain by the composite after short exposure times in D_2O. The equilibrium moisture content used in the calculation was determined from the nuclear analysis of the exterior surface. A comparison between the observed and the predicted moisture profiles indicates that the diffusion rate determined from bulk measurement is consistent with the observed data. At some points in the specimen the observed concentration of D_2O is 0.1 percent higher or lower than the predicted level, but this could be due to resin rich or fiber rich regions within the sample.

CONCLUSIONS

The results of this investigation have demonstrated that the nuclear probe is capable of determining moisture pickup in resins and composites with high sensitivity. Using calibration standards we are able to determine absolute deuterium concentrations and to predict the H_2O response in resin based

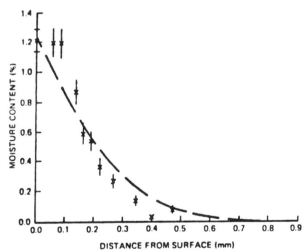

Figure 3. *Measured and predicted moisture profile in AS/3501-5 graphite/epoxy composite conditioned for 2.5 days at 77°C, 75% RH.*

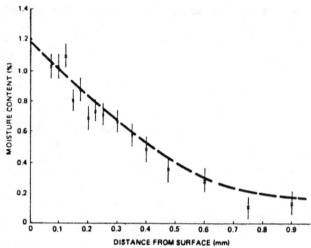

Figure 4. Measured and predicted moisture profile in AS/3501-5 graphite/epoxy composite conditioned for 8.7 days at 77°C, 75% RH.

materials through simulation with D_2O. Moisture profiles with resolution of 12 μm in the direction of the gradient have been measured in unidirectional graphite/epoxy composites and found to be consistent with classical diffusion theory.

REFERENCES

1. DeIasi, R. and Whiteside, J. B., "Effect of Moisture on Epoxy Resins and Composites," in *Advanced Composite Materials—Environmental Effects*, J. R. Vinson, ed. ASTM STP 658, American Society for Testing and Materials, 2–20 (1978).
2. DeIasi, R., Whiteside, J., and Wolter, W., "Effects of Varying Hygrothermal Environments on Moisture Absorption in Epoxy Composites," presented at Fourth Conference on Fibrous Composites in Structural Design, San Diego, CA (Nov. 1978).
3. TRW Inc., Systems Group, "Analysis of Moisture in Polymers and Composites," NASA Lewis Research Center, Contract NAS 320406.
4. Kaelble, D. H., "Wetometer for Measurement of Moisture in Composites," Progress in Quantitative NDE, Third Annual Report, AFML, Contract F33615-74-C-5180 (1977).
5. Kays, A. O., "Determination of Moisture Content in Composites by Dielectric Measurements," Second Interim Technical Report, AFFDL Contract F33615-78-C-3216 (December 1978).
6. Shen, C. H. and Springer, G. S., "Moisture Absorption and Desorption of Composite Materials," *J. Composite Materials, 10*, 2 (1976).

13

Model for Predicting the Mechanical Properties of Composites at Elevated Temperatures

G. S. Springer

1. INTRODUCTION

THE MECHANICAL PROPERTIES OF FIBER REINFORCED COMPOSITES DE-crease when the material is exposed to elevated temperatures. For some period of exposure the strengths and the moduli may remain above their allowable limits. However, after some time, the strengths and the moduli may become so low that the material cannot sustain the imposed loads or maintain the prescribed allowable deflections. Therefore, for practical reasons, it is important to know the decrease in strengths and moduli with exposure time and the time of failure, i.e. the length of exposure time at which the strengths and moduli reach their minimum allowable values.

Although the problem is of great importance, relatively little is known of the response of organic matrix composites to elevated temperatures. Most of the available information is in the form of data. These data can be used to determine the behavior of a given material under a given condition, but cannot be used to predict the behavior of different types of materials under exposure to different temperatures.

The aforementioned shortcomings of purely empirical approaches could be overcome by the use of analytical models. In this paper such a model is outlined. The model has first been developed for predicting the changes in strengths and moduli of wood during fire [1–3]. Here the model is adapted to fiber reinforced organic matrix composites.

2. PROBLEM STATEMENT AND APPROACH

We consider an organic matrix composite constructed of N plies of either unidirectional prepreg tape or fabric (Figure 1). The cross-sectional area of the composite as well as the fiber orientations in the plies are arbitrary. In-

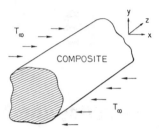

Figure 1. Description of the problem.

itially ($t < 0$) the temperature of the composite T_i, the strength S^o, and the modulus E^o are known. At time $t = 0$ the composite is exposed to an environment in which the temperature T_∞ is different than the initial temperature T_i. The temperature around the surface may vary both with position and time. Either the temperature or the heat flux must be known at each point on the surface. It is desired to determine the changes in strengths and moduli with respect to time and the time of failure.

The tensile, compressive, and shear strengths and moduli normal and parallel to the fibers change at different rates. Therefore, the changes in each of these properties must be known unless, in the given problem, one or more of these properties are negligible. The model presented in this paper applies to changes in each of these properties. However, to simplify the discussions, the analysis is described in general terms, i.e. expressions are developed only for changes in one of the strengths S/S° and one of the moduli E/E°. It is understood that the resulting expressions represent tensile, compressive or shear properties either along or normal to the fibers.

The calculation includes the following major steps:

1. The temperature T as a function of time is calculated at every point inside the material.
2. The mass of vapor formed and released Δm up to time t is calculated inside the material.
3. The changes in strength S/S° and modulus E/E° are related to the mass loss Δm through empirical relationships.
4. The stress ratio R is calculated as a function of time

$$R = \frac{strength\ of\ material}{applied\ stress} \qquad (1)$$

The strength and, therefore, the stress ratio R change with time. The material fails when R reaches a prescribed minimum value R_{min}

$$R > R_{min} \quad no\ failure$$

$$R \leqslant R_{min} \quad failure \qquad (2)$$

Details of the calculation procedures are presented in the next sections. For clarity, the method is given for a material with rectangular geometry. The temperature is taken to be uniform along the z axis (Figures 1,2) (two dimensional problem). Note, however, that the model is general and may be applied to arbitrary geometries and to arbitrary temperature distributions.

3. TEMPERATURE DISTRIBUTION AND MASS LOSS

A rectangular slab of thickness a and width b is considered (Figure 2). Initially (time < 0) the temperature T_i at every point inside the slab is known. At time t = 0, the slab is suddenly exposed to a specified temperature T_∞. It is desired to calculate the temperature inside the slab as a function of position and time and the mass of vapor released at each point due to the increase in material temperature. The slab is assumed to be well ventilated so that the vapors formed during heating can escape directly from the material.

The law of conservation of energy for a small volume element dv inside the slab can be expressed as:

| rate of change of energy in dv | = | net rate of energy transferred into dv by conduction | + | net rate of energy transferred into dv by convection | + | rate of energy liberated (or absorbed) in dv |

The heat capacity of the released vapor is small compared to the heat capacity of the solid. The heat liberated by the vaporization of any water present in the material is small compared to the heat of vaporization of the composite. Finally, the heat transfer through the material by convection is small compared to the heat transfer by conduction. Accordingly, the law of conservation of energy may be written as

$$\frac{\delta(m_s C_s T)}{\delta t} = \frac{\delta}{\delta x} K_x \frac{\delta T}{\delta x} + \frac{\delta}{\delta y} K_y \frac{\delta T}{\delta y} + \frac{\delta m_a}{\delta t} L \qquad (3)$$

where T is the absolute temperature inside the material at the x, y coordinate, C_s is the specific heat (defined below), L is the heat of reaction, K_x and K_y are

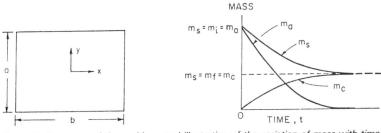

Figure 2. *Geometry of the problem, and illustration of the variation of mass with time.*

the thermal conductivities in the x and y directions. m_s is the mass per unit volume of the composite comprised of virgin ("active") material (mass per unit volume m_a) and of material which cannot pyrolyze further at the given temperature ("char," mass per unit volume m_c)

$$m_s = m_a + m_c \tag{4}$$

m_s, m_a and m_c vary with time as illustrated in Figure 2. Any water which may be present in the material is not included in m_s.

The specific heat and the thermal conductivities are approximated by the expressions

$$C_s = \frac{m_a}{m_s} C_a + \frac{m_c}{m_s} C_c \tag{5}$$

$$K_{x,y} = \frac{m_a}{m_i} K_{ax,y} + \frac{m_c}{m_f} K_{cx,y} \tag{6}$$

where C_a, C_c and K_a, K_c are the heat capacities and thermal conductivities of the active material and the char, respectively. m_i and m_f are the masses at the beginning and at the end of the pyrolysis

$$\left.\begin{array}{l} m_s = m_a = m_i = \varrho_a \\ \\ m_c = 0 \end{array}\right\} t < 0 \qquad \left.\begin{array}{l} m_s = m_f = m_c \\ \\ m_a = 0 \end{array}\right\} t \to \infty \tag{7}$$

ϱ_a is the density of the dry, active material. Note that ϱ_a is a constant for a given type of material. The mass m_f depends upon the temperature. The value of m_f decreases with increasing temperatures. The mass m_s is related to m_i and m_f by the expression

$$m_s = \left(1 - \frac{m_f}{m_i}\right) m_a + m_f \tag{8}$$

It is assumed that the volatile formation may be represented by a single step Arrhenius bulk reaction

$$\frac{\delta m_a}{\delta t} = - m_a \, k_a \, e^{-E_a/RT} \tag{9}$$

k_a is the frequency factor, E_a is the activation energy, and R is the universal gas constant. The reactions occurring during the pyrolysis are complex. Therefore, k_a and E_a may be taken to be constants only within narrow temperature ranges. Values of k_a and E_a appropriate to the temperature must be used in the calculations.

The vapor mass lost during time t is

$$\Delta m = \int_o^t \left(\frac{\delta m_a}{\delta t} \right) dt \qquad (10)$$

In order to obtain solutions to equations (5)–(10) the initial and boundary conditions must be specified. Initially (time t < 0) the temperature T_i and the density ϱ_a, must be given

$$m_s = m_i = m_a = \varrho_a \qquad T = T_i \qquad (11)$$

$$m_c = 0$$

During exposure to elevated temperatures (time t ≥ 0) either the temperature or the heat flux must be known at every point on the surface.

4. CHANGES IN STRENGTH AND MODULUS

The model presented below is for predicting changes in the tensile, compressive, and shear strengths and moduli for composites exposed to high temperatures. The hypothesis is made that degradation in the mechanical properties is related to the mass loss. Accordingly, the strength and the modulus at any point (designated by the coordinates x,y, Figure 3) are expressed as

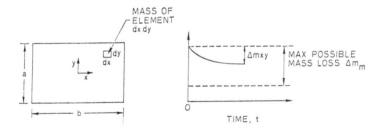

INITIAL STRENGTH (t = 0): S_0

STRENGTH OF ELEMENT
dxdy AT TIME t   : $S_{xy} = S_0[1 - (\Delta m_{xy}/\Delta m_m)^e]$

STRENGTH OF TOTAL
CROSS SECTION AT
TIME t   : $S = \frac{1}{A} \iint_A S_{xy} dA = \frac{1}{(a)(b)} \int_0^a \int_0^b S_0[1 - (\Delta m_{xy}/\Delta m_m)^e] dx dy$

Figure 3. Description of the strength model.

$$\frac{S_{xy}}{S^\circ} = 1 - \left(\frac{\Delta m_{xy}}{\Delta m_m}\right)^e \tag{12}$$

$$\frac{E_{xy}}{E^\circ} = 1 - \left(\frac{\Delta m_{xy}}{\Delta m_m}\right)^g \tag{13}$$

S_{xy} and E_{xy} are the strength and the modulus at time t at the x,y position. S° and E° are the strength and modulus at a reference temperature (room temperature, say). Δm_{xy} is the mass loss at position x,y at the time t. Δm_{xy} is the maximum mass loss. Δm_{xy} and Δm_m are mass losses resulting from volatilization only, and do not include mass losses due to water vaporization. In equations (12) and (13) the exponents e and g are constants which depend on the material but which, by definition, are independent of geometry and temperature. The values of these constants must be determined experimentally, as will be discussed subsequently.

By assuming that the strain ε is constant across the cross section A, the maximum load

$$F = \int_A \int S_{xy} dA \tag{14}$$

where F is the load at which failure occurs. The total ("overall") strength at time t is

$$S = \frac{F}{A} = \frac{1}{A} \int_A \int S_{xy} da \tag{15}$$

The modulus of elasticity can be determined by similar reasoning. It is assumed that up to the point of failure the relationship between the strength and the modulus is linear and can be represented by Hooke's law

$$S_{xy} = E_{xy}\, \varepsilon \tag{16}$$

where the strain ε is taken to be uniform across the cross section. Equations (14) and (16) yield

$$F = \int_A \int E_{xy}\varepsilon dA = \varepsilon \int_A \int E_{xy} dA \tag{17}$$

Equation (17) can be written as

$$\frac{F}{(A)(\varepsilon)} = \frac{1}{A} \int_A \int E_{xy} dA \tag{18}$$

The "overall" modulus of elasticity at time t is

$$E = \frac{S}{\varepsilon} = \frac{1}{A} \iint_A E_{xy} dA \qquad (19)$$

Equations (15) and (19) provide the strength and the modulus as a function of exposure time. It is noted again that these equations apply to tensile, compressive, and shear properties, but the values of the constants e and g are different for the different types of properties.

Using the expressions developed in the foregoing the strength S and the modulus E can be calculated by the following steps:

1. The mass loss at every point $\Delta m_{xy}/\Delta m_m$ is calculated as a function of time by the method given in Section 3.
2. The strength S_{xy} and the modulus E_{xy} at every point are calculated as a function of time using equations (12) and (13).
3. The "overall" strength S and modulus E as a function of time are calculated by equations (15) and (19).

Once the surface temperature or the heat flux to the surface is specified (both of these parameters may vary with position and time) the strength and the modulus can readily be calculated by the procedure outlined above.

5. STRESS RATIO AND FAILURE TIME

The stress ratio R is defined as

$$R = \frac{strength\ of\ material}{applied\ stress} \qquad (1)$$

For combined stresses, the stress ratio may be calculated, for example, by the Tsai-Wu failure criterion which, for the two dimensional problem under consideration, is [4]

$$(F_{11}\ \sigma_1^2 + 2\ F_{12}\ \sigma_1\ \sigma_2 + F_{22}\ \sigma_2^2 + F_{66}\ \sigma_6^2)\ R^2 + (F_1\ \sigma_1^2 + F_2\ \sigma_2)\ R = 1 \quad (20)$$

Here σ_1 and σ_2 are the applied stresses in the directions normal and parallel to the fibers, σ_6 is the applied shear stress in the coordinate system (1–2 are the coordinates in each ply normal and parallel to the fibers). The parameters F are related to the strengths of the material [4]

$$F_{11} = \frac{1}{XX'} \qquad\qquad F_1 = \frac{1}{X} - \frac{1}{X'}$$

$$F_{22} = \frac{1}{YY'} \qquad\qquad F_2 = \frac{1}{Y} - \frac{1}{Y'} \qquad (21)$$

$$F_{66} = \frac{1}{(Sh)^2} \qquad\qquad F_{12} = -\frac{1}{2}\sqrt{F_{11}F_{22}}$$

where X, X' and Y, Y' are tensile and compressive strengths parallel and normal to the fibers. Sh is the longitudinal shear strength. These strengths (X, X', Y, Y', Sh) all vary with exposure time. The changes in the strengths can be calculated by the method described in Section 4. Since the strengths change with time, the stress ratio R changes with time too, as illustrated in Figure 4. The material will fail when R reaches a certain minimum value. Generally this value is taken to be unity. In other words, the failure time is reached ($t = t_f$) when $R = R_f$. By calculating R as a function of time the failure time can be determined (Figure 4).

For a simply supported beam the above procedure can be simplified. According to the flexure formula, the beam is safe when the following inequality is satisfied

$$M \leqslant S_b C \tag{22}$$

where M is the maximum applied bending moment, S_b is the maximum allowable extreme fiber stress in bending, and C is the section modulus. For a rectangular beam of height d and width b the section modulus is $C = bd^2/6$.

The stress ratio is defined as

$$R = \frac{S_b}{M/C} \tag{23}$$

During exposure to elevated temperatures the strength and, consequently, the value of S_b decrease. This results in a decrease in the value of R with exposure time. The beam will fail at the time t_f at which the value of R reaches unity. S_b is now expressed as

$$S_b = S_b^\circ \frac{S_b}{S_b^\circ} \tag{24}$$

where S_b° is the maximum allowable extreme fiber stress in bending of the material prior to exposure to the elevated temperature. Equations (23) and (24) may be combined to yield

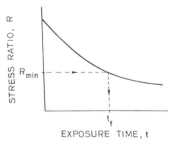

Figure 4. Illustration of the variation of stress ratio with exposure time.

$$R = \frac{(S_b/S_b^\circ)}{[M/(C \times S_b^\circ)]} \qquad (25)$$

The above form of R is convenient because strength loss S_b/S_b° can be calculated, as described in Section 4.

6. SOLUTION PROCEDURE

Solutions to the problem described in Section 3–5 can be obtained by numerical methods. A "user friendly" computer code suitable for generating numerical results was developed and can be obtained from the author.

The input parameters required for the calculations are listed in Table 1. The output given by the computer code is also included in this table. All the required input parameters can be measured directly. The density ϱ may be obtained simply by measuring the weight of a sample of known volume. The specific heats C_a and C_c and thermal conductivities K_a and K_c can be determined using standard calorimetry and thermal conductivity apparatus. The final mass m_f, the heat of vapor generation L, the frequency factor K_a and the activation energy E_a can be measured by a thermogravimetric analyzer.

The constants e and g required in the strength calculations can be obtained by the following procedure [2]. Test specimens are to be exposed to constant

Table 1. Input parameters required by the model and the output parameters provided by the model.

Input parameters

1) Density of composite, ϱ_a
2) Final mass, m_f
3) Specific heat of composite C_a
4) Specific heat of char, C_c
5) Thermal conductivity of composite, K_a
6) Thermal conductivity of char, K_c
7) Heat of vapor generation, L
8) Frequency factor of composite, k_a
9) Activation energy of composite, E_a
10) Constant, e
11) Constant, g
12) Strengths prior to exposure, X, X', Y, Y', Sh

Output parameters

1) Temperature as a function of position and time
2) Changes in strengths as a function of time
3) Changes in moduli as a function of time
4) Change in strength ratio as a function of time
5) Time of failure

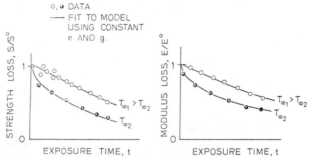

Figure 5. *Illustration of the procedure used to determine the values of the constants e and g.*

temperatures for specified lengths of time. The residual strengths and moduli of the specimens are then measured. Plots of strengths and moduli versus exposure time are constructed, as illustrated in Figure 5. The strengths and the moduli are then calculated for different assumed values of e and g, and the numerical results are compared to the data. The e and g values giving the "best fit" are taken to be the values of these parameters.

It is emphasized again, that e and g will have different values for different properties (tensile, compressive, shear) different orientations (normal and parallel to the fibers) and different materials. The values of e and g should be independent of temperature, but may vary if the temperature range is very wide.

7. CONCLUDING REMARKS

The validity of the first part of the model (temperature distribution and mass loss, Section 3) has been evaluated by comparing the results of the model to date generated using both an organic matrix composite (graphite-epoxy Fiberite T300/1034) and wood [5,1]. The measured and calculated temperature and mass losses agreed well, supporting the validity of this first part of the model. The results given by the second and third parts of the model (changes in strengths and moduli and failure time) have only been compared to data obtained with wood [2,3]. Again, reasonable agreements were found between the data and the results of the model warranting extension of the model to organic matrix composites. Comparisons of the second and third parts of the model to measurements made with composites must await the generation of appropriate data.

ACKNOWLEDGMENTS

This work was supported by the Mechanics and Surface Interactions Branch, Nonmetallic Materials Division, Materials Laboratory, Wright Patterson Air Force Base, Dayton, Ohio.

REFERENCES

1. Springer, G. S., "Mass Loss of and Temperature Distribution in Southern Pine and Douglas Fir in the Range 100 to 800°C," *J. of Fire Sciences, 1,* 271-284 (1983).
2. Do, M. H. and Springer, G. S., "Model for Predicting Changes in the Strengths and Moduli of Timber Exposed to Elevated Temperatures," *J. of Fire Sciences, 1,* 285-296 (1983).
3. Do, M. H. and Springer, G. S., "Failure Time of Loaded Wooden Beams During Fire," *J. of Fire Sciences, 1,* 297-303 (1983).
4. Tsai, S. W. and Hahn, H. T. *Introduction to Composite Materials.* Lancaster, PA: Technomic Publishing Co., Inc. (1980).
5. Pering, G. A., Farrell, P. V., and Springer, G. S., "Degradation of Tensile and Shear Properties of Composite Exposed to Fire or High Temperature," *J. of Composite Materials, 14,* 54-68 (1980).

14

Multi-Material Model Moisture Analysis for Steady-State Boundary Conditions

R. K. Miller

SCOPE

THE THEORY FOR MOISTURE ABSORPTION AND DESORPTION FOR STEADY-state boundary conditions in a single material system was presented in 1976 by Shen and Springer [1]. In aerospace applications today there is a need to analyze multi-material systems for both steady-state and transient boundary conditions. The "W8GAIN" computer code mentioned in Reference [2] performs both analyses but at costs dictated by computer facilities, desired output accuracy, and the complexity of the model being analyzed.

The results presented here define a closed-form solution for analyzing multi-material models with constant boundary conditions. The final form of the solution can be easily programmed on a programmable desk calculator.

CONCLUSIONS AND SIGNIFICANCE

For a two-material system (shown in Figure 1) exposed to constant humidity and constant temperature, the total weight of moisture (per unit area) as a function of time is:

$$mII(t) = m_1(t) + m_2(t) = \rho_1 \frac{Mm_1}{50} \left\{ h_1 - \frac{4S_2}{B\pi^2} \sum_{K=1,3,5...}^{\infty} \frac{1}{K^2} \right.$$

$$\left. \exp\left[\frac{-K^2\pi^2 D_2 t}{S_2{}^2}\right] \left(1 - \cos\left[\frac{K\pi Bh_1}{S_2}\right]\right) \right\} + \rho_2 \frac{Mm_2}{50} h_2$$

162

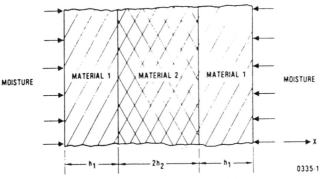

Figure 1. *A two-material system exposed to constant humidity and temperature. The initial moisture concentration is zero.*

$$- \frac{4S_2}{\pi^2} \sum_{K=1,3,5\ldots}^{\infty} \frac{1}{K_2} \exp\left[\frac{-K^2\pi^2 D_2 t}{S_2{}^2}\right] \cos\left[\frac{K\pi B h_1}{S_2}\right] \Bigg\} \quad (1)$$

Mm_i, $m_i(t)$, ϱ_i and h_i, respectively, are the maximum percent weight gained of moisture in material i, the total moisture gained per unit at time t in material i, the density of material i, and one half of the total thickness of material i as shown in Figure 1 ($i = 1$ or 2). The constant B is used to adjust the thickness, h_i, when converting Material 1 into Material 2. B is defined as

$$B = \sqrt{D_2/D_1} \quad (2)$$

Where D_1 and D_2 are the diffusion coefficients for Material 1 and Material 2, respectively. S_2 represents the total thickness of the system after it has been converted into Material 2.

$$S_2 = 2(Bh_1 + h_2) \quad (3)$$

The percent weight gained of moisture for the two-material system is given as

$$MII(t) = \frac{mII(t)}{\rho_1 h_1 + \rho_2 h_2} \quad (4)$$

Similarly, for an N-material system, the total weight gained of moisture per unit area is

$$mN(t) = m_1(t) + m_2(t) + \ldots + m_{N-1}(t) + m_N(t) =$$

$$\rho_1 \frac{Mm_1}{50} \left\{ h_1 - \frac{4S}{B_1 \pi^2} \sum_{K=1,3,5,\ldots}^{\infty} \frac{1}{K^2} \exp \left[\frac{-K^2 \pi^2 Dt}{S^2} \right] \right.$$

$$\left. \left(1 - \cos \left[\frac{K\pi B_1 h_1}{S} \right] \right) \right\} + \rho_2 \frac{Mm_2}{50} \left\{ h_2 - \frac{4S}{B_2 \pi^2} \sum_{K=1,3,5,\ldots}^{\infty} \right.$$

$$\left. \frac{1}{K^2} \exp \left[\frac{-K^2 \pi^2 Dt}{S^2} \right] \left(\cos \left[\frac{K\pi B_1 h_1}{S} \right] - \cos \left[\frac{K\pi}{S} (B_2 h_2 + B_1 h_1) \right] \right) \right\}$$

$$+ \rho_{N-1} \frac{Mm}{50} N-1 \left\{ h_{N-1} - \frac{4S}{B_{N-1} \pi^2} \sum_{K=1,3,5,\ldots}^{\infty} \frac{1}{K^2} \exp \left[\frac{-K^2 \pi^2 Dt}{S^2} \right] \right.$$

$$\tag{5}$$

$$\left. \left(\cos \left[\frac{K\pi}{S} \sum_{j=1}^{N-2} B_j H_j \right] - \cos \left[\frac{K\pi}{S} \sum_{j=1}^{N-1} B_j H_j \right] \right) \right\} + \rho_N \frac{Mm_N}{50}$$

$$\left\{ h_N - \frac{4S}{B_N \pi^2} \sum_{K=1,3,5,\ldots}^{\infty} \frac{1}{K^2} \exp \left[\frac{-K^2 \pi^2 DT}{S^2} \right] \cos \left[\frac{K\pi}{S}^{N-1} B_j H_j \right] \right\}$$

$$B_i = \sqrt{D/D_i} \quad (i = 1, 2, \ldots, N-1, N) \tag{6}$$

$$S = 2(B_1 h_1 + B_2 h_2 + \ldots + B_{N-1} h_{N-1} + B_N h_N) \tag{7}$$

In equations (5) and (6), D is an arbitrarily chosen value. The percent weight gained of moisture for the N-material system becomes

$$MN(t) = \frac{mN(t)}{\rho_1 h_1 + \rho_2 h_2 + \ldots + \rho_{N-1} h_{N-1} + \rho_N h_n} \times 100 \tag{8}$$

NUMERICAL ILLUSTRATION

DeIasi and Whiteside [3] investigated the effects of moisture on boron/epoxy (B/5055), graphite/epoxy (AS/3501-5), and a hybrid boron-graphite/epoxy laminate with a $[0_{3B}/\pm 45_G/90_G/\pm 45_G/0_{3B}]$ stacking sequence. The results presented in their study include the diffusion coefficients at different temperatures for the three previously mentioned systems. Also studied was the equilibrium moisture content for various relative humidities in water vapor. For a steady-state boundary condition of 23C and 100 percent relative humidity, the three systems will have the moisture diffusion parameters shown in Table 1. Using the properties of boron/epoxy and graphite/epoxy and equations (1),(2),(3), and (4), a "theroetical" curve was generated for percent weight gained versus square root time. Treating the boron-graphite/epoxy hybrid as a single material system an "experimental" curve was generated using Shen and Springer's [1] theory. In addition, a W8GAIN computer analysis [2] was performed and the results compared with the two previously mentioned curves. The comparison is shown in Figure 2 indicating an excellent agreement between W8GAIN and the "experimental" curve and a very good agreement for the "theoretical" curve.

SUPPLEMENTARY INFORMATION

For a single-material system, Shen and Springer [1] found that the moisture concentration through a material at any time, $C(x,t)$ can be expressed as

$$C^* = \frac{C(x,t) - C_o}{C_m - C_o} = 1 - \frac{4}{\pi} \sum_{K = 1,3,5,\ldots}^{\infty} \frac{1}{K} \sin \left[\frac{K\pi x}{S} \right] \exp \left[\frac{-K^2 \pi^2 D t}{S^2} \right] \quad (9)$$

For further derivations the initial concentration, C_o, will be dropped so that equation (9) can be expressed as:

$$C^* = C(x,t)/Cm \quad (9)$$

For a two-material system such as the one shown in Figure 1, the distribution of C^* through the model will appear as shown in Figure 3. At both boundaries $C^* = 1$ indicates that the surface is saturated at the boundary conditions. At $X = h_1$ and $X = h_1 + 2h_2$, $C^* = C^*_1$ for both materials. This preserves the mass transfer condition defined by Springer [2]. Also in the midplane of this model, $X = h_1 + h_2$, $C^* = C^*_2$ and the slope of the normalized concentration curve is zero. The layer of Material 2 can be replaced with a layer of Material 1 with thickness $2Ah_2$ and still reproduce the same C^* distribution as before. Similarly, the layers of Materials 1 can be replaced by

Table 1. Moisture diffusion parameters at 23C and 100% relative humidity for B/5505, AS/3501-5, and B/5505-AS/3501-5.

System	$D_X(mm^2/s)$	$M_{MAX}(\%)$
B 5505	2.31×10^{-8}	1.95
AS/3501-5	1.06×10^{-8}	1.80
B 5505-AS/3501-5	3.845×10^{-8}	1.70

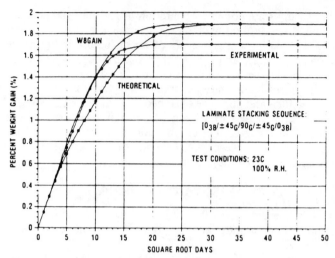

Figure 2. Comparison of a two material model with experimental results for a single hybrid material system.

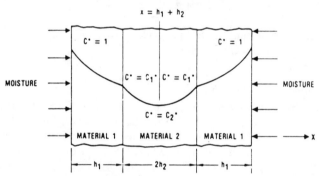

Figure 3. The normalized concentration distribution through a two material system. The initial concentration is zero.

layers of Material 2 with adjusted thicknesses of Bh_1. The results of these two conversions are illustrated in Figures 4a and 4b. Evaluating equation (9) at $X = h_1 + Ah_2$ for the model in Figures 4a and at $X = Bh_1 + h_2$ for the model in Figure 4b yields

$$C_2^* = 1 - \frac{4}{\pi} \sum_{K = 1,3,5,\ldots}^{\infty} \frac{(-1)^{\frac{K-1}{2}}}{K} \exp\left[\frac{-K^2\pi^2 D_1 t}{S_1^2}\right] \quad (10a)$$

$$= 1 - \frac{4}{\pi} \sum_{K = 1,3,5,\ldots}^{\infty} \frac{(-1)^{\frac{K-1}{2}}}{K} \exp\left[\frac{-K^2\pi^2 D_2 t}{S_2^2}\right] \quad (10b)$$

or

$$\frac{D_1}{S_1{}^2} = \frac{D_2}{S_2{}^2}$$ (10c)

Evaluating $C_1{}^*$, for the same two models yields

$$C_1{}^* = 1 - \frac{4}{\pi} \sum_{K=1,3,5,\ldots}^{\infty} \frac{1}{K} \sin\left[\frac{K\pi h_1}{S_1}\right] \exp\left[\frac{-K^2\pi^2 D_1 t}{S_1^2}\right]$$ (11a)

$$= 1 - \frac{4}{\pi} \sum_{K=1,3,5,\ldots}^{\infty} \frac{1}{K} \sin\left[\frac{K\pi B h_1}{S_2}\right] \exp\left[\frac{-K^2\pi^2 D_2 t}{S_2^2}\right]$$ (11b)

Combining equations (10c), (11a), and (11b) gives

$$B = \frac{S_2}{S_1} = \frac{D_2}{D_1} = \frac{1}{A}$$ (12)

Referring to Figure 3, the weight of the moisture per unit area in this system is defined as

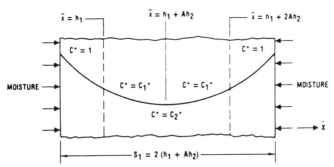

Figure 4a. *The two-material model converted into a single layer of Material 1.*

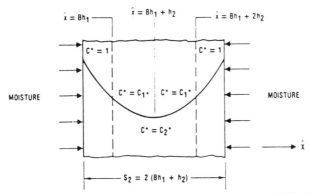

Figure 4b. *The two-material model converted into a single layer of Material 2.*

$$mII(t) = \int_0^{2(h_1 + h_2)} C(x)dx = 2\int_0^{h_1} C(x)dx + 2\int_{h_1}^{h_1 + h_2} C(x)dx$$

$$= 2Cm_1 \int_0^{h_1} C^*(x)dx + 2Cm_2 \int_{h_1}^{h_1 + h_2} C^*(x)dx \tag{13}$$

By converting Material 1 into Material 2, equation (13) becomes

$$mII(t) = \frac{2Cm_1}{B} \int_0^{Bh_1} C^*(\hat{x})d\hat{x} + 2Cm_2 \int_{Bh_1}^{Bh_1 + h_2} C^*(\hat{x})d\hat{x} \tag{14}$$

By integrating equation (9) with $D = D_2$ and $S = S_2$ and evaluating at the limits shown in equation (14) a closed-form solution for $mII(t)$ is generated. The results have already been shown in equation (1).

This same approach can be used to derive the equations for an N-material system. The most important items to remember are that the limits on the integration of the moisture concentration must be adapted to the "converted" model and the same integrals are divided by the conversion factor, B_1. In equation form, the total weight of moisture gained for an N-material system is given as

$$mN(t) = \frac{2Cm_1}{B_1} \int_0^{B_1h_1} C^*(\hat{x})d\hat{x} + \frac{2Cm_2}{B_2} \int_{B_1h_1}^{B_1h_1 + B_2h_2} C^*(\hat{x})d\hat{x} \tag{15}$$

$$+ \ldots + \frac{2Cm_{N-1}}{B_{N-1}} \int_{B_1h_1 + \ldots + B_{N-2}h_{N-2}}^{B_1h_1 + \ldots + B_{N-1}h_{N-1}} C^*(\hat{x})d\hat{x} + \frac{2Cm_N}{B_N} \int_{B_1h_1 + \ldots + B_{N-1}h_{N-1}}^{B_1h_1 + \ldots + B_Nh_N} C^*(\hat{x})d\hat{x}$$

ACKNOWLEDGMENTS

The author wishes to thank Dr. Keith Kedward and Mr. H. McCutchen for their advice and support of this work.

NOMENCLATURE

C_o Initial Moisture Concentration (g mm^{-3})
C_m Maximum Moisture Concentration (g mm^{-3})
$C(x,t)$ Moisture Concentration in the Material (g mm^{-3})
D Mass Diffusivity (mm^2 s^{-1})

h	Thickness of Material (mm)
m	Weight of Moisture per Unit Area (g mm²)
M	Percent Moisture Concentration in the Material (Dimensionless)
Mm	Equilibrium Moisture Content (Dimensionless)
t	Time(s)
ϱ	Density of Material (g mm⁻³)

REFERENCES

1. Shen, C. H. and Springer, G. S., "Moisture Absorption and Desorption of Composite Materials," *J. Composite Materials, 10,* 2 (1976).
2. Springer, G. S., "Moisture Content of Composites Under Transient Conditions," *J. Composite Materials, 11,* 107 (1977).
3. DeIasi, R. and Whiteside, J. B., "Effect of Moisture on Epoxy Resins and Composites," in *Advanced Composite Materials—Environmental Effects,* J.R. Vinson, ed. ASTM STP 658, American Society for Testing and Materials, 2-20 (1978).

15

A Rapidly Convergent Scheme to Compute Moisture Profiles in Composite Materials Under Fluctuating Ambient Conditions

Y. WEITSMAN

BASIC CONSIDERATIONS

CONSIDER A MOISTURE SORPTION PROCESS THAT IS DESCRIBED BY THE classical diffusion laws. In the one dimensional case we have

$$\frac{\partial m}{\partial t} = D \frac{\partial^2 m}{\partial x^2} \tag{1}$$

to which we must attach initial and boundary conditions.

In (1) $m = m(x,t)$ is moisture content, x is the spatial coordinate, t is time, and D is the coefficient of moisture diffusion.

It has been observed [1,2] that the equilibrium moisture content depends on the ambient relative humidity, and we shall also assume that the boundary conditions are determined by the same quantity. Furthermore, the moisture diffusivity was found to be the most sensitive to temperature [3,4]. Several empirical relationships were proposed, and we shall employ

$$M = Cr^\alpha \tag{2}$$

$$D(T) = D_R \exp(A/T_R - A/T) \tag{3}$$

In (2) and (3) M is the equilibrium moisture content, r the ambient relative humidity, T the temperature, T_R the reference temperature, and A, C, α the material constants.

In accordance with previous analyses [5,6] we can uncouple the process of heat diffusion from all other time-dependent material processes, e.g., moisture-diffusion or stress-relaxation. This simplification is justified because for all practical temperature fluctuations and geometrical dimensions the time required to reach thermal equilibrium is several orders of magnitude shorter than the time-scales for moisture diffusion or for relaxation response. Consequently, we consider spatial uniform temperature profiles, namely $T = T(t)$ as prescribed by the fluctuations in ambient temperature, when analyzing transient moisture diffusion.

SYMMETRIC EXPOSURE

Consider an infinite plate of thickness $2L$. Let $-L \leqslant x \leqslant L$ and assume an initial uniform moisture distribution m_o. When the plate is exposed to an elevated ambient relative humidity, the boundary moisture is given by μ, namely $m(x = \pm L, t) = \mu$. Due to the symmetry of the present problem it suffices to analyze only the region $0 \leqslant x \leqslant L$.

For constant μ the moisture content $m(x, t)$ is given by well known expressions [7,8]. Since we aim at extending those expressions to the case of fluctuating $\mu(t)$ and temperature $T(t)$ we choose to represent them in the following form

$$m(x,t) - m_o I_o(x,t) = \mu I(x,t) \tag{4}$$

The functions $I_o(x, t)$ and $I(x, t)$ take two alternate forms

$$I_o(x,t) = \left\{ \begin{array}{c} C(x,t) \\ \text{or} \\ 1 - E(x,t) \end{array} \right. , \quad I(x,t) = \left\{ \begin{array}{c} 1 - C(x,t) \\ \text{or} \\ E(x,t) \end{array} \right. \tag{5}$$

In (5)

$$C(x,t) = -2 \sum_{n=1}^{\infty} (-1)^{n+1} \cos(p_n x/L) \exp(-p_n^2 t^*) \tag{6}$$

and

$$E(x,t) = \sum_{n=1}^{\infty} (-1)^{n+1} \left[\text{erfc}\left(\frac{2n-1-x/L}{2\sqrt{t^*}}\right) \right. \tag{7}$$

$$\left. + \text{erfc}\left(\frac{2n-1+x/L}{2\sqrt{t^*}}\right) \right]$$

$$\text{with } p_n = (2n-1)\pi/2 \tag{8}$$

$$\text{and } t^* = Dt/L^2$$

The complementary error function erfc(z) decays rapidly with z. Its asymptotic value is given by [9] erfc $z \sim (\sqrt{\pi}z)^{-1}\exp(-z^2)$ consequently, for computational precision of $0(10^{-16})$—as obtained in "double precision" routines in digital computers—we can set erfc $z = 0$ for $z > 5.877$. In the sequel we shall designate this "cut-off" number by λ.

The rapid decay of erfc z implies that series (7) converges rapidly for short times. On the other hand it is obvious that series (6) converges rapidly for long times. To achieve computational efficiency we should therefore switch among the two forms of equations (5).

Straightforward arithmetic yields that accuracy of $0(10^{-16})$ is maintained by the following set of rules

$$\text{for} \quad \left(\frac{i-1}{2\lambda}\right)^2 < t^* < \left(\frac{i}{2\lambda}\right)^2 \quad \text{use } i \text{ terms in series (7)} \qquad (9a)$$

$$\text{for} \quad \left(\frac{5}{2\lambda}\right)^2 < t^* < \left(\frac{Q}{9^2}\right) \quad \text{use five terms in series (6)} \qquad (9b)$$

$$\text{for} \quad \frac{Q}{(2i+1)^2} < t < \frac{Q}{(2i-1)^2} \quad \text{use } i \text{ terms in series (6)} \qquad (9c)$$

In (9a) and (9c) $i = 1,2,3,4$. Also $Q = 14.93$ and $\lambda = 5.877$.

For $t^* > Q$ the moisture distribution is uniform to within $0(10^{-16})$.

It follows from (9a) that we never need more than the four following terms in series (7):

$$\text{erfc}\left(\frac{1 - x/L}{2\sqrt{t^*}}\right) + \text{erfc}\left(\frac{1 + x/L}{2\sqrt{t^*}}\right)$$

$$- \text{erfc}\left(\frac{3 - x/L}{2\sqrt{t^*}}\right) - \text{erfc}\left(\frac{3 + x/L}{2\sqrt{t^*}}\right)$$

It can be noted that the form of expressions (9) remains valid for *any* desired accuracy, ε, except that λ and Q depend on ε. Obviously for a lesser accuracy we require even fewer terms in (6) and (7).

Consider now the case of fluctuating temperatures, $T = T(t)$. In view of (3) the diffusivity D is now time dependent and the non-dimensional time t^* in (8) becomes a complicated function of real-time t. However, if we consider $t^* = D_R t/L^2$ at the reference temperature $T = T_R$ then in analogy with thermoviscoelasticity [10] we can replace t^* with the reduced dimensionless time ξ^* whenever $T = T(t)$ as follows

$$\xi^* = \frac{D_R}{L^2} \int_0^t \exp[A/T_R - A/T(s)]ds \qquad (10)$$

The moisture distribution under fluctuating temperatures is given by (4) with t^* replaced by ξ^*. In view of the single-valuedness of $\xi^* = \xi^*(t)$ it is always possible to convert the results back to real time t.

Consider next the case of fluctuating ambient relative humidity $r = r(t)$. By equation (2) this implies $\mu = \mu(t)$. When $\mu = \mu(t)$ and $T = T(t)$ equation (4) yields, upon employment of the superposition integral

$$m(x,t) - m_o I_o(x,\xi^*) = \int_0^t \Pi[x,\xi^*(t) - \xi^*(\tau)] \, \mu'(\tau)d\tau \qquad (11)$$

Equation (11) must of course be evaluated numerically.

NON-SYMMETRIC EXPOSURE

Consider now an infinite plate of thickness L whose faces $x = 0$ and $x = L$ are exposed to different relative humidities which fluctuate independently of each other. We still assume that all temperature fluctuations are spatially uniform within the entire plate.

The solution to differing, but constant boundary conditions $m(o,t) = \mu^o$ and $m(L,t) = \mu^L$ with zero initial moisture $m(x,o) = o$ can be expressed as follows

$$m(x,t) = \mu^o \, H_o(x,t^*) + \mu^L \, H_L(x,t^*) \qquad (12)$$

with

$$H_o(x,t^*) = \begin{cases} S_o(x,t^*) \\ \text{or} \\ U_o(x,t^*) \end{cases} \qquad H_L(x,t^*) = \begin{cases} S_L(x,t^*) \\ \text{or} \\ U_L(x,t^*) \end{cases} \qquad (13)$$

In (13)

$$S_o(x,t^*) = 1 - \frac{x}{L} - \frac{2}{\pi} \sum_{n=1}^{\infty} \frac{1}{n} \sin \frac{q_n x}{L} \exp(-n^2\pi^2 t^*) \qquad (14)$$

$$S_L(x,t^*) = \frac{x}{L} + \frac{2}{\pi} \sum_{n=1}^{\infty} \frac{\cos n\pi}{n} \sin \frac{q_n x}{L} \exp(-n^2\pi^2 t^*)$$

$$U_o(x,t^*) = \text{erfc}\left(\frac{x/L}{2\sqrt{t^*}}\right) + V_o(x,t^*) - W_o(x,t^*)$$

$$\tag{15}$$

$$U_L(x,t^*) = \text{erfc}\left(\frac{1 - x/L}{2\sqrt{t^*}}\right) + V_L(x,t^*) - W_L(x,t^*)$$

The functions V and W in (15) represent the following infinite series

$$V_o(x,t^*) = \sum_{n=1}^{\infty} \text{erfc}\left(\frac{2n + x/L}{2\sqrt{t^*}}\right)$$

$$W_o(x,t^*) = \sum_{n=0}^{\infty} \text{erfc}\left(\frac{2(n + 1) - x/L}{2\sqrt{t^*}}\right)$$

$$\tag{16}$$

$$V_L(x,t^*) = \sum_{n=1}^{\infty} \text{erfc}\left(\frac{2n + 1 - x/L}{2\sqrt{t^*}}\right)$$

$$W_L(x,t^*) = \sum_{n=0}^{\infty} \text{erfc}\left(\frac{2n + 1 + x/L}{2\sqrt{t^*}}\right)$$

In (12–16) $t^* = Dt/L^2$ and $q_n = n\pi$.

Expressions (14) are available in the literature [8], while (15) are obtained by means of a straightforward Laplace transform and inversion method.

Maximal efficiency in evaluating H_o and H_L is again obtained by switching between their alternate forms given in (13) and detailed in (14–16), because (14) is efficient for long times and (15) is advantageous for short times. For instance, for an accuracy of $O(10^{-16})$ we never need more than four terms in each of U_o, U_L, S_o, and S_L as listed in Table 1. For a lesser accuracy, the number of terms is of course smaller.

For fluctuating boundary conditions $\mu^o(t)$ and $\mu^L(t)$, and with varying temperature $T(t)$ we employ the reduced time ξ^* given in (10) and a superposition integral analogous to (11) to get

$$m(x,t) = \int_0^t \left\{ H_o\left[x, \xi^*(t) - \xi^*(\tau)\right] \frac{d\mu^o(\tau)}{d\tau} \right.$$

Table 1. Number of terms required in various truncated series to attain accuracy of $O(10^{-16})$ in moisture profile ($\lambda = 5{,}877$, $R = 16/\pi^2 \log e$).

Range	Largest Number of Terms in Each Series						Total Number Terms in (12)
	V_o	V_L	W_o	W_L	S_o	S_L	
$o < t^* < (1/\lambda)^2$	0	0	0	0	0	0	2
$(1/\lambda)^2 < t^* < (3/2\lambda)^2$	1	1	0	0	0	0	4
$(3/2\lambda)^2 < t^* < (2/\lambda)^2$	1	1	1	1	0	0	6
$(2/\lambda)^2 < t^*\ R/5^2$	2	2	1	1	0	0	8
$R/(i + 1)^2 < t^* < R/i^2$ $(i = 4,3,2,1,0)$	0	0	0	0	i	i	$2i$

$$+ H_L \left[x, \xi^*(t) - \xi^*(\tau)\right] \left. \frac{d\mu^L (\tau)}{d\tau} \right\} \ d\tau \qquad (17)$$

THE NUMERICAL SCHEME

To compute the moisture $m(x, t)$ we divide the time-span of interest t_f into n, not necessarily equal, sub-intervals. These intervals $\Delta_i = t_i - t_{i-1}(i = 1, 2, \ldots n)$ with $t_o = 0$ and $t_n = t_f$ should be selected in a manner that *both* the ambient moistures $\mu^o(t)$ and $\mu^L(t)$ as well as the temperature $T(t)$ are represented to within a satisfactory approximation by the "staircase" functions*

$$\mu^o(t) = \mu_i^o, \qquad \mu^L(t) = \mu_i^L, \qquad T(t) = T_i$$

for $\qquad\qquad\qquad\qquad\qquad\qquad\qquad\qquad\qquad\qquad\qquad\qquad\qquad$ (18)

$$t_{i-1} < t < t_i \ \ (i = 1, 2, \ldots n)$$

Note that in the symmetric case $\mu_o = m_o$.

Denote $g_i = \exp(A/T_R - A/T_i)$ then (10) yields

$$\xi_i^* = \frac{D_R}{L^2} \sum_{k=1}^{i} (t_k - t_{k-1}) g_K \ \ (i = 1, 2, \ldots n) \qquad (19)$$

---

*Obviously, only one ambient moisture $\mu(t)$ is involved in the symmetric case.

Thereby

$$\zeta_{ij}^* = \xi_i^* - \xi_j^* = \frac{D_R}{L^2} \sum_{k=j+1}^{i} (t_k - t_{k-1}) g_k \tag{20}$$

$$(j = 0, 1, \dots i-1, i = 1, 2, \dots n)$$

The integrals (11) and (17) are now represented respectively by the sums

$$m(x,t_i) = m_o I_o(x, \xi_i^*) + \sum_{j=1}^{i} (\mu_o - \mu_{j-1}) I(x, \zeta_{ij}^*) \tag{21}$$

and

$$m(x,t_i) = \sum_{j=1}^{i} [(\mu_j^\circ - \mu_{j-1}^\circ) H_o(x, \zeta_{ij}^*) + (\mu_j^L - \mu_{j-1}^L) H_L(x, \zeta_{ij}^*)] \tag{22}$$

Expressions (21) and (22) remain valid for any intermediate time $\hat{t}_i$ where $t_{i-1} < \hat{t}_i < t_i$, provided we substitute the value of $\hat{t}_i$ in place of t_i in (19) and (20) as well as in (21) and (22).

Computational efficiency is achieved by switching between the two alternate forms given in (5) and (13) which is accomplished by testing the ranges of ξ_i^* and ζ_{ij}^* according to rules (9a–9c) or in Table 1, respectively. Obviously, ξ_i^* and ξ_{ij}^* must replace t^* in equations (9) and Table 1.

2A NUMERICAL EXAMPLE FOR THE SYMMETRIC CASE

To illustrate the method we consider the case of a sixteen ply 5208/T300 graphite/epoxy laminate with $L = 0.04''$. For this material $D_R = 1.5019 \times 10^{-8}$ in²/min and $A = 6340$.

The composite laminate was considered to be exposed to fluctuating ambient relative humidity, which is reflected as a fluctuating boundary moisture μ, and to fluctuating temperatures.

Specifically, we considered the situation in which the ambient moisture level switched between ½ percent and 1 percent every 5000 minutes, while the temperature varied between two fixed levels of 350°K and 297°K also every 5000 minutes. Computations were performed for moisture and temperature fluctuating in phase (case 1) and out-of-phase (case 2). Case 1 is shown by solid lines and case 2 is marked by dashed lines in Figure 1.

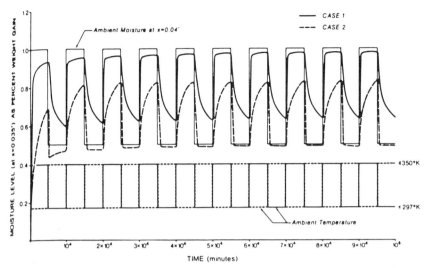

Figure 1. *Moisture levels at x = 0.035″ vs. time in a 0.08″ thick 5208/T300 graphite/epoxy laminate that is exposed symmetrically to two cases of fluctuating ambient relative-humidity and temperature. Case 1: R.H In-plane with temperature. Case 2: R.H. out-of-phase with temperature.*

The results, exhibited in Figure 1, show the variation of moisture level with time at a station located at $x = 0.035″$.

Note that sharp slopes in $m(x,t)$ vs. t occur during the high-temperature time intervals. Consequently, the in-phase case approaches the saturation level of $m(x,t) = 1$ percent during the "wet" intervals while for the out-of-phase case the moisture level at $x = 0.035″$ approaches 0.5 percent during the "dry" intervals. The details are shown by the heavy lines in Figure 1.

ACKNOWLEDGMENTS

This work was conducted, in part, under Contract F33615-79-C-5517 from the Air Force Material Laboratory (AFML/AFWAL) and in part under Contract F 49620-78-C-003 from the Air Force Office of Scientific Research (AFOSR). This support is gratefully acknowledged.

REFERENCES

1. Shen, C. H. and Springer, G. S., "Moisture Adsorption and Desorption of Composite Materials," *Journal Comp. Mat., 10*, 36–54 (Jan. 1976).
2. DeIasi, R. and Whiteside, J. B., "Effect of Moisture on Epoxy Resins and Composites," in *Advanced Composite Materials—Environmental Effects,* J. R. Vinson, ed. STP 658 (ASTM), 2–20 (1978).
3. Shirrel, C. D., "Diffusion of Water Vapor in Graphite/Epoxy Laminates," in *Advanced Composite Materials—Environmental Effects,* J. R. Vinson, ed. STP 658 (ASTM), 21–42 (1978).

178   Y. Weitsman

 4. McKague, E. L. and Halkias, J. E., Private Communication.
 5. Weitsman, Y., "Diffusion With Time-Varying Diffusivity With Application to Moisture-Sorption in Composites," *Journal Compt. Mat., 10,* 193–204 (July 1976).
 6. Douglass, D. A. and Weitsman, Y., "Stresses Due to Environmental Conditioning of Cross-Ply Graphite/Epoxy Laminates," In *Proc. Third International Conference on Composite Materials (ICCM3),* Vol. 1, A.R. Bunsell, et al., ed. Paris, France: Pergamon Press, 529–542 (1980).
 7. Luikov, A. V. *Analytical Heat Diffusion Theory.* Academic Press, 97–114 (1968).
 8. Crank, J. *The Mathematical Theory of Diffusion.* 2nd ed., Oxford University Press, 47, 104–105 (1975).
 9. Abramowitz, M. and Stegun, L. A., "Handbook of Mathematical Functions," National Bureau of Standards, 298 (June 1964).
10. Morland, L. W. and Lee, E. H., "Stress Analysis for Linear Viscoelastic Materials With Temperature Variation," *Transactions Society of Rheology, IV,* 233–263 (1960).

16

A Fickian Diffusion Model for Permeable Fibre Polymer Composites

R. M. V. G. K. Rao, M. Chanda, and N. Balasubramanian

INTRODUCTION

SHEN AND SPRINGER [1,2] PRESENTED EXTENSIVE WORK ON THE MOISTURE absorption behaviour of graphite-epoxy composites, based on a Fickian diffusion model. Romanenkov and Machavariani [3] confirmed the applicability of such a diffusion model for glass-polyester composites and reported that the composite diffusion co-efficient could be considered concentration independent, in void-free, compact composites. Bonniau and Bunsell [4] applied two different diffusion models (single and two phase models) for glass-epoxy composites and concluded that the diamine cured composite obeyed the Fickian diffusion model. Mehta et al [5] reported that glass-ribbon reinforced cellulose acetate films behaved anomalously due to an internal stress phenomenon. They discussed the significance of a structure factor in impermeable fibre based polymer composites.

All the above studies confirm the validity of a Fickian diffusion model for such composites, with the implication of a well-fabricated void-free composite. No attempts, however, have been made so far to verify the adequacy of a Fickian model for describing the moisture absorption in polymer composites containing highly permeable phases such as the natural fibres, except for the only reference of the work by Mehta et al [6], who reported on cheese cloth filled cellulose acetate film composites. They considered molecular diffusion to occur in that part of the unfilled film, while the diffusion through the low resistant cheese cloth was approximated by a convective flow controlled by a pressure gradient across the film. Their morphological studies, however, indicated that the cellulose acetate matrix didn't penetrate into the porous cheese cloth, which can possibly be the reason for a non-Fickian diffusion operating under these conditions.

179

For the first time Rao et al [7] developed modified expressions for jute-epoxy (permeable fibre composite) composites to calculate the composite diffusion coefficient in terms of the volume fraction of a resin impregnated fibre phase (V_f') and its diffusion coefficient (D_f'). This was a consequence of their significant observation that the virgin jute fibre lost its identity in a composite. Rao [8] subsequently pursued moisture diffusion studies on jute and glass-epoxy composites and fully characterised the behaviour of these permeable and impermeable fibre composites relative to each other.

In this paper, the authors present correlations between experimental data and the analytical diffusion plot based on a modified Fickian equation. Very good agreements are noticed between experiments and theory, even under the influence of varied internal (fibre volume fraction) and external (ambient temperature) factors.

THEORY

The one-dimensional diffusion process taking place through two large opposite faces of a solid slab material (Figure 1) can be represented by the Fick's II-law, as

$$\frac{\partial C_A}{\partial t} = D_x \frac{\partial^2 C_A}{\partial x^2} \tag{1}$$

Equation (1) can be solved for the following boundary equations,

$$c = c_i \quad 0 < x < h \quad t \leqslant 0 \tag{2.a}$$

$$c = c_a \quad x = 0, x = h \quad t > 0 \tag{2.b}$$

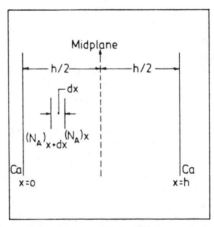

Figure 1. Mass transfer through a slab by molecular diffusion under unsteady state conditions.

The solution of equation (1) with the above boundary conditions, as given by Jost, is

$$\frac{C - C_i}{C_a - C_i} = 1 - \frac{4}{\pi} \sum_{j=0}^{\infty} \frac{1}{(2j + 1)} Sin \frac{(2j + 1)\pi x}{h} exp \frac{[-(2j + 1)\pi^2 D_x t / h^2]}{} \tag{3}$$

Now considering a bone dry specimen ($M_i = 0$) and noting that the moisture content (M_t) can be expressed as,

$$M_t = A \int_0^h c \, dx \tag{4}$$

equation (3) takes the form

$$\frac{M_t}{M_m} = 1 - \frac{8}{\pi^2} \sum_{j=0}^{\infty} exp \frac{[-(2j + 1)^2 \pi^2 (D_x t / h^2)]}{(2j + 1)^2} \tag{5}$$

A plot of the dimensionless absorption parameter (M_t / M_m) against the dimensionless diffusion parameter ($D_x t / h^2$) represents the Fickian diffusion curve for the composite. A good fit of the experimental data with this curve, therefore confirms the applicability of a Fickian model for the material considered.

Modified Diffusion Plot for Permeable Fibre Polymer Composite

Equation (5) can be rewritten for the case of composites containing a highly permeable fibre phase, as

$$F_s = \frac{M_t}{M_m'} = 1 - \frac{8}{\pi^2} exp^{-(D_c' t / h^2) \pi^2}$$

$$for \ D_c' t / h^2 > 0.05 \tag{6.a}$$

$$= \frac{4}{\pi} (D_c' t / h^2)^{1/2}$$

$$for \ D_c' t / h^2 < 0.05 \tag{6.b}$$

As reported by Bonniau and Bunsell, all the plots represented by equations (5), (6.a) and (6.b) merge into one line during the initial part of the diffusion process, so that this initial linear part of the diffusion curve can well be used to evaluate the concentration independent composite diffusion coefficient.

Now, equation (6.b) can be rewritten as,

$$F_s = 1 - \frac{8}{\pi^2} \left[\frac{1}{\underset{exp}{ln(D_c't/h^2)\,\pi^2}} \right] \quad (7)$$

$$exp$$

The diffusion curve as represented by equation (7) is very convenient to use in the analysis of permeable fibre composites, owing to their very high diffusivities. A plot of F_s against $ln(D_c't/h^2)$ will represent the analytical curve for the Fickain diffusion process in such composites. The extent to which the experimental data correlates with such a curve then determines the applicability of a Fickian model for the type of composites under consideration.

Calculation of Composite Diffusion Coefficient (D_c')

Rewriting equation (6.b) for two different instants of exposure (t_1, t_2) and rearranging,

$$D_c' = \pi \left(\frac{h}{4M_m'} \right)^2 \left(\frac{M_2 - M_1}{\sqrt{t_2} - \sqrt{t_1}} \right)^2 \quad (8)$$

EXPERIMENT DETAILS

Materials

The jute fibres of commercial grade (supplied by the Indian Jute Fibre Manufacturer's Association, Calcutta) were chosen together with a hot curing epoxy resin system (LY 556 and HT 972 combination of Ciba Geigy (I) Ltd, Bombay).

Specimen Preparation and Exposure

Unidirectional laminates of 300mm × 300mm × 3mm were fabricated using a filament winding technique and the same cured at 150 °C for 3 hours at 100 Psi pressure. Test specimens of 25mm × 25mm × 3mm were cut off the laminates, using a high speed diamond cutter.

Specimens of varied fibre volume fractions (0.6, 0.7, 0.82) were achieved using 5, 7 and 9 layers of winding (Figure 2).

Specimens of 0.7 volume fraction were then edge coated and immersed in water baths maintained at 298 °K, 313 °K, and 333 °K, while those with the three different volume fractions were immersed in water at 313 °K.

The weight gain by the specimens was carefully monitored as a function of time of immersion (t) using a precision single pan electronic balance, ensuring that the surface moisture of all the specimens was carefully wiped off each time. The weight gain was expressed as,

Figure 2. Moisture absorption specimens: Top: L-R: pure epoxy; jute-epoxy composite, $V_f' = 0.6$; jute-epoxy composite, $V_f' = 0.71$; jute-epoxy composite, $V_f' = 0.82$; bottom: jute fibre specimens.

$$\% M_t = \frac{W_2 - W_1}{W_1} \times 100$$

Where W_2 is the instant weight of the moistened specimen and W_1 that of a bone dry specimen.

The percent weight gains were then plotted against $\sqrt{t}$ to generate the moisture absorption curves necessary to calculate the composite diffusion coefficients.

Calculation of Resin Impregnated Fibre Volume Fraction (V_f')

The calculation of an unmodified fibre volume fraction (V_f) as applicable to impermeable fibre composites is straightforward. In case of a permeable fibre based composite as considered in these studies, the procedure was rendered complicated due to the fact that the jute fibre got converted into a resin impregnated fibre in the composite (Figure 3). A knowledge of the amount of resin absorbed within the fibre itself had become necessary to calculate the volume fraction of such fibre. A resin absorption curve of the jute fibre under typical curing conditions is shown in Figure 4, which shows that the fibres absorbed 88% by weight of resin before setting into a composite. A sample calculation of the volume fraction of the resin impregnated jute fibre is presented in the appendix.

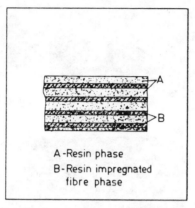

Figure 3. Schematic diagram of a permeable fibre polymer composite.

RESULTS AND DISCUSSION

Figures 5 and 6 show the moisture absorption curves for the jute-epoxy composites at different temperatures and fibre volume fractions, respectively. The composite diffusion coefficients for all the cases have been calculated using equation (8).

Plots of dimensionless absorption (F_s) and diffusion ($D_c't/h^2$) parameters are shown in Figures 7 and 8 at different ambient temperatures and fibre volume fractions. These clearly indicate a very good fit of the experimental

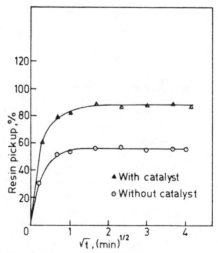

Figure 4. Percent resin absorption in jute fibres.

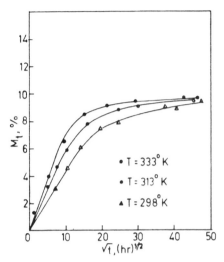

Figure 5. Moisture absorption curves of jute-epoxy composite at different temperatures (V_f' = 0.70, immersion).

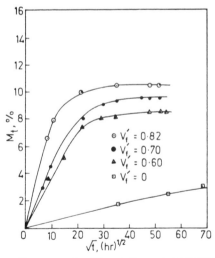

Figure 6. Moisture absorption curves for different fibre volume fractions (V_f') for jute epoxy composite (T = 313°K, immersion).

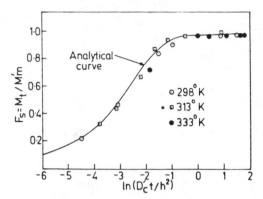

Figure 7. Comparison of analytical (Equation 7) and measured F_s' values for unidirectional jute epoxy composite (V_f' = 0.70) immersion-correlation-I.

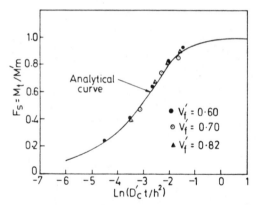

Figure 8. Comparison of analytical (Equation 7) and measured 'F_s' values for unidirectional jute epoxy composite (313°K, immersion)-correlation-II.

data with the analytical curve, as given by equation (7), confirming that a Fickian diffusion model is adequate to represent the moisture absorption phenomenon in polymer composites containing a highly permeable fibre phase.

CONCLUSIONS

Moisture absorption in permeable fibre polymer composites such as jute-epoxy composites can be characterised by a Fickian diffusion model, provided the composite is fabricated under carefully controlled conditions to eliminate any voids or other structural defects. The volume fraction of a resin impregnated fibre and not its superficial fibre volume fraction as applicable to impermeable fibre based composites, should be considered in the moisture absorption analysis of composites containing fibres that are highly permeable to both the matrix resin and the ambient moisture. Systematic moisture absorption characterisation of such composites leads to a better practical utilisation of the natural fibre composites.

APPENDIX

Calculation of the Resin Impregnated Fibre Volume Fraction (V_f')

From Figure 4, the resin absorption of the dry jute fibre under typical curing conditions was found to be 88%.

The weight fraction of the resin impregnated fibres (W_f') is then calculated using the formula,

$$W_f' = \frac{1.88 \ W_j}{W_c'}$$

where W_j = weight of jute fibres

W_c = weight of composite

The volume fraction of the resin impregnated fibres (V_f) is then calculated using the relationship,

$$V_f' = \frac{1}{1 + \dfrac{\varrho_f'}{\varrho_c'} \ \dfrac{1 - W_f'}{W_f'}}$$

where ϱ_f', the specific gravity of the resin impregnated jute fibres, is calculated using the following formula,

$$\varrho_f' = \frac{1.88 \ W_j}{\dfrac{W_j}{\varrho_f^o} + \dfrac{0.88 \ W_j}{\varrho_r}}$$

$$= \frac{1.88}{\dfrac{1}{\varrho_f^o} + \dfrac{0.88}{\varrho_r}}$$

here $\varrho_f^o = 1.34$ for jute fibres

$\varrho_r = 1.15$ for the resin

Therefore $\varrho_f' = 1.25$

Sample Calculation for a 3-layered Composite

$$W_j = 65 \text{ g} \qquad \text{Therefore } W_f' = 1.88 \times 65/235$$
$$= 0.52$$
$$W_c' = 235 \text{ g} \qquad \varrho_c' = 1.3 \text{ from experiments}$$

$$V_f' = 1/[1 + ([1.25/1.3][(1 - 0.52)/0.52])] = 0.60$$

LIST OF SYMBOLS

A surface area of diffusion
C_a diffusant concentration at the surface
C_i diffusant concentration at any time
D_x diffusion coefficient in x-direction
D_c' overall diffusion coefficient of a permeable fibre composite
h specimen thickness
M_m equilibrium moisture content in a composite
M_m' equilibrium moisture content as referred to a permeable fibre composite
M_t moisture content in a composite at any time
t time of exposure
W_1 weight of specimen before exposure
W_2 weight of specimen after exposure
F_s dimensionless absorption parameter
$\dfrac{D_c' t}{h^2}$ dimensionless diffusion parameter

ACKNOWLEDGMENTS

The authors are very thankful to Dr. S. R. Valluri, Director National Aeronautical Laboratory and Dr. A. K. Singh, Head, Materials Sciences Division, National Aeronautical Laboratory, for their support to these investigations. The assistance provided by Mr. S. Gururaja, in typing the manuscripts is highly appreciated.

REFERENCES

1. Shen, Chi-Hung and Springer, George S., "Moisture Absorption and Desorption of Composite Materials," *J. Composite Materials, 10*, 2–20 (1976).
2. Springer, George S., "Moisture Contents of Composites under Transient Conditions," *J. Composite Materials, 11*, 107–122 (1977).
3. Romanenkov, I. G. and Machavariani, Z. P., "Water Absorption of GRPS," *Soviet Plastics,* 49–51 (June 1966).
4. Bonniau, P. and Bunsell, A. R., "A Comparative Study of Water Absorption Theories Applied to Glass-Epoxy Composites," *J. Composite Materials, 15*, 272–93 (May 1981).
5. Mehta B. S., Dibenedetto, A. T., and Kardos, J. L., "Sorption and Diffusion of Water in Glass-ribbon Reinforced Composites," *J. Applied Polymer Science, 21*, 3111–3127 (1977).
6. Ibid. "Diffusion and Permeation of Gases in Fibre Reinforced Composites." *Intern. J. Polymeric Mater., 5*, 147–161 (1976).

7. Rao, R. M. V. G. K., Balasubramanian, N., and Chanda, Manas, "Moisture Absorption Phenomenon in Permeable Fibre Polymer Composites," *J. Appl. Poly. Sci., 26,* 4069–4079 (1981).
8. Rao, R. M. V. G. K., "Diffusion Phenomenon in Polymer Composites, Permeable and Impermeable Fibre Composites," Ph.D. Thesis, Dept. of Chemical Engineering, I.I.Sc, Bangalore (Dec. 1982).

17

Irreversible Hygrothermomechanical Behavior and Numerical Analysis in Anisotropic Materials

T. J. CHUNG AND R. J. BRADSHAW

INTRODUCTION

EPOXY COMPOSITE STRUCTURES REINFORCED WITH GLASS, BORON, OR graphite fibers have been in wide use because of their superior performance characteristics. It has also been known that the adverse effect of combined moisture and thermal environments upon the strength of the composites are significant enough that they should be accounted for in design. The moisture absorption in a composite laminate results in a dilation of the matrix and reduction in the resin softening temperature or glass transition temperature. Furthermore, it is possible that the concentration gradients of the moisture diffusion would cause unequal swelling stresses and result in the formation of misocracks as well as predictable dimensional changes in both resin and laminates.

When the hygroscopic environment is coupled with high temperatures the matrix-fiber composites show significant degradation according to the test results reported in [1-4]. If the mechanical loading is combined with long-term hygrothermal environments, however, one faces insurmountable difficulties in laboratory measurements. Thus, a customary procedure is to first subject a specimen to the hygrothermal loadings and next to the mechanical load independently to determine the degradation. Under the service conditions, however, we would require a complete coupling of hygroscopic, thermal, and mechanical loadings because the effect of one on another must be recognized in their simultaneous actions [5-8].

Suppose the moisture is in contact with the surface of a matrix-laminate. We may postulate that fluid particles wander in a random manner through the lattice, but tend to get trapped or delayed at certain fixed sites. It is further postulated that the probability that the active traps capture a hydrogen

190

atom is proportional to the concentration of diffusing hydrogen. Note also that an additional variable to consider is the fraction of occupied traps which may release the hydrogen atom. Thus, on one hand, we have free phase associated with the diffusing fluid and on the other hand, there is a bound phase containing the trapped fluid particles. Under such conditions, the conventional Fick's law will no longer be valid and an additional relationship describing the time rate of change of the trapped phase must be incorporated [9–12]. In addition, the mass flux for the free phase is assumed to be a function of the gradients of temperature and strain as well as the moisture concentration, which may be called the modified Fick's law. We derive the consistent governing equations for two phase diffusion in the context of chemical potential which is the rate of change of the free energy with respect to the moisture concentration. The temperature distribution is governed by the modified Fourier law in which the heat flux is considered a function of the moisture as well as the temperature gradients. It is shown that the hygrothermal equations are the consequence of the second law of thermodynamics with modification for the moisture diffusion.

The mechanics of deformation of the system of matrix-laminate is described by the first law of thermodynamics coupling completely with the hygrothermal behavior. The constitutive equation includes, among other aspects, the viscoelastic behavior and the irreversible process combined with the second law of thermodynamics. The resulting equations are rigorous in theory and convenient for implementation to numerical analysis such as finite elements.

The purpose of the present paper is to present the latest development in the theoretical formulations and the prediction of responses of the matrix laminate subjected to hygrothermomechanical loadings. A comparison is made between the viscoelastic and the elastic behavior for a simple case. It is concluded that the proposed approach can be used successfully to provide design criteria which will keep the material from failure.

BALANCE LAWS AND CONSTITUTIVE STRUCTURE

The most consistent approach to derive a constitutive theory stems from adherence to the balance laws—conservation of mass, momentum, and energy. Toward this end, it is necessary to choose an adequate coordinate system either Lagrangian or Eulerian. The conservation of mass requires

$$\rho_0 = \rho\sqrt{G} \qquad \text{for Lagrangian coordinates} \qquad (1)$$

$$\frac{\partial \rho}{\partial t} + \nabla \cdot \rho V = 0 \qquad \text{for Eulerian coordinates} \qquad (2)$$

where ϱ is the density, V is the velocity vector, and G is the determinant of the metric tensor. Since no solid or fluid particles are expected to undergo large motions, we need not resort to the Eulerian coordinate system. Thus, the logical choice would be the Lagrangian coordinates. For small deformations, the metric tensor G is a unity and the density ϱ is equal to the initial density ϱ_0. The total mass consists of contributions from matrix, fibers, the free and bound phase of moisture,

$$\rho = \sum_k \rho_k \quad (k = 1,2,3,4)$$

We consider a motion governed by the following functions: body force f_i, stress σ_{ij}, thermal flux $q_i^{(T)}$, moisture flux $q_i^{(M)}$, heat supply $h^{(T)}$, moisture supply $h^{(M)}$, entropy due to temperature $\eta^{(T)}$, and the chemical potential $\eta^{(M)}$ which may also be called the entropy due to moisture. The combined thermal and hygroscopic quantitites are $q_i = q_i^{(T)} + q_i^{(M)}$, $h = h^{(T)} + h^{(M)}$.

The first law of thermodynamics is given by

$$\dot{K} + \dot{U} = R + Q \tag{3}$$

where the kinetic energy K, internal energy U, mechanical power R, and the heat sources Q are defined as

$$K = \frac{1}{2} \int_\Omega \rho V_i V_i d\Omega \tag{4}$$

$$U = \int_\Omega \rho \varepsilon d\Omega \tag{5}$$

$$R = \int_\Omega \rho f_i V_i d\Omega + \int_\Gamma \sigma_{ij} n_j V_i d\Gamma \tag{6}$$

$$Q = \int_\Omega \rho h d\Omega + \int_\Gamma q_i n_i d\Gamma \tag{7}$$

where Ω is the interior domain, Γ is the boundary surface, and n_i is the component of a vector normal to the boundary surface. It can be shown that the consequence of the first law of thermodynamics assumes the conservation of mass and the balance of momentum and energy such that

$$\rho \ddot{u}_i - \sigma_{ij,j} - \rho f_i = 0 \tag{8}$$

$$\rho \dot{\varepsilon} = \sigma_{ij} \dot{\gamma}_{ij} + q_{i,i} + \rho h \tag{9}$$

where u_i and γ_{ij} are the components of the displacement and strain tensor, respectively.

ENTROPIES AND IRREVERSIBLE PROCESS

The second law of thermodynamics in local form is given by

$$d\dot{\eta}^{(T)} - \frac{dQ^{(T)}}{\theta} \geqslant 0 \tag{10a}$$

where θ is the absolute temperature, $\theta = T_0 + T$, with T_0 and T being the initial temperature and the temperature change, respectively. The global form of (10a) for an irreversible process is

$$\int_\Omega \rho\dot{\eta}^{(T)}d\Omega - \int_\Omega \frac{\rho h^{(T)}}{\theta}d\Omega - \int_\Gamma \frac{q_i^{(T)}n_i}{\theta}d\Gamma \geqslant 0 \tag{10b}$$

Using the similar argument, we write $M = M_o + M$ where M, M_o, and M are the total moisture content, initial moisture content, and the change in moisture. The chemical potential $\eta^{(M)}$ has a unit of (lb–in)/(unit mass) as compared with (lb–in)/(unit mass- °C) for the entropy due to temperature. We assert that the irreversibility arises also from moisture infiltration resulting in the condition [13],

$$d\dot{\eta}^{(M)} - \frac{dQ^{(M)}}{M} \geqslant 0 \tag{11a}$$

and

$$\int_\Omega \rho\dot{\eta}^{(M)}d\Omega - \int_\Omega \frac{\rho h^{(M)}}{M}d\Omega - \int_\Gamma \frac{q_i^{(M)}n_i}{M}d\Gamma \geqslant 0 \tag{11b}$$

Using the Green-Gauss theorem, the expressions (10b) and (11b) are written as, respectively,

$$\int_\Omega \left(\rho\theta\dot{\eta}^{(T)} - q_{i,i}^{(T)} - \rho h^{(T)} + \frac{1}{\theta}q_i^{(T)}\theta_{,i}\right)d\Omega \geqslant 0 \tag{12a}$$

and

$$\int_\Omega \left(\rho M\dot{\eta}^{(M)} - q_{i,i}^{(M)} - \rho h^{(M)} + \frac{1}{M}q_i^{(M)}M_{,i}\right)d\Omega \geqslant 0 \tag{12b}$$

For the above expressions to be valid for all arbitrary volumes, we require the integrands to be of the form,

$$D^{(T)} + \frac{1}{\theta} q_i^{(T)} \theta_{,i} \geq 0, \quad D^{(M)} + \frac{1}{M} q_i^{(M)} M_{,i} \geq 0 \qquad (13)$$

where the total internal dissipation D is given by

$$D = D^{(T)} + D^{(M)} \geq 0 \qquad (14)$$

for an irreversible process, and the internal dissipations due to temperature and moisture are, respectively,

$$D^{(T)} = \rho\theta\dot{\eta}^{(T)} - q_{i,i}^{(T)} - \rho h^{(T)} \qquad (15a)$$

$$D^{(M)} = \rho M \dot{\eta}^{(M)} - q_{i,i}^{(M)} - \rho h^{(M)} \qquad (15b)$$

The expressions (15a) and (15b) are the equations of heat conduction and diffusion, respectively. To satisfy (14) in which the total internal dissipation is the sum of dissipation from both temperature and moisture we introduce a parameter ϕ such that

$$0 \leq \phi \leq 1$$

We propose to write the internal dissipations due to temperature and moisture with respect to the total internal dissipation, respectively

$$D^{(T)} = \phi D \qquad (16)$$

$$D^{(M)} = (1 - \phi)D \qquad (17)$$

Here the parameter ϕ signifies the portion of dissipation arising from temperature only. Thus, we assure that, for an irreversible process, $D^{(T)}>0$ and $D^{(M)}>0$. If the process is reversible, then, we have $D^{(T)} = 0$ and $D^{(M)} = 0$. The notion of dissipation due to moisture infiltration is important because not only the classical diffusion equation results as a special case but also we are now able to take into account the very reasonable assertion that the moisture infiltration does crate an irreversible state of stress and strain and, consequently, the dissipative energy.

The governing equations for the heat conduction and moisture diffusion completely coupled with deformation are, from (15a) and (15b), respectively,

$$\rho\theta\dot{\eta}^{(T)} - q_{i,i}^{(T)} - \rho h^{(T)} - D^{(T)} = 0 \qquad (18)$$

$$\rho M \dot{\eta}^{(M)} - q_{i,i}^{(M)} - \rho h^{(M)} - D^{(M)} = 0 \qquad (19)$$

Note that the expressions (18) and (19) are the consequence of energy equation (9) and the second law of thermodynamics (11).

We now require the constitutive equations for the stress (σ_{ij}), entropies ($\eta^{(T)}$, $\eta^{(M)}$), fluxes ($q_i^{(T)}$, $q_i^{(M)}$), and internal dissipations ($D^{(T)}$, $D^{(M)}$). We begin with the Helmoholtz free energy which is assumed to be dependent on the strain (γ_{ij}), absolute temperature (θ), moisture concentration (M), and the internal state variables ($\alpha_{ij}^{(r)}$) [13],

$$\psi = \psi\,(\gamma_{ij},\theta,M,\alpha_{ij}^{(r)}) = \varepsilon - \theta\eta^{(T)} - M\eta^{(M)} \tag{20}$$

Here $\alpha_{ij}^{(r)}$ ($r = 1,2 \ldots$) implies a possible damping effect (viscoelastic) which may lead to an irreversible process and may be given by a generalized Maxwell-type kernal [14]

$$\alpha_{ij}^{(r)} = \int\limits_{0}^{t} \exp\left[\frac{(t - \tau)}{T_{(r)}}\right] \frac{\partial\gamma_{ij}}{\partial\tau}\,d\tau \tag{21}$$

where $T_{(r)}$ is the relaxation time and τ is the time variable.

FREE AND BOUND PHASES

The moisture diffusion is considered to be associated with a free phase ($M^{(f)}$) subjected to diffusion through the matrix. It is possible that some of the moisture particles (hydrogen) are attraced to the surfaces of the solid material (matrix-laminate) thus forming a bound phase ($M^{(b)}$). We further assume that $M^{(f)}$ is a function of both space and time whereas $M^{(b)}$ is a function of only time. This because once a hydrogen atom is trapped it remains immobile or it does not change with respect to the spatial position but changes only with respect to time as the hydrogen may be freed sooner or later. The time rate of change of $M^{(b)}$ may be given by [9–12]

$$\frac{\partial M^{(b)}}{\partial t} = C_1 M^{(f)} - C_2 M^{(b)} \tag{22a}$$

where C_1 and C_2 denote constants associated with probabilities of hydrogen being captured and freed, respectively. The total moisture concentration M, from the assumptions given above, is related as follows:

$$M = M^{(f)} + M^{(b)} \tag{22b}$$

$$\dot{M} = \dot{M}^{(f)} + \dot{M}^{(b)} \tag{22c}$$

$$M_{,i} = M_{,i}^{(f)} \tag{22d}$$

The expressions (22a–d) will be incorporated into the two-phase diffusion equations.

CONSTITUTIVE EQUATION

The time rate of change of free energy per unit volume in (20) combined with (14) and (9) yields, for an irreversible process,

$$D = D^{(T)} + D^{(M)} = \sigma_{ij}\dot{\gamma}_{ij} - \rho\left(\frac{\partial\psi}{\partial\gamma_{ij}}\dot{\gamma}_{ij} + \frac{\partial\psi}{\partial\theta}\dot{\theta} + \frac{\partial\psi}{\partial M}\dot{M}\right.$$

$$\left. + \frac{\partial\psi}{\partial\alpha_{ij}^{(r)}}\dot{\alpha}_{ij}^{(r)}\right) - \rho\eta^{(T)}\dot{\theta} - \rho\eta^{(M)}\dot{M} \geqslant 0 \tag{23}$$

which provides the constitutive relations,

$$\sigma_{ij} = \rho\,\frac{\partial\psi}{\partial\gamma_{ij}} \tag{24}$$

$$\rho\eta^{(T)} = -\rho\,\frac{\partial\psi}{\partial\theta} \tag{25}$$

$$\rho\eta^{(M)} = -\rho\,\frac{\partial\psi}{\partial M} \tag{26}$$

$$D^{(M)} = -\phi\sum_{r=1}^{n}\frac{\partial\psi}{\partial\alpha_{ij}^{(r)}}\alpha_{ij}^{(r)} \geqslant 0 \tag{27}$$

$$D^{(T)} = -(1-\phi)\sum_{r=1}^{n}\frac{\partial\psi}{\partial\alpha_{ij}^{(r)}}\dot{\alpha}_{ij}^{(r)} \geqslant 0 \tag{28}$$

It is interesting to note that the hygroscopic entropy assumes an identical form of the conventional chemical potential which represents the derivative of the Helmholtz energy with respect to concentration. In view of the functional representation (20) and small strain assumption it is appropriate to expand the free energy in a quadratic Taylor series form of the variables γ_{ij}, T, M, and $\alpha_{ij}^{(r)}$. This gives

$$\rho\psi = \frac{1}{2}E_{ijkl}\,\gamma_{ij}\gamma_{kl} - \beta_{ij}^{(T)}\,T\gamma_{ij} - \beta_{ij}^{(M)}\,M\gamma_{ij}$$

$$-\frac{1}{2}\frac{c}{T_0}T^2 - \frac{1}{2}\frac{b}{M_o}M_0 - \chi TM - \sum_{r=1}^{n} B_{ij}^{(T)}\,(r)\,T\alpha_{ij}^{(r)}$$

$$\text{(29)}$$

$$-\sum_{r=1}^{n} B_{ij}^{(M)}\,(r)\,M\alpha_{ij}^{(r)} + \frac{1}{2}\sum_{r=1}^{n} \xi_{ijkl}^{(r)}\,\alpha_{ij}^{(r)}\,\alpha_{kl}^{(r)} + \sum_{r=1}^{n} \xi_{ijkl}^{(r)}\alpha_{ij}^{(r)}\gamma_{kl}$$

where E_{ijkl} denotes the fourth order tensor of anisotropic elastic moduli, $\beta_{ij}^{(T)}$ and $\beta_{ij}^{(M)}$ are the second-order tensors of anisotropoic thermoelastic moduli and anisotropic hygroelastic moduli, respectively. The constants c and b are the heat capacity and hygroscopic capacity, respectively. The symbol χ is defined as the hygrothermal coefficient. $B_{ij}^{(T)}\,(r)$ and $B_{ij}^{(M)}\,(r)$ are the second order tensors of anisotropic dissipative thermoelastic and hygroelastic constants respectively. $\xi_{ijkl}^{(r)}$ denotes the fourth order tensor of dissipative elastic constants.

It follows from (24)–(28) that

$$o_{ij} = E_{ijkl}\gamma_{kl}\,\beta_{ij}^{(T)}T - \beta_{ij}^{(M)}M + \sum_{r=1}^{n} \xi_{ijkl}^{(r)}\,\alpha_{kl}^{(r)} \qquad \text{(30)}$$

$$\rho\eta^{(T)} = \beta_{ij}^{(T)}\gamma_{ij} + \frac{c}{T_0}T + \chi M + \sum_{r=1}^{n} B_{ij}^{(T)}\,(r)\alpha_{ij}^{(r)} \qquad \text{(31)}$$

$$\rho\eta^{(M)} = \beta_{ij}^{(M)}\gamma_{ij} + \frac{b}{M_o}M + \chi T + \sum_{r=1}^{n} B_{ij}^{(M)}(r)\,\alpha_{ij}^{(r)} \qquad \text{(32)}$$

Furthermore, adequate forms of heat flux and mass flux must be described. To this end, we invoke the diffusion-thermal effect or "Duffour effect," $q_{ij}M_{,j}$, to the heat flux and the thermal-diffusion effect or "Soret effect," $Q_{ij}T_{,j}G_{ijkl}\gamma_{kl,j}$, to the mass flux such that [15]

$$q_i^{(T)} = k_{ij}T_{,j} + q_{ij}M_{,j}^{(f)} \qquad \text{(33)}$$

$$q_i^{(M)} = D_{ij}M_{,j} + Q_{ij}T_{,j} - G_{ijkl}\gamma_{kl,j} \tag{34}$$

where k_{ij} and D_{ij} are the standard coefficients of thermal conductivity and diffusion, respectively, q_{ij} is the diffusion-thermal coefficient, Q_{ij} is the thermal-diffusion coefficient, and G_{ijkl} is the hygroelastic modulus. Note that all of these constants refer to an anisotropic media and assume appropriate forms of isotropic tensors when the material is isotropic. For example, $G_{ijkl}\gamma_{\chi\lambda,o}$ becomes $\zeta\gamma_{kk,i} = \zeta o_{kk,i} = \zeta P_{,i}$ so that the fourth order tensor G_{ijkl} is changed into a scalar and the equivalent of the gradient of pressure P would result. It is well known that the diffusion coefficient D_{ij} may be dependent on temperature so that [16]

$$D_{ij} = D_{ij}^0 \exp\left[\frac{-e}{R\theta}\right] \tag{35}$$

where D_{ij}^0 is the frequency factor or permeability index, e is the activation energy, and R is the universal gas content.

GOVERNING EQUATIONS

In view of (8),(18),(19), and (22) the final forms of our governing equations for an irreversible process are written as

Momentum Equation

$$\rho\ddot{u}_i - E_{ijkl}u_{k,lj} + \beta_{ij}^{(T)}T_{,j} + \beta_{ij}^{(M)}M_{,j} - \sum_{r=1}^{n}\xi_{ijkl}^{(r)}\alpha_{kl}^{(r)}\ \rho f_i = 0 \tag{36}$$

Heat Conduction Equation

$$\frac{\theta}{T_0}\ c\dot{T} + \theta\left(\beta_{ij}^{(T)}\dot{\gamma}_{ij} + \sum_{r=1}^{n}B_{ij}^{(T)}(r)\dot{\alpha}_{ij}^{(r)} + \chi\dot{M}\right) - (K_{ij}T_{,j})_{,i}$$
$$- (q_{ij}M_{,j})_{,j} - \rho h^{(T)} - D^{(T)} = 0 \tag{37}$$

Two-Phase Diffusion Equations

$$\frac{M}{M_0}\ b\dot{M} + M\ \beta_{ij}^{(M)}\gamma_{ij} + \sum_{r=1}^{n}B_{ij}^{(M)}(r)\dot{\alpha}_{ij}^{(r)} + \chi T\ - (D_{ij}M_{,j})_{,j}$$
$$- (Q_{ij}T_{,j})_{,j} + (G_{ijkl}\gamma_{kl,j})_{,i} - \rho h^{(M)} - D^{(M)} = 0 \tag{38}$$

$$\dot{M}^{(b)} = C_1 \dot{M}^{(f)} - C_2 \dot{M}^{(b)} \qquad (39)$$

$$M = M^{(f)} + M^{(b)} \qquad (40)$$

It should be noted that, for transient analysis, the temperature change is small in comparison with the reference temperature. Thus, we set $\theta/T_0 \cong 1. \theta \cong T_0$. Now the total number of variables to be solved are the displacement u_i, temperature change T, total moisture M, free phase moisture content $M^{(f)}$, and bound phase moisture content $M^{(b)}$ with the expression (40) being applied as a constraint condition. The calculations may be simplified by eliminating $M^{(b)}$ such that (39) and (40) result in a single equation

$$\dot{M} - \dot{M}^{(f)} - (C_1 - C_2)M^{(f)} + C_2 M = 0 \qquad (41)$$

If the initial moisture content is zero, then the first two terms in (38) should be dropped. It is seen that the simultaneous solution of (36), (37), (38), and (41) takes into account a complete coupling of deformation field with heat and mass transfer of moisture consistent with the first and second laws of thermodynamics, and observed physical phenomena cited in the literature.

Note that a number of simplifications can be made and the governing equations become equivalent to special cases reported by other investigators. For example, we may pursue the following simplifications:

1. The process is reversible; the $D^{(T)} = D^{(M)} = 0$.
2. No viscoelastic effect is present; then all terms with the internal state variables α_{ij} are set equal to zero.
3. The bound phase is negligible; then the expressions (39)–(41) are disregarded ($M^{(b)} = 0$, $M = M^{(f)}$).
4. The Duffour effect is negligible; then set the moisture gradient term in (33) equal to zero.
5. The Soret effect is negligible; then the terms of temperature gradient and strain gradient in (34) are disregarded.
6. The hygrothermal coupling effect is negligible ($\chi = 0$); then the terms involved with χ in (37) and (38) are set equal to zero.
7. Thermoelastic coupling is negligible; then the term associated with $\beta_{ij}^{(T)}$ (37) is disregarded.
8. Hygroelastic coupling is negligible; then the term associated with $\beta_{ij}^{(M)}$ in (38) is disregarded.

It is seen that, with these simplifications, the decoupled equations of motion, heat conduction, and diffusion result from (36), (37), and (38) respectively.

The effects of high temperature and moisture infiltration into the matrix-laminates are to degrade the material properties through the relation of the type

$$E = \hat{E}\,(1 - e^{-d}) \qquad (42)$$

where d is the total degradation factor given by

$$d = f[\theta, M, \delta^{(T)}, \delta^{(M)}] \qquad (43)$$

Here, $\delta^{(T)}$ and $\delta^{(M)}$ are the degradation coefficients due to temperature and moisture, respectively, and $\hat{E}$ is the Young's modulus for a one-dimensional matrix fiber composite specimen given by

$$\hat{E} = v_f \hat{E}_f + v_m \hat{E}_m \qquad (44)$$

with v_m and v_f denoting the volume contents of matrix and fiber, respectively, and E_m and E_f the Young's modulus for the matrix and fiber, respectively. It should be noted that the degradation factor may theoretically be between zero and ∞. The practical range, however, should be quite narrow. For example, if the material suffers a loss of strength to between 50 and 95 percent, then the corresponding degradation factor should be between 0.7 and 3. The functional relationship (43) can be determined by means of controlled laboratory experiments. A provision should be made such that the degradation factor be dependent of the current temperature and moisture concentration, but that the influence be allowed to disappear at the reference temperature and the zero moisture infiltration.

APPLICATIONS

The governing equations derived in (36)–(41) are based on the first and second laws thermodynamics cast in the Lagrangian coordinates, chemical thermodynamics, and the principle of heat and mass transfer. The assumption incorporated in the equations are so general that some of the effects built into the theory may be disregarded by the experimental evidences. Thus, the appropriate simplifications (eight of them mentioned in the previous section) can be suggested according to the types of material and other practical engineering judgments. The linear isotropic finite elements are used in this analysis.

Pending thorough investigations of effects of all terms included in (36)–(41) the present paper is concentrated on the damping and irreversibility due to temperature and moisture on the two-dimensional glass fiber reinforced composites. The Duffour and Soret effects are disregarded. Also, we consider the bound phase for diffusion is negligible. Thus, our governing equations for the present example are of the form

$$\rho u_i - E_{ijkl}\, U_{k,lj} + \beta_{ij}^{(T)} T_{,j} + \beta_{ij}^{(M)} M_{,j} - \sum_{r=1}^{n} \xi_{ijkl}^{(r)}\, \alpha_{kl,j}^{(r)} - \rho f_i = 0$$

$$(45)$$

$$C\dot{T} + T_o \left(\beta_{ij}^{(T)} \gamma_{ij} + \sum_{r=1}^{n} B_{ij}^{(T)}(r) \, \dot{\alpha}_{ij}^{(r)} + \chi\dot{M} \right) + k_{ij}T,_{ji} - \rho h^{(T)} - D^{(T)} =$$

$$(46)$$

$$\frac{M}{M_o} bM + M \left(\beta_{ij}^{(M)} \dot{\gamma}_{ij} + \sum_{r=1}^{n} B_{ij}^{(M)}(r) \, \dot{\alpha}_{ij}^{(r)} + \chi\dot{T} \right) - D_{ij}T,_{ij}$$

$$(47)$$

$$- \rho h^{(M)} - D^{(M)}$$

Furthermore, we consider $D_{ij} = D_{ij}^o$ in (35) and $\dot{E} = E$ in (42).

A composite material in a state of plane strain is analyzed. The composite consists of four layers of continuous graphite fibers in an epoxy matrix. Figure 1a shows a cross section of the composite under study. The filaments in layers 1 and 4 have a wrap angle (θ) of $90°$. Layers 2 and 3 are made of two plies each. One ply has a wrap angle of $\pm45°$ and the other ply has a wrap angle of $-45°$. Figure 1b shows how the wrap angle is measured. Material properties are given in Table 1. Values given correspond to the filament longitudinal and transverse axis system. Material properties are transformed to the composite reference coordinate system. Figure 2 shows the finite element model and the boundary conditions. Initial conditions are listed in Table 2.

Selection of the relaxation time ($T_{(r)}$) for the viscoelastic analysis was found to be sensitive to the time step (Δt). Specifically the ratio of $\Delta t / T_{(r)}$ was limited to approximately 80 for stability. In addition, since the moisture migration into the material takes place over days a $\Delta t = 10,000$ sec was required. This resulted in selection for this study of a relaxation time equal to 125 sec. For the material selected the relaxation time should equal approximately 2.5 $\times$

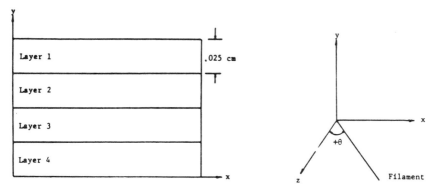

a. Fiber composite cross section   b. Wrap angle

Figure 1. Fiber composites.

Table 1. Material properties.

Material Property	Symbol	Fiber-Reinforced
Fiber element longitudinal elastic modulus (N/cm^2)	E_L	20.68×10^6
Fiber element transverse elastic modulus (N/cm^2)	E_T	2.1×10^6
Fiber shear modulus in transverse-longitudinal plane (N/cm^2)	G_{TL}	8.96×10^5
Fiber shear modulus in transverse-transverse plane (N/cm^2)	G_{TT}	6.89×10^5
Poisson's ratio in fiber element transverse-transverse plane	ν_{TT}	0.4
Poisson's ratio in fiber element transverse-longitudinal plane	ν_{TL}	0.262
Heat capacity [(N-cm)/(cm^3-$^\circ$C)]	c	0.207
Fiber transverse conductivity [(N-cm)/(cm-sec-$^\circ$c)]	k_T	1.33
Fiber longitudinal conductivity [(N-cm)/(cm-sec-$^\circ$c)]	k_L	0.667
Hygroscopic capacity [(N-cm)/cm^3]	b	1379.
Fiber transverse diffusivity [(N-cm)/(cm-sec)]	d_T	2.39×10^{-6}
Fiber longitudinal diffusivity [(N-cm)/(cm-sec)]	d_L	2.39×10^{-7}
Hygrothermal coefficient [(N-cm)/(cm-$^\circ$c)]	χ	1.241
Fiber element longitudinal thermal expansion coefficient (1/$^\circ$c)	$\alpha_L^{(T)}$	6.5×10^{-5}
Fiber element transverse thermal expansion coefficient (1/$^\circ$c)	$\alpha_T^{(T)}$	1.3×10^{-5}
Fiber longitudinal moisture expansion coefficient	$\alpha_T^{(M)}$	6.67×10^{-3}
Fiber transverse moisture expansion coefficient	$\alpha_L^{(M)}$	3.33×10^{-3}
Relaxation time (sec)	$T_{(r)}$	125

10^{-4} sec. This results, however, in a Δt equal to 0.02 sec. A Δt this small would require a prohibitive amount of computer time to carry the calculation out to several days. For the purposes of this study, therefore, a Δt equal to 10,000 sec and a relaxation time equal to 125 sec were selected to study the effect of the viscoelastic behavior.

Figure 3 shows the deformed shape of the model after 11.6 days. It appears from this figure that the dilation caused by the temperature and moisture loading is predominating. Plus and minus signs in each element indicate the direction of σ_x, σ_y, and $_{xy}$ respectively. σ_x is compressive in each element, supporting the conclusion that the temperature and moisture loading are

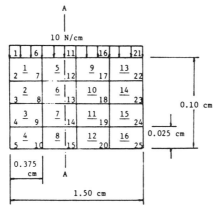

Figure 2. Finite element approximation used to compare with (7), $T_{(r)}$ = 125 sec, Δt = 10,000 sec. (Temperature loading on top surface = 100°C and moisture loading of 10% on the top surface).

Table 2. Initial boundary conditions for temperature and moisture.

Reference Moisture Concentration	0.02
Reference Temperature (°K)	300.0
Change in Moisture Concentration-Upper Surface	0.10
Temperature Change (°C)-Upper Surface	100.0

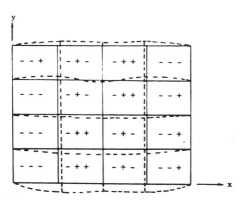

Figure 3. Deformed shape after 11.6 days, $T_{(r)}$ = 125 sec, Δt = 10,000 sec.

203

predominating. σ_y is compressive in the elements adjacent to the fixed boundaries and tensile in the central elements. Table 3 compares the elastic and viscoelastic displacements after 11.6 days. As shown by this table the percent difference between the two formulations is insignificant. Table 4 is a comparison of the stresses predicted by the elastic and viscoelastic formulations after 11.6 days. As shown in this table the two formulations result in essentially the same stresses. The greatest difference takes place between the σ_y stresses. σ_{xy} is identical for both formulations. Table 5 compares the change in moisture concentration. In general, the viscoelastic formulation predicts a lower change in moisture concentration at nodes 2–5 and 7–10 which corresponds to higher viscoelastic σ_y stresses for elements 1–4. Change in moisture concentration is computed to be higher than the elastic prediction for nodes 12–15 which corresponds to lower viscoelastic σ_y stresses for elements 5–8.

This seems to indicate the viscoelastic effect increases stresses near the fixed boundary and decreases them in the center. Consideration of the viscoelastic effect seems to result in faster absorption of the moisture in regions of the material remote from the fixed boundaries accompanied by a reduction in stresses. Figure 4 shows this effect at nodes 11–15 in the center of the

Table 3. Comparison of elastic and viscoelastic displacements (11.6 days).

Node Number	Elastic	Viscoelastic	% Difference
X-Displacement (X10⁻⁴ cm)			

Node Number	Elastic	Viscoelastic	% Difference
X-Displacement ($\times 10^{-4}$ cm)			
6	0.17585	0.17571	0.080
7	0.19316	0.19295	0.109
8	0.18724	0.18696	0.150
9	0.13765	0.13734	0.225
10	0.0562	0.0559	0.534
11	0.0	0.0	0.0
12	0.0	0.0	0.0
13	0.0	0.0	0.0
14	0.0	0.0	0.0
15	0.0	0.0	0.0
Y-Displacement ($\times 10^{-4}$ cm)			
6	1.31369	1.31295	0.056
7	0.35331	0.35266	0.184
8	−0.52142	−0.52180	0.073
9	−1.33560	−1.33561	0.001
10	−2.12226	−2.12185	0.019
11	0.44454	0.44349	0.236
12	−0.21051	−0.21147	0.454
13	−0.81584	−0.81656	0.088
14	−1.37926	−1.37964	0.028
15	−1.91398	−1.91400	0.001

Table 4. Comparison of elastic and viscoelastic stresses after 11.6 days,
$T_{(r)}$ = 125 sec., Δt = 10,000 sec.

Element Number	Elastic	Viscoelastic	% Difference
		$\sigma_x (N/cm^2)$	
1	−20571	−20572	0.005
2	−12075	−12075	0.0
3	−11267	−11267	0.0
4	−17138	−17137	0.006
5	−21425	−21424	0.005
6	−11615	−11616	0.009
7	−10783	−10783	0.0
8	−17228	−17227	0.006
		$\sigma_y (N/cm^2)$	
1	−2689	−2690	0.037
2	−2473	−2473	0.0
3	−2305	−2306	0.044
4	−2210	−2211	0.045
5	517	516	0.193
6	475	473	0.421
7	446	443	0.673
8	435	431	0.920
		$\sigma_{xy} (N/cm^2)$	
1	168	168	0.0
2	−8	−8	0.0
3	118	−118	0.0
4	−267	−267	0.0
5	−202	−202	0.0
6	−81	−81	0.0
7	43	43	0.0
8	166	166	0.0

material. The trend is reversed for the nodes approaching the fixed boundary.

In general, the viscoelastic effect results in reduced magnitude of stress in both horizontal and vertical directions. Such reduction is more significant for the horizontal stress in sections closer to the top and bottom surfaces, whereas the vertical stresses are reduced more uniformly through the cross section due to viscoelastic effects.

Figure 5 shows a comparison of the x and y displacements, temperature, and change in moisture concentration for the viscoelastic and elastic formulations over time at node 10. Viscoelastic behavior reduces the deflections. This indicates that the increased σ_y stresses in elements near the fixed boundary are due to the viscoelastic behavior and not greater deflections. Temperature

Table 5. Comparison of elastic and viscoelastic change in moisture concentration (11.6 days).

Node Number	Elastic	Viscoelastic	% Difference
	Change in Moisture Concentration		
2	.073763	.073743	0.0271
3	.052050	.052012	0.0730
4	.038042	.037991	0.1341
5	.033248	.033191	0.1714
7	.073763	.073755	0.0108
8	.052050	.052034	0.0307
9	.038042	.038019	0.0604
10	.033248	.033222	0.0782
12	.073763	.073768	0.0068
13	.052050	.052057	0.0134
14	.038042	.038048	0.0158
15	.033248	.033252	0.0120

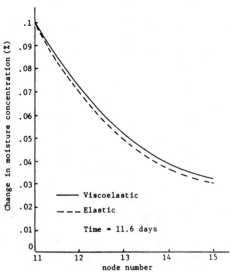

Figure 4. Comparision of elastic and viscoelastic change in moisture concentration, $\Delta t = 10,000$ sec.

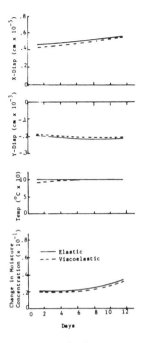

Figure 5. *Displacements, temperature and moisture vs. time at node 10.*

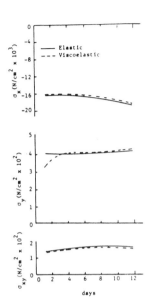

Figure 6. *Stresses vs. time for element 8.*

takes approximately 3.5 days to stabilize when the viscoelastic effect is considered. The change in moisture concentration is lower when viscoelasticity is considered. This is consistent with behavior at nodes closer to the fixed boundary observed earlier.

Figure 6 shows a comparison of the stresses over time in element 8. In general σ_x and σ_{xy} are reduced by the viscoelastic effect. σ_ξ on the other hand exhibits slightly more erratic variations. In general, σ_y is reduced over time by viscoelastic effects. Between 11 and 11.6 days σ_y shows a trend whereby the viscoelastic formulation predicts higher stresses as observed earlier. This seems to indicate the viscoelastic influence increases with time.

CONCLUSION

A composite structure subjected to hygrothermomechanic loadings has been investigated. The constitutive theory and numerical applications have been presented. From the viewpoint of possible physical phenomena, the governing equations are considered to be rigorous. Various simplifications are suggested, depending on experimental evidences.

The example problem demonstrated herein included irreversible processes in which the viscoelastic behavior is taken into account. For a range of relaxation time considered in this example, the effect of damping or irreversibility appears to be negligible. The effect of temperature and moisture is found to be significant in the deformation and stress field.

The future study will involve the evaluation of degradation factors from experimental results and incorporate them in the analysis. In addition, the Duffour and Soret effects and two-phase diffusion model must be investigated numerically and justification for inclusion of these effects in the analysis should be established. With additional data available from these results, it is now possible to evaluate the design criteria in order to avoid failure from exposure to temperature and moisture infiltration.

REFERENCES

1. Fried, N., "Degradation of Composite Materials: The Effect of Water on Glass-Reinforced Plastics," Proc. Fifth Symp. Naval Structural Mechanics, Philadelphia (May 8–10, 1967).
2. Hertz, Jules, "High Temperature Strength Degradation of Advanced Composites," in *Space Shuttle Materials*. Vol. 3, Society of Aerospace Material and Process Engineers, 9–16 (Oct. 1970).
3. McKaug, E. L., Reynolds, J. D., and Halkias, J. E., "Moisture Diffusion in Fiber Reinforced Plastics," Transactions of ASME, *Journal of Engineering Materials and Technology,* 92–95 (January 1976).
4. Unger, D. J. and Aifantis, E. C., "Solutions of Some Diffusion Equations Related to Stress Corrosion Cracking."
5. Aifantis, E. C. and Gerberich, W. W., "Gaseous Diffusion in a Stressed-Thermoelastic Solid, Part II: Thermodynamic Structure and Transport Theory," *Acta Mechanica, 28,* 25–47 (1977).
6. Chung, T. J. and Prater, J. L., "A Constitutive Theory for Anisotropic Hygrothermoelasticity with Finite Element Applications," *Journal of Thermal Stresses, 3,* 435–452 (1980).
7. Prater, J. L., "Finite Element Analysis in Two-Dimensional Composites Subjected to Hygrothermoelastic Loadings," M.S. Thesis, The University of Alabama in Huntsville (1980).
8. Nowacki, W., "Certain Problems of Thermodiffusion in Solids," *Arch. Mech., 23* (6), 731–735 (1971).
9. McNabb, A. and Foster, P. K., "A New Analysis of the Diffusion of Hydrogen in Iron and Ferritic Steels," *Transactions of The Metallurgical Society of AIME, 227,* 618–627 (1963).
10. Caskey, G. R. and Pillinger, W. L., "Effect of Trapping on Hydrogen Permeation," *Metallurgical Transactions, 6A,* k67–476 (1975).
11. Carter, H. G. and Kibler, K. G., "Langmeir-Type Model for Anomalous Moisture Diffusion in Composite Resins," *J. Composite Materials, 22,* 118–131 (1978).
12. Gurtin, M. E. and Yatomi, C., "On a Model for Two Phase Diffusion in Composite Materials," Vol. 13, 126–130 (1979).
13. Vincenti, W. G. and Kruger, C. J. *Introduction to Physical Gas Dynamics.* New York: John Wiley and Sons, Inc. (1965).
14. Chung, T. J., "Thermomechanical Response of Inelastic Fiber Composites," *Int. Journal of Numerical Methods in Engineering, 9,* 169–185 (1975).
15. Eckert, R. G. and Drake, R. M., ed. *Analysis of Heat and Mass Transfer.* McGraw-Hill Book Col. (1972).
16. Van Amerongen, G. J., "Diffusion in Elastomers," *Rubber Chemistry and Technology, 37* (5), 1067–1074 (1964).

18

A Comparative Study of Water Absorption Theories Applied to Glass Epoxy Composites

P. BONNIAU AND A. R. BUNSELL

INTRODUCTION

G LASS FIBER REINFORCED EPOXY RESIN IS INCREASINGLY BEING USED FOR primary structures. This is because of the relatively low cost of the fiber compared to other reinforcements and the superior electrical, mechanical and chemical properties of epoxy resin when compared to other reinforcements and the superior electrical, mechanical, and chemical properties of epoxy resin when compared to polyester resin which is more often associated with glass fiber reinforced composites. Previous studies have shown however that, like all organic materials, the properties of epoxy matrix composites are influenced by the absorption of water and by the surrounding environments [1,2]. Epoxy resin absorbs water from the atmosphere with the surface layer reaching equilibrium with the surrounding environment very quickly followed by diffusion of the water into all of the material. The water absorbed is not usually in the liquid form but consists of molecules or groups of molecules linked by hydrogen bonds to the polymer. In addition, liquid water can be absorbed by capillary action along any cracks which may be present or along the fiber-matrix interface [3,4,5,6].

As the properties of any composite depend on the behavior of the matrix, fibers, and the interface, it seems essential to know the rate of water absorption in glass epoxy composites in order to predict their long-term behavior. In many cases water absorption obeys Fick's law and diffusion is driven by the water concentration gradient between the environment and the material producing continuous absorption until saturation is reached [7]. In other cases however, this model is seen not to be applicable or to break down under certain conditions of humidity and temperature or during cycles of exposure to water vapour followed by drying which can initiate other mechanisms such as

209

cracking or chemical attack [4,8]. In this study we have applied two diffusion models to the interpretation of the absorption of water by glass epoxy composites under both vapor and liquid conditions.

THE DIFFUSION MODELS

Two diffusion theories have been studied. The first was the classical single free phase model of absorption in which the water molecules are not combined with the matrix and the second, the Langmuir two phase model considers a, free diffusion phase and a second combined phase which does not involve diffusion. Both these models are based on diffusion theory and Fick's law which consider that the driving force of diffusion is the water concentration gradient.

In order to simplify the analysis of the diffusion equations the following hypotheses have been made:

—the diffusivity D of the free phase is independent of the concentration.
—we consider the planar phase case so that diffusion occurs in one direction only, normal to the plane.

$$\frac{dc}{dt} = D \frac{d^2c}{dx^2} \tag{1}$$

in which C is the water concentration at time t at a point a distance x from the surface and D is the diffusivity.

In this study we have considered the case in which there is a maximum of water concentration at the surface and this does not change during absorption. In addition we have tested plates of thickness h which were sufficiently big for edge effects to be minimal and the plates were initially dried before being put into an environment of constant temperature and humidity.

SINGLE PHASE ABSORPTION MODEL

It has been shown [7,9] that with single phase diffusion the weight gain due to absorption can be expressed in terms of two parameters, the diffusivity D and the weight gain at saturation $Mm\%$ (see Figure 1).

$$M\% = Mm\square . \left[1 - \frac{8}{\pi^2} \sum_{n=0}^{\infty} \frac{1}{(2n+1)^2} \exp\left[-\frac{D.t}{h^2} \pi^2(2n+1)^2 \right] \right] \tag{2}$$

For $\frac{Dt}{h^2} > 0,05$, this relationship reduces to

$$M\% = Mm\% . \left[1 - \frac{8}{\pi^2} \exp\left[-\frac{D.t}{h^2} . \pi^2 \right] \right] \tag{3}$$

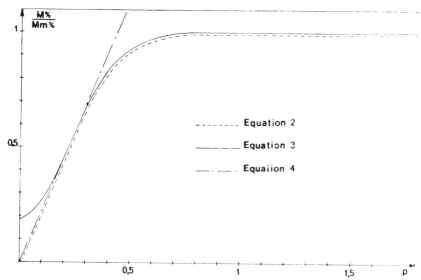

Figure 1. Single phase absorption model. Theoretical curves.

For $\dfrac{Dt}{h^2} > 0,05$, the relationship describing absorption is given by:

$$M\% = Mm\% \cdot \frac{4}{\pi} \cdot \sqrt{\frac{D t}{h^2}} \tag{4}$$

TWO PHASE MODEL

With this model the weight gain $M\%$ as a function of time t is written in terms of four parameters, the diffusivity D, weight gain at saturation $Mm\%$, the probability a of a molecule of water passing from a combined state to the free phase and the probability β of a molecule of water passing from the free to the combined phase [9,10,11,] (see Figure 2).

For

$$a << \frac{D\pi^2}{h^2} \quad \text{and} \quad \beta << \frac{D\pi^2}{h^2}$$

We can write:

$$M\% = Mm \cdot \left[1 - \frac{\beta}{a + \beta} \exp\left[-a.t \right] \right] - \frac{a}{a + \beta} \cdot \frac{8}{\pi^2} \sum_{n=0}^{\infty} \frac{1}{(2n+1)^2}$$

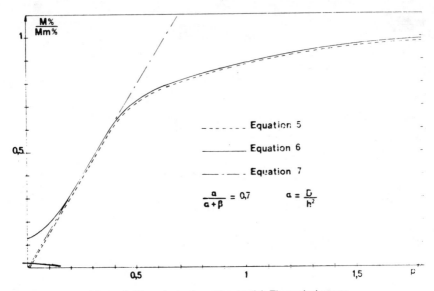

Figure 2. *Two phase absorption model. Theoretical curves.*

For $\dfrac{Dt}{h^2} > 0{,}05$ this reduces to

$$M\% = Mm\% . \left[1 - \frac{\beta}{a+\beta} \exp\left[-a.t \right] - \frac{a}{a+\beta} \cdot \frac{8}{\pi^2} \right.$$

$$\left. \exp\left[-\frac{D.t}{h^2} . \pi^2 \right] \right] \tag{6}$$

For $\dfrac{Dt}{h^2} < 0{,}05$

$$M\% = Mm\% . \frac{a}{a+\beta} \cdot \frac{4}{\pi} \cdot \sqrt{\frac{Dt}{h^2}} \tag{7}$$

It can be see the two phase model reduces to the single phase case when $a = 1$, $\beta = 0$.

TREATMENT OF THE CURVES

The calculation of the parameters involved in diffusion was made using equations (3) and (6) depending on the model considered and for points for which $Dt/h^2 > 0,05$. A curve fitting technique was used based on minimum variances which was developed for curves for an exponential form [12]. The calculations were made by successive approximations with the aid of a T.I. 59 programmable calculator and the parameters were obtained with a precision of 0,001 in ten minutes for the single phase model and in one hour for the two phase model. Essentially the approach adopted was as follows. Equations (3) and (6) were of the form:

$$M = f(t, Mm, D).$$

This can be represented by a Taylor's series so that,

$$f(t, Mm, D = f(t, Mm_o, D_o) + \frac{df}{dMm} (t, Mm_o, D_o)\Delta Mm + \frac{df}{dD} (t, Mm_o, D_o)\Delta D$$

(8)

Now, for the single phase model, we have:

$$\frac{df}{dMm} = \left[1 - \frac{8}{\pi^2} \exp \left[- \frac{D.t}{h^2} .\pi^2 \right] \right]$$

(9)

and

$$\frac{df}{dD} = Mm. \frac{8}{h^2} .t.exp \left[- \frac{D.t}{h^2} \right] .\pi^2$$

(10)

The variance V is given by,

$$V = \sum \left[Mi - f(ti) \right]^2$$

Where Mi is the weight gain measured and $f(ti)$ is the calculated value so that:

$$V = \sum \left[Mi - f(ti, Mm_o, D_o) - \frac{df}{dMm} (ti, Mm_o, D_o)\Delta Mm - \frac{df}{dD} (ti, Mm_o, D_o)\Delta D \right]^2$$

(11)

Minimum variance is given by:

$$\frac{dV}{d(\Delta Mm)} = 0$$

(12)

$$\frac{dV}{d(\Delta D)} = 0$$

(13)

214 P. BONNIAU AND A. R. BUNSELL

which results in two simultaneous equations and two unknowns, ΔMm, and ΔD.

Taking a value to begin with of Mm_o in the range of experimental scatter and D_o obtained from equation (4), we could calculate a curve which fitted these values so that Mm_o is replaced by Mm_1 such that:

$$Mm_1 = Mm_o + \Delta Mm \tag{14}$$

and D_o by D_1

$$D_1 = D_o + \Delta D \tag{15}$$

An acceptable fit is defined as:

$$\left| \frac{\Delta Mm}{Mm_1} \right| + \left| \frac{\Delta D}{D_1} \right| < 0.001 \tag{16}$$

For the two phase model it is necessary to consider the two additional parameters a and β so that the approach outlined above leads to four simultaneous equations instead of two.

EXPERIMENTAL DETAILS

The three materials which have been studied were fabricated in the form of plates by UDD-FIM and are primarily used as insulators in electric motors. All three consisted of a $0,90°$ woven E-type glass fiber cloth in Bisphenol A epoxy resin. The fiber volume fraction was around 50 percent and the quality of the material was excellent with no included bubbles and a high reproducibility between different specimens. Three types of material were produced by using different resin hardeners, diamine, dicyandiamide, and anhydride. Table 1 gives more details of the specimens which were tested.

The dimensions of the specimens were chosen in order to render edge negligible and to assure that diffusion occurred in one direction only normal to the fibers as is presumed in the analysis. The plates had a thickness h of 1 $\pm$ 0,16 mm, width n of 210 $\pm$ 0,4 mm, and length l of 300 $\pm$ 0,4 mm. In the case of the first material, diamine hardener, specimens of different thicknesses were also examined ($h = 0,2$-$0,5$-1-2 mm).

Tests were conducted in ten environmental chambers which allowed relative humidities in the range 0 to 100 percent to be achieved plus liquid immersion in distilled water. The range of temperature was 25 to 90°C. A schematic view of a chamber is shown in Figure 3. Relative humidity was controlled by using saturated salt solutions, the vapor pressure of which when in equilibrium with the air gave very steady values ($\pm$3% RH). The relative humidities produced by the six salt solutions are shown in Figure 4 as a func-

Table 1. *Characteristics of the three studied materials.*

		Material 1	Material 2	Material 3
Type of hardener used with Bisphenol A resin		Diamine	Dicyandiamide	Anhydride
Glass fibre cloth	Filament type	E.C. 9.68	E.C. 9.68	E.C. 9.68
	No. of threads in warp/cm	17	17	17
	No. of threads in weft/cm	13	13	13
Composite	No. of cloth layers/mm	7	6	6
	Fibre volume faction	56%	48%	48%
	Density gr/cm³	1,90	1,80	1,80
Usual Upper Service Temperature		155°C	140°C	140°C

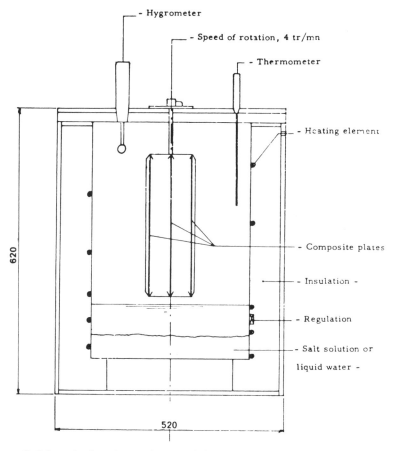

Figure 3. *Schematic view of an environmental chamber: —Hygrometer—Speed of rotation, 4 tr/mn — Thermometer—Heating element—Composite plates—Insulation—Regulation—Salt solution or liquid water.*

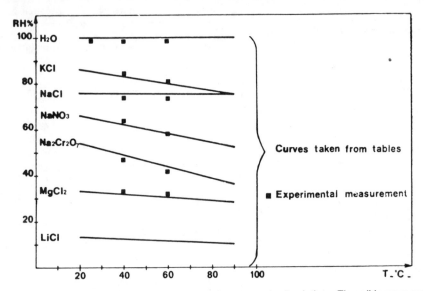

Figure 4. *Relative humidity versus temperature for saturated salt solutions. The solid curves are the mean values given in References 13 and 14. The plotted symbols are the values obtained in this study.*

tion of temperature [13,14]. Relative humidity measurements were made using a lithium chloride hygrometer made by Richard-Pekly type DMS 100. Temperature measurements were made with a numerical thermometer made by A.O.I.P. type PN 25 and an electronic thermometer made by National-semiconductors type LX-5600 which controlled the heating element.

A Sartorius balance type 1205 MP was used to weigh the specimen to within 1 mg. The balance was placed inside a Faraday cage in order to eliminate the influence of static electricity on the plates and was also useful in avoiding fluctuations due to draughts. All experimental points shown on the curves represent the mean of measurements taken from two or three specimens primarily dried for two weeks at 100°C.

RESULTS AND DISCUSSION

In Water Vapor

Diamine hardener—As Figure 5 shows, the behavior of the first material subjected to humid conditions showed excellent correspondence with the classical diffusion model: The initial linear relationship between $M\%$ and $\sqrt{t}$ was clearly observed followed by saturation. The results obtained with plates of different thicknesses showed that the water is uniformly distributed

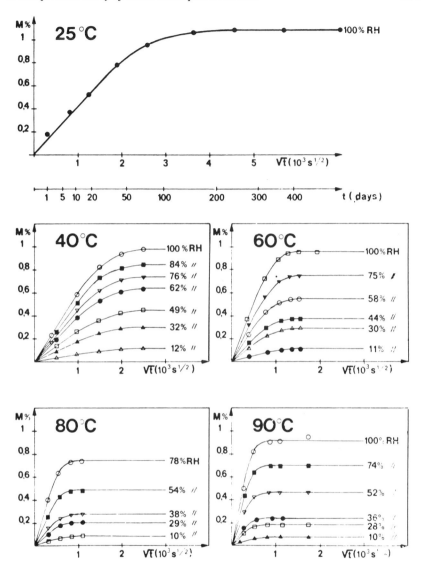

Figure 5. First material absorption curves. The solid curves are theoretical and obtained from consideration of the single free phase diffusion model.

throughout the plates at saturation, see Figure 6. From Figure 7, it is clear that the mechanisms involved are reversible. For the diamine material, it was therefore possible to obtain the two parameters D and Mm for the diffusion and confirm the model described by Springer [7].

The diffusivity D was only a function of temperature and followed a rela-

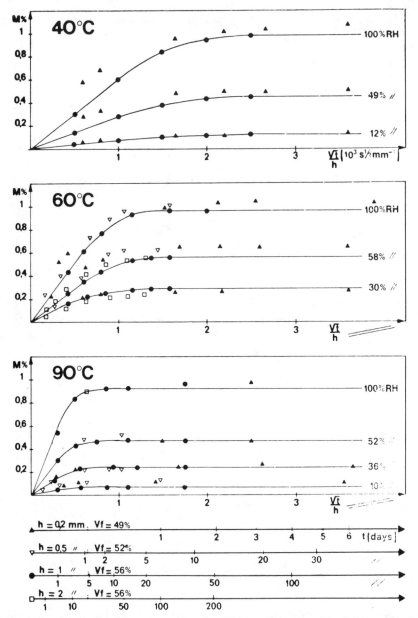

Figure 6. First material absorption curves for different thicknesses. The solid curves are theoretical.

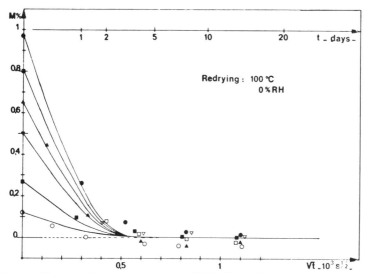

Figure 7. *First material desorption curves at 100°C. The solid curves are theoretical.*

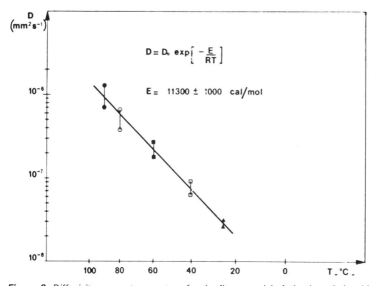

Figure 8. *Diffusivity versus temperature for the first material. Arrhenius relationship.*

tionship of the Arrhenius type, as seen in Figure 8. The saturation limit was a linear function of the relative humidity as shown in Figure 9. This first material was seen therefore to come into equilibrium with the surrounding environment. The observed behavior indicated that the water was absorbed by the resin and existed in the form of molecules or groups of molecules linked

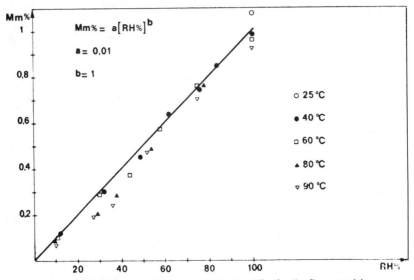

Figure 9. *Saturation limit versus relative humidity for the first material.*

to the resin by hydrogen bonds. In this case it was not surprising that the material appeared not to have been damaged by the presence of the water over the greater part of the range of relative humidity. Damage did occur however under the most severe conditions (90°C—100% RH) and for exposure times above two weeks, and was seen as a progressive whitening of the material although no extra weight gain was detected. It seems most probable that this damage was due to microcracking of the resin surface produced by partial condensation of the water vapor on to the plates.

DICYANDIAMIDE HARDENER

The water update of this second material in the presence of water vapor was qualitatively similar at first sight to the behavior seen with the first material in that absorption proceeded until saturation. The shapes of the absorption curves were however different with the second material showing a break in the continuous smooth water update curve, as can be seen in Figure 10. The single phase model was not sufficient to describe this behavior and we applied the Langmuir two phase model with success to describe the results obtained. Using equation (6) it has been possible to obtain values for all four parameters. The diffusivity D was again dependent only on the temperature whilst the saturation level was a function of the relative humidity, as shown in Figures 11 and 12. In addition, it can be seen from Figure 13 that the probabilities a and β increased with temperature and that the water in the free phase was a constant fraction of the total water absorbed as $a/a + \beta$ was about 0,7 over a wide range of temperature as seen in Figure 14.

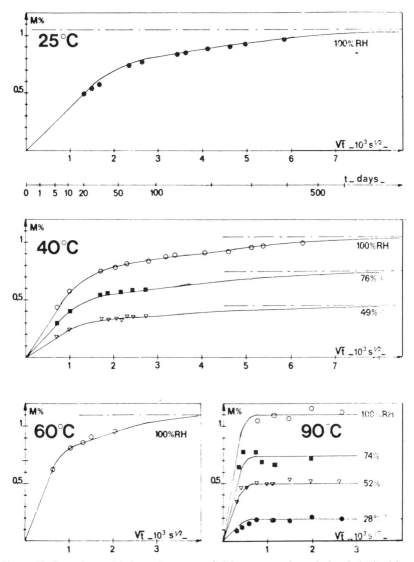

Figure 10. *Second material absorption curves. Solid curves are theoretical and obtained from consideration of the two phase model.*

It is important to note that much longer exposure times were necessary to achieve saturation with this second material due to the influence of the two additional phases which considerably modify the absorption curves. Application of the single phase model led to very imprecise values of the parameters D and Mm (see Figures 11,12). As with the first material then the second was

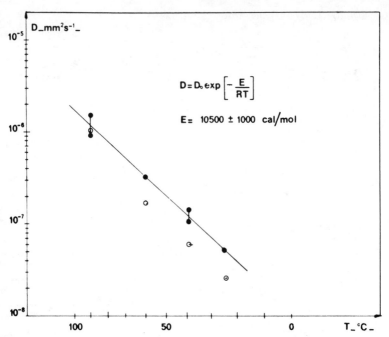

Figure 11. Diffusivity versus temperature for the second material. Open and filled symbols represent the first and the second model respectively.

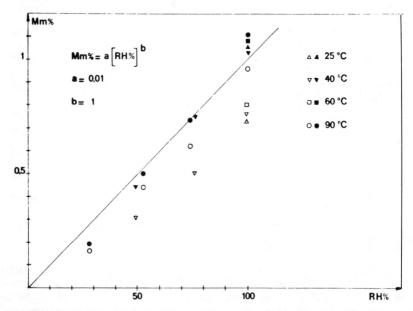

Figure 12. Saturation limit versus relative humidity for the second material. Open and filled symbols represent the first and the second model respectively.

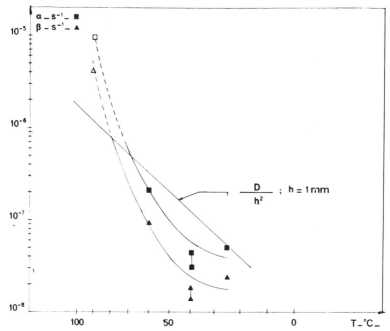

Figure 13. *Probabilities a and β for the second material versus temperature. The dotted line corresponds to the limit of validity for the two phase model a and β are too high.*

seen to come into equilibrium with its surroundings. The mechanisms seemed to be uncombined free diffusion as well as a combined phase in which the water molecules were strongly bound to sites in the polymeric molecular structure. The observed two phase absorption was not accompanied by any visible damage to the material except under the most severe conditions in which whitening occurs as described for the first material.

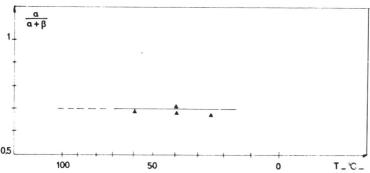

Figure 14. *Approximative fraction of the water concentration in the free phase to the total water absorbed, a/a + β versus temperature.*

ANHIDRIDE HARDENER

Under humid conditions the rate of water update of the third material at first appeared to be similar to that of the other two materials, increasing with relative humidity and temperature but saturation was not observed as shown in Figure 15. Considerable loss of material was observed due principally to erosion of the resin at the surface of the plates at 40 °C and above. The water uptake at 25 °C was similar to that seen with the dicyandiamide hardened resin and indicates a two phase water update but material loss, which was very evident at higher temperatures, excludes the possibility of a general description of water absorption by diffusion theory.

LIQUID ENVIRONMENT

Diamine hardener

In order to normalize results the weight gain experienced by the first material in a liquid environment has been plotted as a function of:

$$p = \frac{D.t}{h}$$

with the values of D those obtained in humid conditions, see Figure 16. On the same graph is plotted the results obtained at 100 percent RH at various temperatures.

The rate of water uptake in the liquid environment at all temperatures was seen to be greater than that in humid conditions. As immersion continued the uptake of water began to increase dramatically and the curve deviated from the curve obtained under 100 percent humidity. This was accompanied by a whitening of the composite with weight gains greater than about 1 percent and around 1,4 percent cracking of surface layers occurred. It appears that in the liquid environment the limit of water concentration for equilibrium to be reached with the surrounding was exceeded at all temperatures although the associated phenomena occurred very slowly at low temperatures.

DICYANDIAMIDE HARDENER

Figure 17 compares the absorption of water in the liquid and humid environments for the second material. The observed behavior was similar to that seen for the first material. Equilibrium was not reached with the outside environment and saturation did not occur indicating that again the water concentration threshold was exceeded at all temperatures in liquid water. The composite was again seen to progressively become white although cracking was not seen which explains the slower absorption of water by this second material at 90 °C when compared to the first.

As was the case with this third material under humid conditions, immersion produced thermally activated behavior with important losses in material

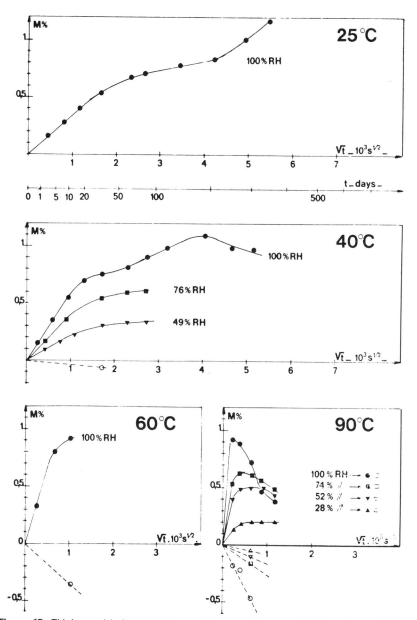

Figure 15. Third material absorption curves. Open symbols represent loss of material after drying.

225

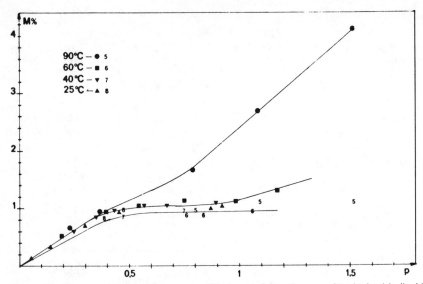

Figure 16. First material absorption curves. Filled symbols are from results obtained in liquid water and numerical symbols from results after exposure to water vapor at 100% RH.

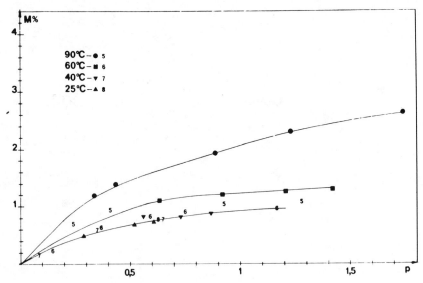

Figure 17. Second material absorption curves. Filled symbols are from results obtained in liquid water and numerical symbols from results after exposure to water vapor at 100% RH.

due to leaching of the resin from the surface of the plates at temperatures of 40 °C and above (see Figure 18).

CONCLUSIONS

It is possible to describe the water absorption of the first two materials (diamine and dicyandiamide hardeners) exposed to water vapor by the use of diffusion theory. The first material for which a diamine hardener was used exhibited single phase absorption in which water diffuses into the epoxy matrix and was weakly bound to the polymer by hydrogen bonding. The second material for which a dicyandiamide hardener was used showed two phase behavior indicating that as well as classical diffusion occurring there was also a phase in which the water molecules became more firmly fixed to sites in the polymer. Both of these materials came into equilibrium with the surrounding environment and saturation occurred under the most severe conditions (100 percent RH, 90 °C) progressive whitening of the material due to erosion of the surface was however observed. These two materials showed different behavior when immersed in liquid water as other phenomena were seen from the beginning of immersion. Both materials became progressively white and opaque and cracking occurred in the first material. These materials do not seem to reach a state of equilibrium with their liquid surroundings although at ambient temperature the rate of damage is extremely low.

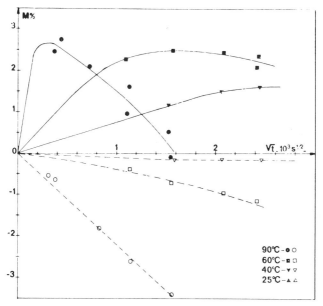

Figure 18. Third material absorption curves in liquid water. Open symbols represent loss of material after drying.

The third material for which the anhydride hardener was used exhibited irreversible damage, in the form of microcracks or erosion of the surface, in both water vapor and when immersed and this resulted in considerable loss of material. This phenomena although occurring at room temperature was however extremely slow.

The behavior of the materials was seen to be different in water vapor than when immersed in liquid water. This was the case even when the relative humidity was very high, around 100 percent RH. The method of accelerating the effect of water vapor on the aging of this type of material by immersion in hot or boiling water was shown to be invalid as different mechanisms are involved in each case.

NOMENCLATURE

D	composite diffusivity	$m^2.s^{-1}$
t	Time	s
h	Thickness of plate	m
$p = \dfrac{Dt}{h}$	Diffusion parameter	dimension less
W	Weight of plate	Kg
Wd	Weight of dry plate	Kg
Wm	Weight of saturated plate	Kg
$M = \dfrac{W-Wd}{Wd}$	Maximum moisture content percentage	%
$Mm = \dfrac{Wm-Wd}{Wd}$	Maximum moisture content percentage	%
Vf	Fiber volume fraction	%
T	Temperature	°C
RH	Relative Humidity	%
a	Probability of a molecule of water passing from the combined phase to the free phase	s^{-1}
β	Probability of a molecule of water passing from the free phase to the combined phase	s^{-1}

REFERENCES

1. Shen, C. H. and Springer, G. S., "Effects of Moisture and Temperature on the Tensile Strength of Composite Materials," *J. Composite Materials, 11*, 2 (Jan. 1977).
2. Shen, C. H. and Springer, G. S., "Environmental Effects on the Elastic Module of Composite Materials," *J. Composite Materials, 11*, 250 (July 1977).
3. DeIasi, R. and Whiteside, J. B., "Effect of Moisture on Epoxy Resins and Composites," *Advanced Composite Materials—Environmental Effects*, J.R. Vinson, ed. ASTM-STP, 658 (1978).
4. Shirrell, C. D., "Diffusion of Water Vapor on Graphite/Epoxy Composites," *Advanced Composite Materials—Environmental Effects*, J.R. Vinson, ed., ASTM-STP, 658 (1978).
5. Carter, H. G., "Fundamental and Operational Glass Transition Temperature of Composite Resins and Adhesives," *Advanced Composite Materials—Environmental Effects*, J.R. Vinson, ed. ASTM-STP, 658 (1978).

6. Hahn, H. T. and Kim, R. Y., "Swelling of Composite Laminates," *Advanced Composite Materials—Environmental Effects,* J.R. Vinson, ed. ASTM-STP, 658 (1978).
7. Shen, C. H. and Springer, G. S., "Moisture Absorption and Desorption of Composite Materials," *J. Composite Materials, 10,* 2 (Jan. 1976).
8. Ishai, O., "Environmental Effects on Deformation, Strength, and Degradation of Unidirectional Glass-Fiber Reinforced Plastics," *Polymer Engineering and Science, 15* (July 1975).
9. Crank, J. *The Mathematics of Diffusion.* Oxford: Clarendon Press.
10. Gurtin, M. E. and Yatomi, C., "On a Model for Two Phase Diffusion in Composite Materials," *J. Composite Materials, 13,* 126 (April 1979).
11. Carter, H. G. and Kibler, K. G., "Langmuir-Type Model for Anomalous Moisture Diffusion in Composite Resins," *J. Composite Materials, 12,* 118 (Apr. 1978).
12. Lewis, David, III, "Curve Fitting Techniques and Ceramics," *Ceramic Bulletin, 57* (1978).
13. "Measurement of Humidity," *HMSO* (1970).
14. O'Brien, F. E. M., *Journal of Scientific Instruments, 25,* 73 (1948).

19

Moisture Gradient Considerations in Environmental Fatigue of CFRP

E. C. EDGE

INTRODUCTION

T HE EARLIEST ATTEMPTS TO CONDITION COMPOSITE MATERIAL IN ORDER
to determine its residual strength in a postulated worst situation often involved soaking at a standard setting for a standard duration. A number of
European Companies, for instance, soaked for 2000 hours at 50°C and 95
percent R.H., as was done at British Aerospace in order to obtain a number
of "degraded" properties from CFRP coupon specimens [1,2]. Material of 2
mm thickness, such as most of these coupons were, would become virtually
saturated, and test results obtained after such conditioning would inevitably
be pessimistic particularly in the hot wet condition. On the other hand, the
treatment was often applied irrespective of material thickness, and very often
no attempt was made to monitor moisture uptake.

When results from natural weathering and in-service experiments became
available [3–6] it could be seen that moisture saturation is not approached.
The need to predict the moisture contents to be expected in service and to
judge how these could be achieved in the laboratory [6–7] became apparent.
This aim has proved fairly easy to achieve with help from the development of
mathematical modelling techniques [7–9]. These predict moisture content as a
function of time for input forcing functions describing weather, in-service
and laboratory conditions.

Further development of experimental technique [10] has revealed the
through-the-thickness moisture profiles which typically obtain in the quasi
steady-state condition. The variation in material property with moisture content [11] and the associated resin swelling and hygro stress effects have made
it desirable to reproduce these practical profiles in the laboratory. This problem is now reasonably well understood. In the process the dangers of over-

accelerated conditioning have become apparent [6] and care is now exercised in this direction.

The development of environmental fatigue programs [12] in which specimens are simultaneously subjected to loading and environmental cycling, would appear to present considerably complex problems. Until primary composite structures on actual aircraft have themselves experienced the appropriate number of stress and environmental cycles, it cannot be considered to be completely established that these test regimes satisfactorily simulate in-service conditions. This will not happen for some time. In the meantime a number of observations can be made.

These programs necessarily involve a considerable timescale acceleration, that is, both the loading and environmental input will have component frequencies a degree higher than will occur in the real life situation. In the former case, this is unlikely to matter, because, unless an input frequency is close to a resonance, the fatigue life depends on the number of applied loading cycles, the frequency being of much less significance. However, in the case of environmental cycling, a simple mathematical approach, developed in Section 2, suggests that there is a strong interaction between the frequency and the moisture profile within the material. The depth below the exposed surface which experiences a change of moisture content is inversely proportional to the square root of the frequency. It therefore appears that if the environmental test input frequency is appreciably higher than that occurring naturally, the moisture gradients, and presumably the hygro stress levels, also induced near the surface, will be unrepresentatively high. Conversely, they can also be expected to be too low in the band which the correct frequency would penetrate but higher ones do not. This suggests that experimental workers, in designing this type of program, should carefully consider the frequency content of the applied environmental input. Simulation of the fatigue effect of hygrostress reversals would appear to be especially difficult.

The small penetration of the high frequency inputs makes difficulties in the theoretical analysis. The simple Fickian diffusion equations, whose applicability to non-homogeneous and anisotropic fiber composites is not obvious even under more favorable circumstances, cannot be expected to represent well the diffusion of water into the fillets of resin between fibers when the affected portion, as will be illustrated later, can be well under one ply width.

In discussion of the philosophy of environmental fatigue, the point has been made that the exposed surface should be subjected to the same extremes of relative humidity as obtained in real life, in order to allow the generation of representative moisture gradient reversals. Quite apart from the frequency considerations, there can be no experimental verification as these reversals are so rapid and over such a narrow band of material as to be beyond the current measuring techniques. It is, however, possible, as is attempted in the next sections, by appropriate use of the diffusion equations to carry out a somewhat speculative comparison between the calculated moisture gradients corresponding to certain input conditions which can reasonably be expected to occur in real life, in flight, and in the laboratory.

In Section 2 following, the mathematical model used to calculate the moisture gradients is developed from the Fickian model normally employed to predict moisture content and profile. In Section 3 a theoretical comparison is attempted between four different conditioning regimes. These are natural weathering, a simulated flight condition, and environmental fatigue cycling with and without humidity variation during the test period. Results are given in the Results and Comments section.

MATHEMATICAL MODEL

Moisture Gradients Induced by Outdoor Daily Solar Cycle

During outdoor weathering, particularly in the tropics, the daily solar cycle can include a large variation in effective relative humidity at the surface of exposed material. This variation in relative humidity will, of course, lead to a corresponding variation in the moisture concentration at the surface. United States Air Force data for Woodbridge, Suffolk, U.K., for the summer months suggests that this surface concentration can vary from 0.4 percent to 1.6 percent by weight of composite (carbon fiber) during the 24-hour cycle.

To allow for the variation of diffusivity with temperature, the change of variable described in Section 7.1 and Reference [8] is employed. The time variable t is replaced by T where

$$T = \int_0^t D(t') \, dt' \qquad (1)$$

and $D(t)$ = (time dependent) diffusion co-efficient.

Consider a laminate thickness X exposed to a daily solar cycle so that the surface concentration contains a component given by

$$C_p = C_v \sin(wT + E) \qquad (2)$$

where C_v = amplitude
      w = frequency
      E = constant

The solution used here is taken from Chapter 3.6 of Reference [9] and assumes equal exposure on both sides. However, in this case, interaction between the side is negligible. The $x = 0$ axis was taken to be at the center of the specimen.

The response to the input described in (2) is given by

$$C_p{}^* = C_v{}^* \sin(wT + \phi + E) +$$

$$4\pi \, C_v \sum_{n=0}^{\infty} (-1)^n \, Z \exp(-Y) \cos\frac{(2n+1)\,\pi\,x}{X} \tag{3}$$

where

$$\frac{C_v{}^*}{C_v} = \text{attenuation} = \left(\frac{\cosh 2k\,x + \cos 2kx}{\cosh kX + \cos kX}\right)^{1/2} \tag{4}$$

$$\phi = \text{phase difference} = \arg\left(\frac{\cosh\ \ kx\ (1+i)}{\cosh\ \ \dfrac{kx\ (1+i)}{2}}\right) \tag{5}$$

X = thickness

$$Y = \frac{(2n+1)^2\,\pi^2 T}{X^2} \tag{6}$$

$$Z = (2n+1)\left[\frac{X^2\,w\cos E - (2n+1)^2\,\pi^2\sin E}{4X^2\,w^2 + \pi^2(2n+1)^4}\right] \tag{7}$$

$$k = \left(\frac{w}{Z}\right)^{1/2} \tag{8}$$

$$i = -\sqrt{1} \tag{9}$$

Differentiating (2) and (3) gives

$$\frac{\partial C_p{}^*}{\partial x} \quad \frac{C_v{}^* k\,(\sinh 2kx - \sin 2kx)\,\sin(wT + \phi + E)}{(\cosh 2kx + \cos 2kx)^{1/2}\,(\cosh kX + \cos kX)^{1/2}}$$

$$+ C_v{}^* \, \frac{\partial \phi}{\partial x}\,\cos(wT + \phi + E)$$

$$+4\pi \, C_v \sum_{n=0}^{\infty} (-1)^{n+1}\,\frac{(2n+1)\,\pi Z}{X}\exp(-Y)\sin\frac{(2n+1)\,\pi x}{X} \tag{10}$$

where $\partial\phi/\partial x$ can be obtained by differentiating (5) w.r.t.x. This leads to a rather intractable expression, but the following approximation can be used for all values of the parameters involved which are applicable to this exercise:

$$\frac{\partial \phi}{\partial x} \simeq k \tag{11}$$

Moisture Gradients Induced During High Temperature Flight Phases

Consider an exposed surface when the relative humidity at that surface undergoes a rapid change. During the time range of interest interaction between the ingress from each side of a composite slab exposed on both sides is negligible. A linear change of concentration will be assumed as a step input leads to a singular solution for $T = 0$.

Let C_s be the total change in concentration occurring in an interval of time represented by $T = T_o$. The surface concentration over this time is as illustrated in Figure 1, and a steady state is assumed prior to the ramp input.

Assume an input, superimposed over a steady state condition, given by

$$C(0,T) = C_r T - C_r (T - T_o) H (T - T_o) \tag{12}$$

where

$C(0,T_1)$ = surface concentration at $T = T_1$

$C_r T_o = C_s$

$$H(T) = \text{Heaviside step function} = \begin{array}{l} 0 : T < 0 \\ \frac{1}{2} ; T = 0 \\ 1 ; T > 0 \end{array} \tag{13}$$

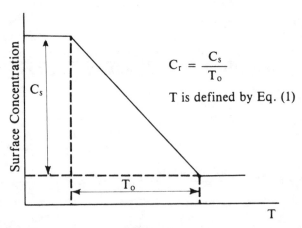

$$C_r = \frac{C_s}{T_o}$$

T is defined by Eq. (1)

Figure 1.

The response to the input (11) is given by Reference [8], equation (3.16)

$$C\,(x,T) = C_r\,(F(T) - F(T - T_o)\,H\,(T - T_o))$$ (14)

where

$$F(T) = \left(T + \frac{x^2}{2}\right)\,\text{erfc}\,\left(\frac{x}{2\sqrt{T}}\right) - x\left(\frac{T}{\pi}\right)^{\frac{1}{2}}\,\exp\,(-x^2/4T)$$ (15)

Differentiating (14) gives

$$\frac{\partial C}{\partial x} = C_r\,\frac{\partial F(T)}{\partial x} - \frac{\partial F(T - T_o)}{\partial x}\,\cdot\,H\,(T - T_o)$$ (16)

where

$$\frac{\partial F(T)}{\partial x} = x\,\text{erfc}\,\left(\frac{x}{2\sqrt{T}}\right) - 2\left(\frac{T}{\pi}\right)^{\frac{1}{2}}\,\exp\,(-x^2/4T)$$ (17)

PRACTICAL EXAMPLE

Natural Weathering

For this example the values of temperature and humidity are taken from United States Air Force data for Woodbridge, Suffolk, U.K., Air Force base. These figures are taken from Reference [13], p. D-7 and are illustrated in Figure 2. Solar effects are considered by Grumman in calculating these figures which are for the upper surface skin ignoring the effects of precipitation for the month of June. Although the effect of precipitation is to increase the daily average humidity, the amplitude of the solar cycle is larger if it is ignored so that slightly higher moisture gradients are calculated. By taking this daily cycle as a steady-state condition, a pessimistic estimate is obtained of the average moisture gradients in summer. On the other hand, even higher moisture gradients will be obtained under tropical conditions.

Simulated Flight Conditions

Owing to the excessive computation costs resulting, only a much simplified model has been employed. On a constant background of 20 °C and 70 percent R.H. the following is superimposed, illustrated in Figure 3.

a) A climb of 2 minutes in which the temperature rises linearly from 20 °C to 120 °C, while the humidity drops from 70 percent to zero. In order to make use of equations (12–17), it is assumed that the change in humidity is proportional to T.

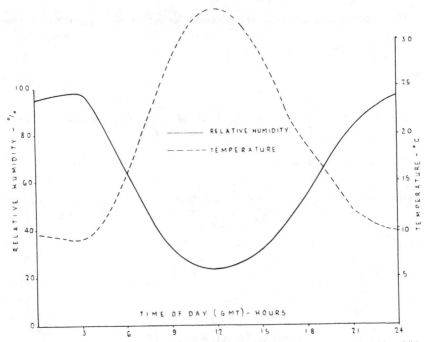

Figure 2. Natural weathering input (from Reference 13) calculated temperatures and humidities at upperskin surface (without precipitation) based on USAF data for Woodbridge, Suffolk, U.K., for month of June.

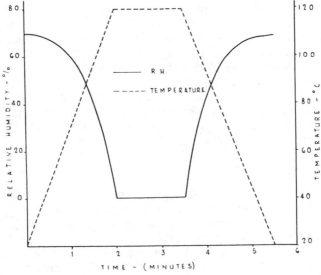

Figure 3. Simulated flight condition assumed temperature and relative humidity distribution.

236

b) 90 seconds at 120° and zero humidity.
c) A descent back to the 20 °C, 70 percent R.H. condition which is the mirror image of (a).

Simplified Environmental Cycling

The simplified environmental cycle considered consisted of a period during the day when testing was assumed to take place in a completely dry environment, with overnight topping up of moisture content at 70 °C and 95 percent R.H. This regime is illustrated in Figure 4.

Environmental Cycling with Humidity Variation During Test

Here only the thermal and humidity cycling during the test period was considered, because of the limit on the costs of computation. The regime used in this part of the calculations is illustrated in Figure 5. This is repeated 35 times during the testing period and it has in fact been taken as a quasi steady state condition, so that the transient terms in the analysis have been ignored. Even with these simplifications it proved impossible within the cost constraints to include enough high frequency terms to adequately simulate the excursions to sub-zero temperature. The actual humidity levels calculated using those terms which it was possible to include are also shown in Figure 5.

Despite the problems encountered it is still felt that the results obtained are

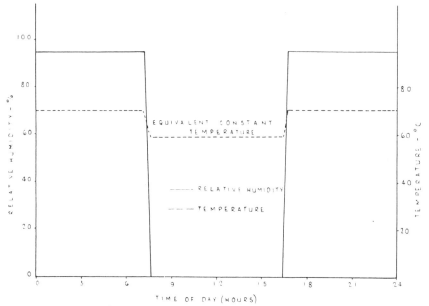

Figure 4. Simplified environmental cycling regime.

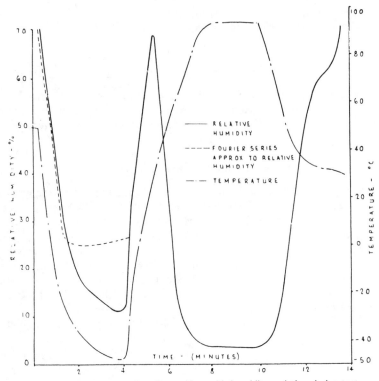

Figure 5. *Environmental cycling regime with humidity variation during test.*

indicative. If it had been possible to include more terms it can be anticipated that still higher moisture gradients would have been estimated for the low temperature portions.

RESULTS AND COMMENTS

Comparison Between Natural Weathering and the Simplified Environmental Cycle

The calculated results obtained using the natural weathering and simplified environmental cycling inputs are illustrated together for comparative purposes. Figure 6 shows the calculated moisture gradients at the surface and Figure 7 those calculated at a depth below the surface of 0.1375 mm, one-tenth of the half thickness of 1.375 mm.

The surface gradient calculated under natural weathering is tolerably sinusoidal, as is to be expected, the maximum value being 300 percent per cm. The simplified environmental cycling, again as expected, gives sharp peaks at the instants of humidification and purge. The maximum gradient calculated

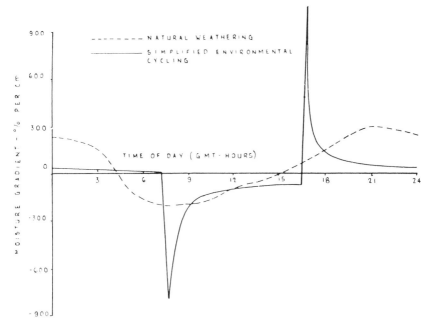

Figure 6. *Natural weathering compared with simplified environmental cycling — surface moisture gradients.*

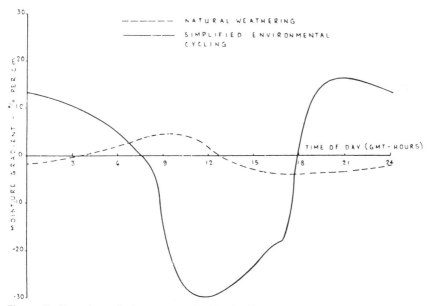

Figure 7. *Natural weathering compared with simplified environmental cycling — moisture gradients at depth of 0.1375 mm.*

239

240   E. C. EDGE

is about 1050 percent per cm, but the level between peaks is lower in general than that obtaining under the natural weathering regime.

In contrast at a depth of 0.1375 mm below the surface the calculated gradients due to the simplified environmental cycle are greater by a factor of about six than those obtained for natural weathering, illustrating the higher frequency of the latter input. In both cases, however, attenuation at this depth is large, the maximum absolute value being less than 30 percent per cm.

Results for Simulated Flight Conditions

The calculated results obtained using the simulated flight conditions are illustrated in Figures 8–10. Figure 8 gives the calculated gradients at the surface for the first 45 minutes from the start of the maneuver. The values obtained are similar in magnitude to those calculated for the simplified environmental cycling, illustrated in Figure 6, but because purge and humidification are separated by only 90 seconds, the peaks inevitably are much closer together in the simulated flight. The rate of humidity change is also somewhat greater resulting in a maximum absolute value of about 1350 percent per cm.

Figure 9 gives the longer term response at the surface, illustrating the gradual decay of the moisture gradients. Figure 10 illustrates the calculated gradients at a depth of 0.1375 mm below the surface. It will be seen that the attenuation is rather greater than in the case of the simplified environmental cycle due to the higher effective frequency of the simulated flight input.

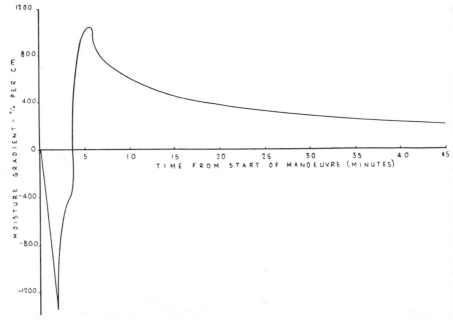

Figure 8. Simulated flight condition short term response at surface.

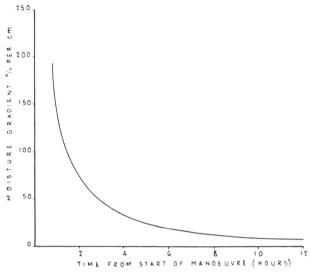

Figure 9. *Simulated flight condition longer term response at surface.*

Results for Environmental Cycling with Humidity Variation During Test

The calculated results obtained using the environmental cycling regime with humidity variation during test are illustrated in Figures 11 and 12. Calcu-

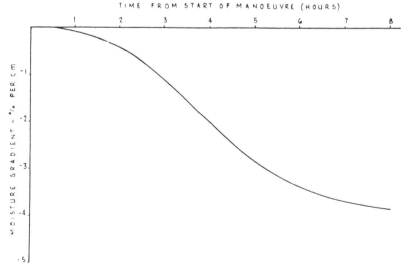

Figure 10. *Simulated flight condition response at depth of 0.1375 mm.*

242 E. C. Edge

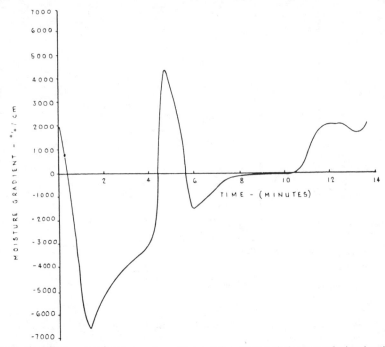

Figure 11. Environmental cycling regime with humidity variation during test calculated moisture gradient at surface.

lated moisture gradients at the surface are given in Figure 11. It will be seen that high values have been deduced, the maximum absolute value exceeding 6000 percent per cm. As already indicated, the regime illustrated in Figure 5 is repeated 35 times consecutively. Although each represents a flight simulation, the unrepresentatively high frequency of application compared to a real service situation is the major contributor to the calculated gradients being considerably greater than for the simulated flight condition. Furthermore, even higher estimates would have been produced had it been possible to include the very high frequency terms required for correct simulation of the sub-zero portions of the cycling regimes.

The high frequency nature of the input is further shown in Figure 12, which illustrates the calculated gradients at a depth below the surface of 0.0275 mm, one fiftieth of the half thickness of 1.375 mm. The highest absolute value obtained is less than 17 percent per cm, and less than that obtained at five times this depth for the simulated flight condition. One ply width is 0.125 mm, so that the penetration of this high frequency input into the material is not much more than 20 percent of a ply width.

These results suggest that a more realistic simulation of real life may be obtained if the daily testing routine were to be performed dry instead of introducing rapid humidity variations of the type in Figure 5.

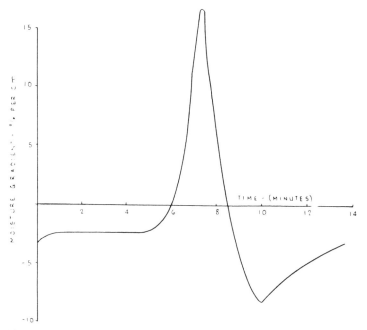

Figure 12. *Environmental cycling regime with humidity variation during test calculated moisture gradient at depth of 0.0275 mm.*

CONCLUSIONS

A theoretical method is presented here which appears to be of general applicability in assessing the realism of accelerated conditioning and testing of CFRP specimens.

The method has been illustrated by applying it to four different conditioning regimes. These are natural weathering, a simulated flight condition, and environmental fatigue cycling without and with humidity variation during the test period.

The results appear to indicate that the inclusion of humidity variation during the test period may, because of the high frequency inputs involved, decrease rather than increase the realism of the test regime when compared with actual service life.

ACKNOWLEDGMENTS

This work has been funded privately by British Aerospace P.L.C. However the author wishes to acknowledge the part played in the evolution of the ideas expressed in this paper by his involvement in the F.A.D.D. environmental fatigue program. This program is being funded by the United Kingdom Ministry of Defense. Reference [13] has been produced under the F.A.D.D.

contract. Reference [1] and [2] have been produced on earlier M.O.D. contracts.

The author wishes to thank the Directors of British Aerospace P.L.C. for giving permission for the publication of this paper.

REFERENCES

1. Eastham, J., BAe Warton Division, Report No. MDR 0182—"Final Report on M.O.D. Contract K/LR32B/2126 in Depth Evaluation of C.F.R.P." (1979).
2. Prestwich, J. D., BAe Manchester Division, Report No. HSA-MSM-R-GEN-0295, "Unidirectional Properties of Carbon Fire Composites" (1978).
3. Trabocco, R. E. and Stander, M., ASTM STP 602, "Effect of Natural Weathering on the Mechanical Properties of Graphite/Epoxy Composite Materials," 67–84 (1976).
4. Roylance, D. and Roylance, M., ASTM STP 602, "Influence of Outdoor Weathering of Dynamic Mechanical Properties of Glass/Epoxy Laminates," 85–94 (1976).
5. Pride, R. A., TM 78716, "Environmental Effects on Composites for Aircraft."
6. Edge, E. C., AGARD CP 288, "The Implications of Laboratory Accelerated Conditioning of Carbon Fibre Composites" (1980).
7. Unman, J. and Tenney, D. R., "Analytical Prediction of Moisture Absorption/Desorption in Resin Matrix Composites Exposed to Aircraft Environments," AIAA Conference, San Diego (March 1977).
8. Crank, J. *The Mathematics of Diffusion.* 2nd Edition, Oxford: Claredon Press (1975).
9. Carsaw, H. S. and Jaeger, J. C. *Conduction of Heat in Solids.* 2nd Edition, Oxford: Claredon Press (1959).
10. Sandorff, P. E. and Tajima, Y. A., "The Experimental Determination of Moisture Distribution in Carbon/Epoxy Laminates," *Composites* (January 1979).
11. Chamis, C. C., Lark, R. J., and Sinclair, J. M., "Effect of Moisture Profiles and Laminate Configuration on the Hygro Stress in Advanced Composites," SAMPE Conference (October 1978).
12. Gerharz, J. J. and Schutz, D., "Fatigue Strength of CFRP under Combined Flight-by-Flight Loading and Flight-by-Flight Temperature Changes," AGARD CP 288 (1980).
13. Dvoskin, N., Jordan, V., and Wolter, W., "Collaborative BAe/GAC Environmental Fatigue Test Programme, Environmental Definition," Grunman A.C. Technical Report AC-79.5 (1979).

20

Thermal Response of Graphite Epoxy Composite Subjected to Rapid Heating

C. A. GRIFFIS, R. A. MASUMURA, AND C. I. CHANG

INTRODUCTION

DUE TO THEIR HIGH STRENGTH TO WEIGHT RATIO AT LOW TEMPERATURES, fiber reinforced organic matrix composites have been successfully utilized in numerous military and space applications. However, intense heating produced by fire or laser irradiation can adversely effect the integrity of composite structures due to the rapid degradation of mechanical properties (tensile and shear strength, elastic moduli, etc.) at temperatures above 200 °C [1]. Severe thermal irradiation may also induce undesirable geometric changes through ablation of critical load carrying members or by generation of high stress concentration due to localized burnthrough [2]. Thus, accurate modeling of thermal response constitutes an important element in assessing the overall reliability of composite structures subjected to rapid, high-intensity heating.

In the present study, a simplified one-dimensional mathematical approach is presented for transient heat transfer analysis of composite plates exposed to surface heating. This model is then employed, in conjunction with an experimental study, to verify recently proposed [3] thermophysical properties of AS/3501-6 graphite epoxy laminate. In a separate investigation, computational results (temperature distributions and ablative characteristics) from the thermal analysis will be coupled with detailed stress analyses to evolve a thermomechanical failure criterion for AS/3501-6 laminates under combined laser irradiation and uniaxial tensile loading.

ANALYTICAL DEVELOPMENT

Heat Transfer Equations for One-Dimensional Model

Since current study is concerned largely with scenarios where a uniform

245

high-intensity energy flux is applied over an area whose characteristic dimensions are significantly greater than the composite thickness, a one-dimensional approach is adopted. Heat transfer is assumed to occur by conduction in the thickness direction, i.e., normal to the plane of the fibers. Thermal decomposition of the organic matrix (e.g., epoxy resin) is approximated [3] by elevation of the bulk heat capacity over an appropriate temperature range; thus, temperature dependent material properties are included in the treatment. In addition, the fiber reinforcing component (e.g., graphite or boron) as well as any residual matrix constituents are assumed to undergo a simple sublimation reaction when the surface temperature reaches a predetermined constant value. Mathematically this surface ablation may be considered equivalent to a classical melting problem [4] in which the melt is instantaneously removed. For subsequent numerical convenience, a convective coordinate system [5,6] is adopted which moves continuously with the ablating fibers as indicated in Figure 1, i.e.,

$$z = x - x_b, \tag{1}$$

where z and x are the convected and global coordinates, respectively, and $x_b(t)$ denotes the instantaneous position of the receding surface. With respect to the moving coordinate, the one-dimensional Fourier conduction equation is given by

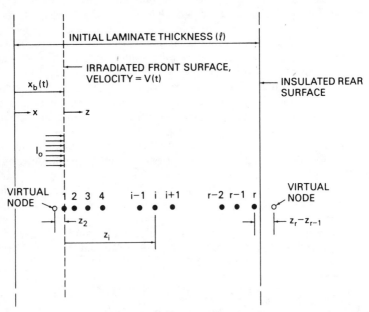

Figure 1. Convected coordinate system and finite difference grid employed in one-dimensional analysis.

$$\frac{\partial}{\partial z}\left(k\,\frac{\partial T}{\partial z}\right) + \rho\,C_p\,V\,\frac{\partial T}{\partial z} = \rho\,C_p\,\frac{\partial T}{\partial t}, \tag{2}$$

where $T(z,t)$, k, ϱ and C_p represent the temperature, thermal conductivity, density, and heat capacity, respectively. The quantity $V \equiv dx_b/dt$ denotes the surface recession rate and is regarded as an unknown function of time.

Subsequent to the onset of ablation ($t<t_0$), boundary conditions at the irradiated surface ($z = 0$) include:

$$T(0,t) = T_s = \text{constant} \tag{3}$$

and a straightforward energy balance expressed by

$$\rho\,H_s\,V = \alpha I_0 + k\,\frac{\partial T}{\partial z} + I_r + I_c, \tag{4}$$

in which H_s is an effective heat of ablation (assumed constant) and I_0 represents the incident heat flux. Surface energy losses due to radiation and convection are I_r and I_c respectively, in equation (4), and the coupling coefficient α denotes the fraction of the incident energy which is absorbed by the receiver.

The radiation loss cited previously is computed according to

$$I_r = \sigma\,\varepsilon\,(T_0^4 - T_s^4), \tag{5}$$

where σ is the Stefan-Boltzmann constant; ε is the surface emissivity; and T_0 represents the surrounding air temperature.

The convective heat loss, I_c, resulting from the flow of high velocity air parallel to the irradiated surface is given by the Newton heat transfer equation, viz.,

$$I_c = h\,(T_r - T_s), \tag{6}$$

where h and T_r are the convection coefficient and air recovery temperature, respectively. In the present study a laminar boundary layer is considered in which case the Pohlhaussen flat plate equation may be used to estimate the convection coefficient. A simplified form [7] of this equation is:

$$h\,(\text{btu}/\text{in}^2 - \text{sec} - {}^\circ\text{F}) = 0.3\,\frac{k_a}{s}\,Re^{1/2}, \tag{7}$$

where k_a is the thermal conductivity of air; s represents the distance from the leading edge of the boundary layer; and Re is the Reynolds Number.*

Due to the low transverse conductivity exhibited by many organic matrix composites at elevated temperatures, a relatively small temperature rise is frequently observed at the unirradiated surface over the exposure times of interest. Consequently, the rear surface is assumed to be insulated, i.e.,

$$\frac{\partial T}{\partial z} = 0 \text{ at } z = l - x_b \tag{8}$$

where l is the initial laminate thickness.

It may be noted that prior to ablation ($t < t_0$, $T(0,t) < T_s$), equations (1) and (2) are applicable with $x_b = V = 0$. Appropriate boundary conditions are then expressed by equations (3–8) with $V = 0$ and the surface temperature T_s regarded as an unknown function of time.

Numerical Solution

Solutions to the nonlinear boundary value problem defined by equations (1–8) were obtained numerically using a modified Crank-Nicholson finite difference method. Figure 1 gives a schematic representation of the one-dimensional array of nodal points employed in the calculations. To efficiently characterize the steep temperature gradient near the irradiated surface, a variable mesh spacing is used in which the nodal separation increases with increasing distance from the front surface. Since equations (1–8) are written with respect to the convected coordinate z, the entire finite difference grid is uniformly translated along with the receding front surface as the solution proceeds. However, because the global coordinate of the unirradiated surface ($x = l$) is fixed, the total number of nodes actually utilized decreases as ablation continues; i.e., once a node has traversed the rear surface it is ignored in subsequent calculations.

With respect to the generic sequence of interior nodal points $i-1$, i, $i+1$ (Figure 1), finite difference approximations for the first and second temperature derivatives at node i are:

$$\left. \frac{\partial T}{\partial z} \right)_{i,j} = \frac{T_{i+1,j} + (r^2 - 1) T_{ij} - r^2 T_{i-1,j}}{r (r + 1) (x_i - z_{i-1})} \tag{9a}$$

and

---

*The semi-empirical expressions given in Reference [7] were utilized to evaluate the dependence of k_a and Re on surface temperature. Also, the thermodynamic analysis in Reference [7] was employed to compute T_r (equation (6)) in terms of the airstream velocity and T_0; this approach assumes a constant value of the recovery factor equal to 0.88.

$$\left.\frac{\partial^2 T}{\partial z^2}\right)_{i,j} = \frac{2[rT_{i-1,j} + T_{i+1,j} - (r+1)\,T_{i,j}]}{r(r+1)\,(z_i - z_{i-1})^2} \tag{9b}$$

where $r \equiv (z_{i+1} - z_i)/(z_i - z_{i-1})$ and the subscripts i and j on T and its derivatives indicate evaluation at coordinate z_i at time t_j. As described in Reference [8] the Crank-Nicholson method entails using in equations (1-8) the following average spatial derivatives over a generic time increment $\Delta t \equiv t_{j+1} - t_j$:

$$\left.\frac{\partial T}{\partial z}\right)_{i,m} = \left[\left.\frac{\partial T}{\partial z}\right)_{i,j+1} + \left.\frac{\partial T}{\partial z}\right)_{i,j}\right]/2 \tag{10a}$$

and

$$\left.\frac{\partial^2 T}{\partial z^2}\right)_{i,m} = \left[\left.\frac{\partial^2 T}{\partial z^2}\right)_{i,j+1} + \left.\frac{\partial^2 T}{\partial z^2}\right)_{i,j}\right]/2, \tag{10b}$$

where the subscript m implies evaluation at the mean time over the increment $t_m = (t_{j+1} + t_j)/2$. The corresponding average time derivative over the interval Δt is represented by the forward difference expression:

$$\left.\frac{\partial T}{\partial t}\right)_{i,m} = \frac{T_{i,j+1} - T_{i,j}}{\Delta t} \tag{11}$$

Substitution of equations (9),(10), and (11) into equation (2) yields the general difference equation to be satisfied at all nodal points, viz.,

$$\left(\frac{-\lambda_1 + \lambda_2 r}{r+1}\right) T_{i-1,j+1} + \left[1 - \lambda_2\left(\frac{r-1}{r}\right) + \frac{\lambda_1}{r}\right] T_{i,j+1}$$

$$- \left[\frac{\lambda_1 + \lambda_2}{r(r+1)}\right] T_{i+1,j+1} = \left(\frac{\lambda_1 - \lambda_2 r}{r+1}\right) T_{i-1,j}$$

$$+ \left[1 + \lambda_2\left(\frac{r-1}{r}\right) - \frac{\lambda_1}{r}\right] T_{i,j} + \left[\frac{\lambda_1 + \lambda_2}{r(r+1)}\right] T_{i+1,j} \tag{12}$$

$$+ \lambda_3 [T_{i+1,j+1} + T_{i+1,j} + (r^2-1)(T_{i,j+1} + T_{i,j}) - r^2(T_{i-1,j+1} + T_{i-1,j})]^2,$$

where

$$\lambda_1 = \frac{k \, \Delta t}{\rho C_p (z_i - z_{i-1})^2}$$

$$\lambda_2 = \frac{V_m \, \Delta t}{2(z_i - z_{i-1})}$$

$$\lambda_3 = \frac{(dk/dT) \, \Delta t}{4r^2(1 + r)^2(z_i - z_{i-1})^2 \rho C_p}$$

The material properties, (ρ, C_p, k, dk/dT) referred to in equation (12) are generally temperature dependent and are to be evaluated at the mean temperature $T_{i,m}$ existing at node i over the time Δt. Similarly, the surface recession rate V_m appearing in λ_2 is an average value associated with time t_m.

According to equation (3) one boundary condition at the irradiated surface (node 1 in Figure 1) is simply the constant temperature condition

$$T(0,t_j) \equiv T_{1,j} = T_s, \tag{13}$$

which is valid for $t > t_0$. The conservation of energy condition, equation (4), is also applicable at node 1 and may be conveniently expressed in difference form by introduction of a virtual node at $z = -z_2$ as indicated in Figure 1. A straightforward substitution of equations (9a) and (10a) into equation (4) then gives

$$T_{2,j+1} + T_{2,j} - T_{0,j+1} - T_{0,j} = \frac{-4z_2}{k} (Q - \rho H_s V_m) \tag{14}$$

in which $Q \equiv \alpha I_0 + I_r + I_c$ is constant by virtue of equation (3) providing I_0 does not vary with time. By applying the general expression, equation (12), at node 1 and combining the result with equation (14), the virtual temperatures ($T_{0,j+1}$ and $T_{0,j}$) are eliminated. This gives

$$2T_s - T_{2,j+1} - T_{2,j} = \left[\frac{\rho H_s V_m - Q}{k}\right]\left[\frac{\rho C_p z_2^2 V_m}{k} - 2z_2 + \frac{(dk/dT) z_2^2}{k^2}\right] \tag{15}$$

where all material properties are evaluated at the surface temperature T_s.

The zero heat flux condition at the unirradiated surface is implemented numerically by again using a virtual node at $z = 2z_r - z_{r-1}$ (see Figure 1). A simple two-term series expansion about node r is employed to estimate the temperature gradient at the insulated surface, viz.,

$$\left.\frac{\partial T}{\partial z}\right)_{surf.,m} = \left.\frac{\partial T}{\partial z}\right)_{r,m} + \left.\frac{\partial^2 T}{\partial z^2}\right)_{r,m} \delta_m = 0, \tag{16}$$

where the second equality follows from equation (8), and δ_m refers to the temporally averaged distance between the node r and the unirradiated surface over the time increment Δt. The quantity δ_m can be expressed by the recursive relation:

$$\delta_m = \frac{\delta_{j+1} + \delta_j}{2} = \delta_j - \frac{V_m \Delta t}{2} \tag{17}$$

in which δ_j is the separation existing at the beginning of the time increment, i.e., at $t = t_j$. Combination of equations (9),(10),(16), and (17) yields:

$$2(T_{r,j+1} + T_{r,j}) \lambda_4 - (T_{r-1,j+1} + T_{r-1,j})(\lambda_4 - 1/2)$$

$$= (T_{r+1,j+1} + T_{r+1,j})(\lambda_4 + 1/2), \tag{18}$$

where

$$\lambda_4 = \frac{2\delta_j - V_m \Delta t}{2(z_r - z_{r-1})} .$$

Finally, the temperatures $T_{r+1,j+1}$ and $T_{r+1,j}$ at the virtual node can be eliminated by invoking the general difference relationship (equation (12)) at node r and combining the result with equation (18). This procedure provides:

$$\left[-\lambda_1 + \lambda_2 + \frac{(\lambda_1 + \lambda_2)(2\lambda_4 - 1)}{2\lambda_4 + 1} \right] T_{r-1,j+1} + \left[2(1 + \lambda_1) \right.$$

$$\left. - \frac{4\lambda_4(\lambda_1 + \lambda_2)}{2\lambda_4 + 1} \right] T_{r,j+1} = \left[\lambda_1 - \lambda_2 - \frac{(\lambda_1 + \lambda_2)(2\lambda_4 - 1)}{2\lambda_4 + 1} \right] T_{r-1,j}$$

$$+ \left[2(1 - \lambda_1) + \frac{4\lambda_4(\lambda_1 + \lambda_2)}{2\lambda_4 + 1} \right] T_{r,j} + \left[\frac{32\lambda_3\lambda_4^2}{(2\lambda_4 + 1)^2} \right] \tag{19}$$

$$(T_{r,j+1} + T_{r,j} - T_{r-1,j+1} - T_{r-1,j})^2 \ .$$

Equations (12),(15), and (19) define an implicit numerical method for solution to the present boundary value problem; given the solution at a generic time t_j, simultaneous solution of these equations provides the temperature distribution and recession rate at t_{j+1}. Since the difference equations are obviously nonlinear, their solution at each time step was obtained iteratively using the method of successive approximations. For a typical time increment $\Delta t = t_{j-1} - t_j$, the following algorithm was adopted. In the first iteration, the quadratic terms appearing in equations (12) and (19) as well as the values of λ_1, λ_2, λ_3, and λ_4 were evaluated using the known nodal temperatures and recession rate existing at t_j. The resulting linear version of equations (12) and (19) were solved simultaneously for the temperatures of nodes 2 through r at t_{j+1}. Then, using equation (15) in conjunction with the recently computed $T_{2,j+1}$, an initial estimate of V_m was calculated. The process was then repeated using the most recent temperatures and V_m to generate a new (updated) linear form of equations (12) and (19). At each time step, the calculations were continued until uniform values of V_m (within two percent) were obtained in successive iterations.

As suggested in the section Heat Transfer Equations for One-Dimensional Model, finite difference expressions analogous to equations (12),(15), and (19) may be developed for pre-ablation regime ($t < t_0$) by noting that the surface temperature $T_{1,j}$ is variable and that $V_m = 0$. The resulting simultaneous difference equations are again nonlinear necessitating use of an iterative procedure. A method similar to that described above was employed in which convergence was based on constancy of surface temperature (rather than V_m) in successive iterations.

EXPERIMENTAL PROCEDURE AND MATERIAL PROPERTIES

Laser Irradiation Tests

To evaluate the analytical mode, laminated composite panels were subjected to rapid heating using the 15 kw, continuous wave, CO_2 laser at NRL. Twenty-ply AS/3501-6 graphite epoxy coupons having in-plane dimensions 5.6 × 10 cm. were fabricated by the NRL Chemistry Division. The panel thickness was 2.54 mm. and the ply layup sequence was

$$[+ 45, - 45, 90, 0, 0, 0, - 45, + 45, 0, 0,]_s,$$

where the subscript denotes symmetry about the laminate midthickness and 0, + 45, - 45, etc. indicate the angular orientation of the lamina fibers with respect to the lengthwise direction of the specimen. To measure the thermal response during irradiation, one thermocouple was embedded in the 11th ply during specimen fabrication and a second was placed at the rear surface (ply 20). In addition, an optical pyrometer was employed to establish the front surface temperature. Testing was conducted at several intensity levels (I_0)

which were generated using a fixed 25 mm beam diameter with variable power output. To simulate aerodynamic cooling effects, a Mach 0.3 airflow as applied parallel to the irradiated surface during each test.**

Thermophysical Properties

The temperature dependence of the material properties of AS/3501-6 which were employed in the numerical calculations are indicated in Figure 2. These properties were obtained from Reference [3] with minor modifications. The density variation $\varrho(T)$ proposed in Reference [3] was increased uniformly

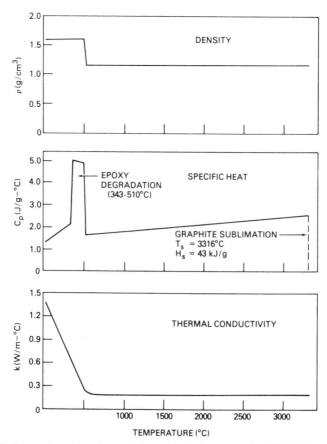

Figure 2. *Temperature dependence of thermophysical properties for AS/3501-6 composite.*

**For these experiments the appropriate value of *s* to be used in the computation of the convection coefficient, equation (7), is 10.48 cm.

by 6 percent based on a higher room temperature value observed for the composite used in the present experiments. The plateau in the heat capacity curve $C_p(T)$ for $343 < T < 510\,°C$ represents additional heat absorbed during resin (matrix) decomposition; effective heat capacity values over this range were computed [3] using a theoretically derived latent heat of 996 J per gram of resin. In the present work the abrupt variations in C_p at 343 and 510 °C suggested in Reference [3] were replaced by more gradual changes occurring over 28 °C intervals. Similarly, a smooth variation in conductivity K (T) and a less severe 28 °C transition in $\varrho(T)$ were introduced at 538 °C. These modest shape changes in the property curves produced no significant change in analytical results and were adopted solely to provide enhanced stability and convergence characteristics in the numerical calculations.

As noted in the $C_p(T)$ curve of Figure 2, the values of H_s and T_s for graphite sublimation were assumed to be 43 kJ/g and 3316 °C, respectively. These properties were also obtained from Reference [3].

DISCUSSION OF RESULTS

The accuracy of the analytical procedure and assumed material properties may be assessed by comparison of the experimental and numerical results shown in Figures 3, 4, and 5 for incident beam intensities (I_0) of 0.232, 1.33, and 2.79 kw/cm² respectively[†]. It is noted that the calculated measured temperatures agree reasonably well, particularly at the lowest intensity level (Figure 3) where the disparity is less than 45 °C over most of the exposure time investigated. Examination of the numerical results for the 2.79 kw/cm² case (Figure 5) reveals that the rapid increase in interior (ply 11) temperature after approximately 4 seconds is due to the close proximity of the advancing irradiated surface to the (global) $x = 1.4$ mm coordinate. The comparable increase in the embedded thermocouple response suggests that the ablation characteristics of the laminate are modeled quite well.

Figures 3–5 also indicate good correspondence between predicted and measured front surface temperatures. The lower temperature, 1860 °C, measured after irradiation for 9.2 sec at 0.232 kw/cm² is reasonably close to the corresponding analytical result of 2132 °C. No ablation is predicted for this test condition. On the other hand, significant ablation is projected for 1.33 and 2.79 kw/cm² beam intensities and the measured surface temperatures, viz. 2938 and 3371 °C, respectively, compare very well with the assumed 3316 °C temperature level necessary to sustain graphite sublimation.

Figure 6 compares the analytical and experimental mass loss as a function of exposure time for material subjected to an *average* intensity of 1.81 kw/cm². The analytical $\Delta m(t)$ was determined from the easily derived expression

---

[†]These intensities refer to maximum values existing at the center of the beam. Also, the surface emissivity (ϵ) and absorption coefficient (a) were both assigned a value of 0.92 in the numerical calculations.

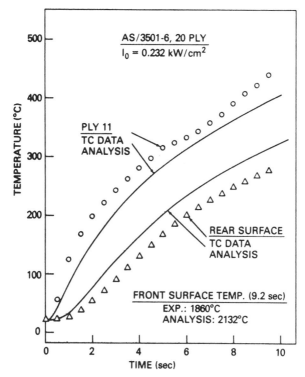

Figure 3. *Comparison of measured and predicted temperature response for AS/3501-6 laminate. Beam intensity = 0.232 kw/cm².*

$$\Delta m(t) = A \int_{x_b(t)}^{l} [\varrho_0 - \varrho(T)]dx,$$

where $\varrho(T)$ is given in Figure 2; ϱ_0 represents the initial (room temperature) density; and A is the total beam area. The temperature distribution $T = T(x,t)$ as well as the instantaneous position of the irradiated surface $x_b(t)$ were obtained from the numerical solution. Although excellent agreement between predicted and measured mass loss is evident, it should be noted that an average rather than peak intensity was employed in the computation of $T(x,t)$ and x_b. Use of the peak flux (2.57 kw/cm²) in the calculations would lead to significantly greater mass loss rates. However, utilization of the average intensity appears reasonable since the experiments themselves represent a spatially average response to irradiation which includes the effects of non-uniform beam intensity.

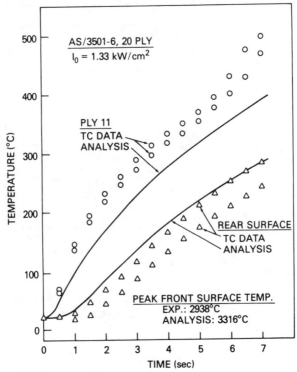

Figure 4. Comparison of measured and predicted temperature response for AS/3501-6 laminate. Beam intensity = 2.79 kw/cm².

Figures 7 and 8 show the computed variation in recession rate V with time for I_0 values of 1.81 and 2.79 kw/cm², respectively. The inset curve in each figure indicates the short-time response. Once the sublimation condition is attained at $t = t_0$, an initial rapid increase in V is apparent followed by a significant period during which the irradiated surface advances at a relatively uniform velocity. Figure 7 also indicates that at longer exposure times V again increases somewhat as the moving front surface approaches the back face. This latter effect simply reflects a less severe temperature gradient ahead of the front surface which is caused by the insulation condition imposed at $x = l$. This reduced gradient implies that the second term on the right hand side of equation (4) becomes larger (less negative) indicating higher values of V.

The nearly constant recession rate displayed over much of the exposure time may be approximated by consideration of the steady state solution of equations (1–7) for a semi-infinite plate. Regarding V and all material properties as constant and requiring that $T \to T_0$ as $z \to \infty$, the solution is

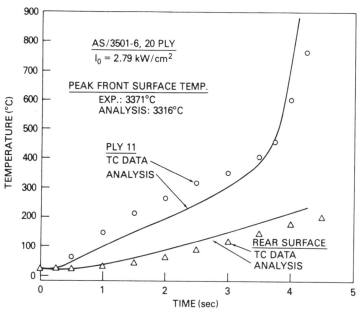

Figure 5. *Comparision of measured and predicted temperature response for AS/3501-6 laminate. Beam intensity = 2.79 kw/cm².*

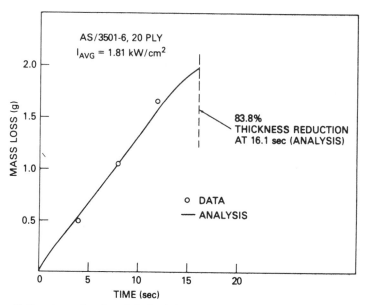

Figure 6. *Experimental and analytically derived mass loss as a function of exposure time. Average beam intensity = 1.81 kw/cm².*

$$T(z) = (T_s - T_0) \exp \left(\frac{-\rho V C_p z}{k} \right) + T_0$$

which may be inserted in equation (4) to give

$$V = \frac{\alpha I_0 + I_r + I_c}{\rho[H_s + C_p(T_s - T_0)]}$$

Evaluation of I_r, I_c, and material properties at $T_s = 3316\,^\circ\text{C}$ provides the following dependence of V on I_0 for the present experiments:

$$V(\text{mm/sec}) = 0.156\, I_0 - 0.150$$

where I_0 has units of kw/cm². As indicated by the dashed lines in Figures 7 and 8, this relationship provides a good estimate of the average velocity over the exposure times indicated and may prove useful in projecting plate burn-through times for AS/3501-6 composite.

Although the numerical treatment presented in this study is adequate for most engineering studies associated with elevated temperature composite behavior, improved modeling is undoubtedly possible. Further examination of the calculated thermal responses shown in Figures 3–5 indicates that the

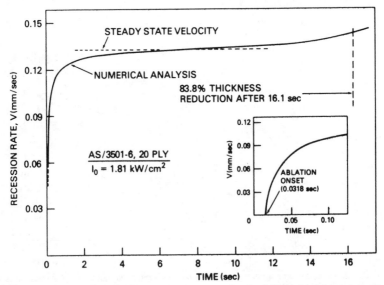

Figure 7. Variation of recession rate with exposure time for 1.81 kw/cm² beam intensity.

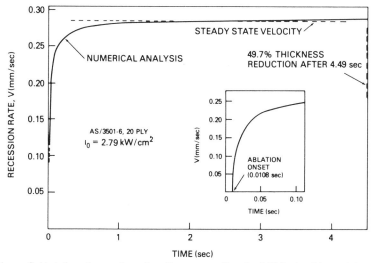

Figure 8. *Variation of recession rate with exposure time for 2.79 kw/cm² beam intensity.*

analysis consistently underestimates the measured interior temperatures and tends to overestimate the rear surface values. Improved correlations at the back surface may result if reduced thermal conductivity values were introduced at low temperatures, and Reference [9] contains experimental conductivity data which would support such an approach. In addition, a recent study [1] of graphite epoxy exposed to flame heating has established that *exothermic* resin decomposition is possible and a kinematic expression quantifying the associated heat generation has been proposed. Inclusion of such a heat source term in the conduction equation should lead to higher prediction temperatures inside the laminate and better agreement with experimental results.

SUMMARY

The primary conclusions from the present study of the thermal response of graphite epoxy composite subjected to rapid heating include:

1. The proposed one-dimensional numerical model coupled with the material properties contained in Reference [3] provides temperature distributions and ablative response which are in good agreement with experimental results from laser irradiation tests. When incorporated with appropriate stress computations, the present analysis provides a practical method for projecting the effects of severe thermal environments on composite plate structures.

2. The present numerical results indicate that a steady state thermal response is approximately achieved over a significant portion of the exposure times

investigated. Analytic expressions for the temperature distribution and recession rate under these conditions have been given.
3. Suggestions for improved modeling include incorporation of exothermic matrix decomposition kinetics and the inclusion of reduced bulk thermal conductivity values at low temperature.

REFERENCES

1. Pering, G. A., Farrell, P. V., and Springer, G. S., "Degradation of Tensile and Shear Properties of Composites Exposed to Fire or High Temperature," *Journal of Composite Materials, 14,* 54 (Jan. 1980).
2. Kibler, K. G., Carter, H. G., and Eisenmann, J. R., "Response of Graphite Composites to Laser Irradiation," TR-77-0706, Air Force Office of Scientific Research, Bolling Air Force Base, DC (March 1977).
3. Menousek, J. F. and Monin, D. L., "Laser Thermal Modeling of Graphite Epoxy," NWC Technical Memorandum Report 3834, Naval Weapons Center, China Lake, CA (June 1979).
4. Boley, B. A. and Weiner, J. H. *Theory of Thermal Stresses.* John Wiley and Sons, Inc., 190 (April 1967).
5. Carslaw, H. S. and Jaegar, J. C. *Conduction of Heat in Solids.* Oxford Univ. Press, 15 (1959).
6. Cohen, M. I., "Melting of a Half-Space Subjected to a Constant Heat Input," *Journal of the Franklin Institute, 283* (4), 271 (April 1967).
7. Hobbs, N. P., Dalton, T. A., and Smiley, R. F., "TRAP2-A Digital Computer Program for Calculating the Response of Mechanically Loaded Structures to Laser Irradiation," KA TR-143 Kamon Avidyne, Burlington, Mass. (October 1977).
8. McCracken, D. D. and Dorn, W. S. *Numerical Methods and Fortran Programming.* John Wiley and Sons, Inc., 384 (1964).
9. Menousek, J. F., Herring, R. L., and Wiggins, E. W., "Thermal Properties of Graphite Epoxy at Elevated Temperatures," MDC-IR0115, McDonnell Aircraft Company, St. Louis, Missouri (Dec. 1978).

21

Some Investigations of Effective Thermal Conductivity of Unidirectional Fiber-Reinforced Composites

W. Gogół and P. Furmański

SCOPE

THIS PAPER CONCERNS THE PROBLEM OF PREDICTING THE EFFECTIVE thermal conductivity of fiber-reinforced materials. Its knowledge is of considerable interest in heat transfer calculations and evaluating the thermoelastic behavior of materials. The effective thermal conductivity of a composite homogeneous on the level of macrostructure is defined as a constant value linking temperature gradient to heat flux density both averaged by volume of fundamental region.

$$<\bar{q}> = - \lambda^{ef} <\overline{grad}\ T>. \tag{1}$$

A well known analogy exists between heat conduction and certain problems of elasticity and electromagnetism. On its basis considerations and calculations dealing with composite thermal conductivity refer as well to effective electric conductivity, dielectric or magnetic permeability, and elastic shear modulus. Shear modulus values of unidirectional composite with symmetrically arranged fibers (square packing array) obtained by finite difference method were presented by Adams and Doner [1]. To make their results more versatile from a practical point of view Hewitt and de Malherbe [9] suggested a formula approximating the values of shear modulus. Utilizing shear loading analogy, Springer and Tsai [2] transferred the results of [1] to composite thermal conductivity and proposed an approximate "thermal model" with the use of which they were able to obtain dependence of λ^{ef} on fiber volume fraction and thermal conductivities of matrix and fibers in closed form expressions. Comparing their values of λ^{ef} to the results of [1] they found that the "thermal model" gave lower values of the composite thermal conductivity than the numerical ones. This conclusion was also confirmed by Zinsmeister and Purohit [4] for a different shape of fibers. Adams and Doner's calculations were repeated by Symm [3] with the use of collocation and least-squares methods. His results turned out to be lower by values than those of [1]. A little earlier Keller [7] derived, on the basis of general properties of Laplace's equation, a following interchange theorem for transverse conductivity of unidirectional composite

$$\lambda_x^{ef} (\lambda_f/\lambda_m) \, \lambda_y^{ef} (\lambda_m/\lambda_f) = \lambda_m \lambda_f. \tag{2}$$

Subscripts x and y in the above formula refer to the principal directions of thermal conductivity tensor.

The object of this paper was to calculate the transverse λ^{ef} of unidirectional fiber-reinforced composite with fibers arranged in rectangular lattice by different theoretical methods and to compare the results obtained to experimental data from a model of composite and the composite itself. This way seemed to us to be an attempt of more general approach to the problem. The more particular goals underlying described objective were as follows:

—to explain existing discrepancies between values of the composite thermal conductivity presented in literature (for example obtained by analogy from [1] and [3])
—to investigate in detail the anisotropic arrangement of fibers (a rectangular lattice) being more representative for laminated composites. The corresponding values of the shear modulus analogical to λ^{ef} were not given in [1]
—to explore if the Keller's theorem is fulfilled in theoretical calculations and experiments
—to compare the results obtained to bounds on effective thermal conductivity of the composite with the assumed arrangement of fibers.

CONCLUSIONS AND SIGNIFICANCE

Good agreement between values of λ^{ef} calculated by different methods and corresponding results of experiments on a model of the composite and composites was obtained. In particular it was found that

—results of Symm [3] were closer to actual values of λ^{ef} than those of Adams and Doner's [1]
—for the whole investigated range of fiber to matrix thermal conductivities ratio and up to fiber volume fraction of about $V_f \cong 0.5$ Hashin and Rosen's formula [8] for "random arrangement" of fibers could be satisfactorily used for calculations of the composite thermal conductivity
—Keller's theorem had been fulfilled in our calculations and experiments with a good accuracy

—bounds on the composite thermal conductivity based on Hashin and Rosen's model were narrower for smaller values of V_f but for higher values of fiber volume fraction bounds obtained with the assumption of one-dimensional temperature or heat flux fields occurred to be better

—for $\lambda_f/\lambda_m>1$ actual values of the effective thermal conductivity came closer to lower bounds for the assumed composite structure while for $\lambda_f/\lambda_m<1$ the reverse was true.

ANALYTICAL BACKGROUND

Formulation of the Problem and Method of Solution

It was assumed, similarly to [2], a) that fibers in the composite were arranged symmetrically in matrix according to a rectangular packing array, b) that fibers were circular, c) that the thermal contact resistance between fibers and matrix was negligible, d) that a temperature field in the composite, by determination of transverse thermal conductivity, was two-dimensional which was due to unidirectional alignment of fibers. The temperature field should then satisfy Laplace equation

$$\frac{\partial^2 T}{\partial x^2} + \frac{\partial^2 T}{\partial y^2} = 0 \tag{3}$$

To find the effective thermal conductivity, fundamental region representative for the assumed structure was considered. A quadrant of the region is shown in Figure 1 together with the appropriate temperature field obtained from theory. According to definition [1] the composite thermal conductivity should be independent of the conditions assumed on boundaries of the fundamental region. One could then express them as

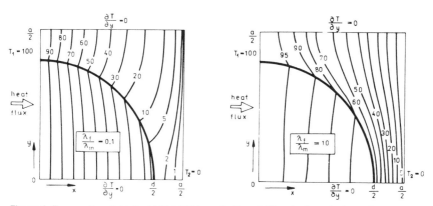

Figure 1. *Temperature fields for $V_f \cong 0.5$ (b/a = 1, d/a = 0.8) and selected values of λ_f/λ_m. Note almost linear change of temperature with x in the fiber region.*

$$T = T_1 = 100 \qquad \text{at } x = 0$$
$$T = T_2 = 0 \qquad \text{at } x = a/2 \qquad (4)$$
$$\frac{\partial T}{\partial y} = 0 \qquad \text{at } y = 0 \text{ and } y = b/2$$

In addition to (4) the conditions

$$T_f = T_m$$

$$\lambda_f \frac{\partial T_f}{\partial n} = \lambda_m \frac{\partial T_m}{\partial n} \qquad (5)$$

should be satisfied at the interface between the fiber and the matrix. In (5) n symbolizes the direction normal to the interface.

To solve the problem given by equations (3) to (5) three different theoretical methods [5] were used, namely: the finite element method, the modified least-squares analytical method, and even the graphical one. Temperature fields in the fundamental region of the composite were obtained for a wide range of fiber to matrix thermal conductivity ratio λ_f/λ_m fiber volume fraction V_f and the dimension ratio b/a of the fundamental region. The effective thermal conductivity of the composite in the direction of x coordinate axis was calculated from formula

$$\lambda^{ef} = \frac{\int_0^{b/a} 9 \times (0, y)\, dy}{b/a\,(T_1 - T_2)} \qquad (6)$$

The form of (6) came directly from definition (1) when the fundamental region and the boundary conditions had been assumed.

Bounds

The composite thermal conductivity calculated by approximate numerical methods is subject to certain inaccuracies which are often hard to estimate. It seems then necessary to compare the results obtained to existing bounds on λ^{ef}. The bounds may be as well used to estimate range of applicability of numerous approximate formulae for the composite thermal conductivity and sometimes even used as the formulae themselves. Two pairs of bounds were used to analyze the results of the solution of the discussed problem. The first pair was derived on the basis of Hashin and Rosen's model [8], [10]. For the considered composite structure and isotropic case it was given in the following form:

— upper bound

$$\frac{(\lambda^{ef})^+}{\lambda_m} = \frac{(\mu + 1)\,\pi/4 + (\pi/2 - 1)\,(\mu - 1)\,V_f}{(\mu + 1)\pi/4 - (\mu - 1)\,V_f} \tag{7}$$

— lower bound

$$\frac{(\lambda^{ef})^-}{\lambda_m} = \frac{(\mu +)\,\pi/4 + (\mu - 1)\,V_f}{(\mu + 1)\,\pi/4 - (\pi/2 - 1)\,(\mu - 1)\,V_f} \tag{8}$$

where $\mu = \lambda_f/\lambda_m$.

The second pair of bounds (for isotropic case) may be obtained by assuming one-dimensional temperature or heat flux fields in the fundamental region [5,6]. This leads to relationships:

— upper bound

$$\frac{(\lambda^{ef})^+}{\lambda_m} = \left[(1 - d/a) + \int_0^{d/2} \frac{dx}{a/2 + (\mu - 1)\sqrt{(d/2)^2 - x^2}} \right]^{-1}, \tag{9}$$

— lower bound

$$\frac{(\lambda^{ef})^-}{\lambda_m} = (1 - d/a) + \int_0^{d/2} \frac{dx}{a/2 - (1 - \mu)^{-1}\sqrt{(d/2)^2 - x^2}} \tag{10}$$

The dependence of λ^{ef}/λ_m on fiber volume fraction in relation to the given pairs of bounds is shown in Figure 2.

EXPERIMENTAL PROCEDURE

Experimental investigations of the composite thermal conductivity were carried out on a model of the fundamental region (a rectangle of dimensions a and b) [5] and on the composite itself.

The model of the composite structure was made from plexiglass plates of dimensions $50 \times 50 \times 10$ mm (Figure 3) in which holes of diameter $d = 6$ mm ($V_f = 0.28$) or $d = 8$ mm ($V_f = 0.5$) were bored. The relative distances between holes for the separate plates were chosen as $b/a = 0.83$, 1.0, 1.25, 1.66, 2.5, and 5.0. In the holes different substances were placed, e.g. polystyrene, polyurethane, water with gelatine, plexiglass, different kinds of ceramics, 1H18N9T steel, copper, etc. having thermal conductivities measured by other methods [11,12].

Specimens of the composite consisted of copper (fibers) and epoxy resin (matrix) with unidirectional alignment of fibers (Figure 4). They differed as

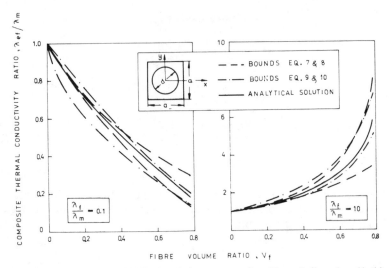

Figure 2. *Calculated values of λ^{ef}/λ_m in relation to two pairs of bounds (based on Hashin and Rosen's model or obtained from assumption of one-dimensional temperature or heat flux fields) as a function of fiber volume fraction V_f. Note change of the spread of bound pairs for different V_f values.*

to their inner structure, namely ratio b/a of average dimensions of the fundamental region, as determined by microscopic observations, was found to be equal at 0.54, 0.85, 1.33 and the fiber diameter to average value of dimension a equal to 0.325.

The measurements of the effective thermal conductivity of composites and models of the composite perpendicular to fibers were made in a plate apparatus in steady state. Thermal contact resistance existing at the interface of heater or cooler and specimen we tried to deminish by covering the places

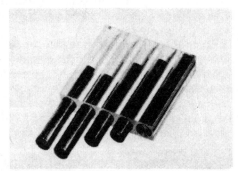

Figure 3. *Model of composite structure (sample used in measurements of λ^{ef}).*

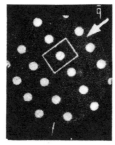

Figure 4. *Inner structure of the copper-epoxy composite.*

with silicon paste. Experimental errors were estimated as about 4 percent. For the composite specimens, thermal conductivity measurements were also carried out for the direction along fibers. The experiments were made in a rod apparatus [12] in steady state by comparative method with precision not exceeding 5 percent.

The results of experiments were presented in Figure 5, Figure 6, and in the accompanying Table.

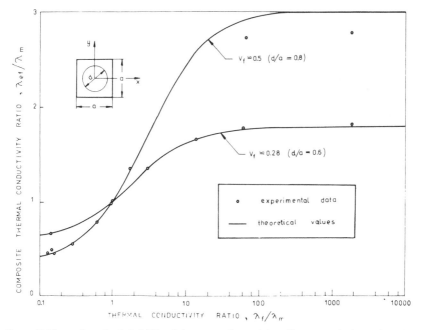

Figure 5. *Thermal conductivity λ^{ef}/λ_m of the composite model vs. fiber to matrix thermal conductivities ratio λ_f/λ_m. Note existence of asymptotes for $\lambda_f/\lambda_m \rightarrow \infty$.*

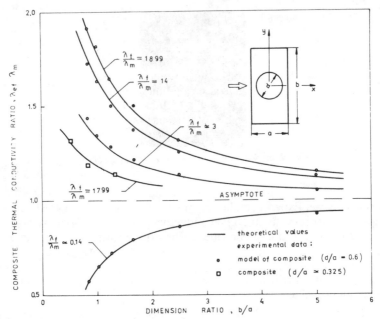

Figure 6. Thermal conductivity λ^{ef}/λ_m vs. dimensions b/a of fundamental region for the model of composite and the copper-epoxy composite.

RESULTS

The values of λ^{ef}/λ_m for the model of the composite determined by different theoretical methods were almost equal for fixed values of λ_f/λ_m and V_f. In the range of the calculations carried out by the separate methods, average discrepancies were not greater than ~ 0.8 percent while maximal ~ 1.5 percent.

Function $\lambda^{ef}/\lambda_m = f(\lambda_f/\lambda_m)$ for two different fiber volume fractions and the average values of experimental data for the model of isotropic composite were presented in Figure 5. The results of calculations came in good agreement with experimental ones except for high fiber to matrix thermal conduc-

Table 1. Experimental data and calculated values of λ^{ef}/λ_m for copper-epoxy composites.

Average Dimension Ratio b/a	Experimental Data		Theoretical Results ⊥ Fibers
	‖ Fibers	⊥ Fibers	
0.54	266.7	1.32	1.31
0.85	186.2	1.19	1.21
1.33	111.8	1.13	1.13

tivities ratio (approximately for $\lambda_f/\lambda_m = 100$) at high fiber volume fraction. The latter was mainly due to the fact that for these V_f values it was difficult to satisfy the proper boundary conditions on the faces of specimens. The calculated values of λ^{ef}/λ_m were about 5 to 10 percent lower than given by Adams and Doner [1]. They fitted almost exactly the results of [3]. It was also verified if the theoretical results and experimental data had satisfied Keller's theorem. The deviation from the theorem didn't exceed 0.6 percent what provided confidence in the accuracy of the results obtained. When looking for a formula approximating numerical results we found that Hashin and Rosen's expression [8] obtained for "random array" gave values of λ^{ef}/λ_m the closest to the exact ones for the whole range of λ_f/λ_m values and up to 50 percent of fiber volume fraction. On the contrary to the function presented in [9] approximating Adams and Doner's results the discussed expression satisfied also Keller's theorem. Dependence of λ^{ef}/λ_m on dimensions ratio b/a of the fundamental region was also studied. Some results of calculations and experimental data are shown in Figure 6. Good agreement between values of the effective thermal conductivity obtained from theory and experiments on the model of the composite and copper-epoxy composites was observed. In the Table the composite thermal conductivity perpendicular to fibers both calculated and experimental is given. In addition, the effective thermal conductivity in the direction along fibers is presented there as well. The latter values satisfied the well known "mixture rule" [5].

The relationship of the calculated values of λ^{ef}/λ_m to the described pairs of bounds (7),(8), and (9),(10) was also analyzed. Some representative results are given in Figure 2 as a function of fiber volume fraction. It was found that the values of the effective thermal conductivity always lay between the upper and lower bounds. Nevertheless they came closer to the lower bound for $\lambda_f/\lambda_m > 1$ and to the upper for $\lambda_f/\lambda_m < 1$. Here we should note that Springer and Tsai's formula [2] based on so called "thermal model" is in fact identical, for $\lambda_f/\lambda_m > 1$ with the lower bound given by expression (10). After studying curves of the kind presented in Figure 3 one could conclude that bounds obtained from Hashin and Rosen's model (7),(8) gave values closer to the exact ones for low fiber volume fraction. The reverse was true for bounds based on the assumed one-dimensional temperature or heat flux fields (isotherms or adiabats) which occurred to be narrower for high values of V_f.

NOMENCLATURE

a,b	dimensions of composite fundamental region
d	fiber diameter
n	direction normal to fiber-matrix interface
$\overline{q}$	heat flux density
T	temperature
V	volume fraction
x,y	cartesian coordinates

μ fiber to matrix thermal conductivity ratio
λ thermal conductivity

Subscripts

ef effective
f refers to fibers
m refers to matrix

REFERENCES

1. Adams, D. F. and Doner, D. R., "Longitudinal Shear Loading of a Unidirectional Composite," *J. Composite Materials, 1,* 4 (1967).
2. Springer, G. S. and Tsai, S. W., "Thermal Conductivities of Unidirectional Materials," *J. Composite Materials, 1,* 166 (1967).
3. Symm, G. T., "The Longitudinal Shear Modulus of a Unidirectional Fibrous Composite," *J. Composite Materials, 4,* 426 (1970).
4. Zinsmeister, G. E. and Purohit, K. S., "Comments on Springer and Tsai's Method of Predicting Effective Thermal Conductivities of Unidirectional Materials," *J. Composite Materials, 4,* 278 (1970).
5. Furmański, P. and Gogól, W., "Investigation of Steady-State Heat Conduction in Two-Dimensional Model of Composite with Symmetrically Arranged Circular Fibers," Archives of Thermodynamics and Combustion, No. 2, Archiwum Termodynamiki i Spalania, in Polish (1979).
6. Stiepanow, S. W., "Zawisimost koefficientov tieploprowodnosti uporiadocziennych dwuchfaznych sistiem ot obiemnoj koncientracji wkliuczieni," Tieplofiziczieskije swojstwa twiordych wieszczestw, Izd. "Nauka," Moskva (1971).
7. Keller, J. B., "A Theorem on the Conductivity of a Composite Medium," *J. Mathematical Physics, 5* (4) (1964).
8. Hashin, Z. and Rosen, B. W., "The Elastic Moduli of Fibre-Reinforced Materials," *J. Applied Mechanics, 31* (1964).
9. Hewitt, R. L. and de Malherbe, M. C., "An Approximation for the Longitudinal Shear Modulus of Continuous Fiber Composties," *J. Composite Materials, 4,* 280 (1970).
10. Donea, J., "Thermal Conductivities Based on Variational Principles," *J. Composite Materials, 6,* 262 (1972).
11. Gogól, W., Gogól, E., and Artecka, E., "Thermal Conductivity Investigations of Moist Soils," (in Polish), Biulethyl Inf. Inst. Techn. Cieplnej, No. 40 (1973).
12. Gogól, E. and Gogól, W., "Apparatus for Measurements of Thermal Conductivity of Metals in Moderate Temperatures by the Comparison Method," (in Polish), Biuletyn Inf. Inst. Techn. Cieplnej, No. 49 (1977).

22

Bounds of Effective Thermal Conductivity of Short-Fiber Composites

S. NOMURA AND T. W. CHOU

INTRODUCTION

SHORT-FIBER COMPOSITES ARE ATTRACTIVE STRUCTURAL MATERIALS because of their ease in fabrication and versatility in properties [1]. The effective elastic properties of short-fiber composites have been examined recently by Chou, Nomura, and Taya [2] and Nomura and Chou [3,4]. The aim of this paper is to derive the bounds of effective physical constants of short-fiber composites.

The important physical constants relevant to the transport properties of a composite include dielectric constants, heat conduction, electrical conduction, magnetic permeability, and diffusion coefficients. Since all these properties are second rank tensors, only the bounds of thermal conductivity are derived in this paper to illustrate the method of analysis. Several approaches have been employed by researchers to examine the effective thermal conductivity of nonhomogeneous materials. Statistical methods were used by Beran [5], and Beran and Molyneux [6]. Beran [5] applied the concept of the theory of turbulence to this subject, and by solving the moment correlation equation with a perturbation procedure and a variational method, derived the upper and lower bounds of effective thermal conductivity for statistically homogeneous and isotropic materials. Hori and Yonezawa [7] used a cumulant approximation of the many point correlation function and obtained an approximate expression of effective thermal conductivity for a perfectly disordered material, the concept of which was first introduced by Kröner [8]. A self-consistent model approach, originally developed by Budiansky [9] and Hill [10] in elasticity problems, has been adopted by Hashin [11] and extended recently by Willis [12] to treat the problem of thermal conductivity. The second phase material assumes the shape of spheres in Hashin's work [11] and

271

short fibers in Willis' approach [12]. The variational method approach was pioneered by Hashin and Shtrikman [13], who derived the upper and lower bounds of effective thermal conductivity for a matrix reinforced with spherical inclusions. Their bounds were improved by Brown [14] and Beran [5] (see Beran [15] for a review). Beran's work takes into account the correlation of thermal conductivity up to the third order and has given the best bounds known so far. But due to the complexity of the moment correlation equation, extension of Beran's approach to statistically anisotropic medium and to derive the correlation function beyond the third order term seem to be very difficult (see Hori [16]).

The above reviewed works, with the exception of Reference [12] are restricted to a statistically homogeneous and isotropic material. In this paper, we examine the effective thermal conductivity of unidirectional short-fiber composite with transverse isotropy. To provide the background of the present approach it is necessary to recapitulate the work of Dederichs and Zeller [17], Zeller and Dederichs [18], Kröner [8,19], and more recently, Nomura and Chou [3] for deriving the effective elastic moduli of short-fiber composites. The work of Dederichs and Zeller established a method for analyzing the elastic behavior of disordered materials. They treated a polycrystalline aggregate without texture in which the crystallographic orientations of different grains are not correlated and the orientations are independent of the shape of the grains. They have shown that the basic equation for a disordered material has the form of the Lippmann-Schwinger's equation in statistical quantum scattering theory and made use of various results of scattering theory to seek an approximate solution of the basic equation. This basic equation was previously also examined by Kroner [8] using a different approach. Kroner was interested in materials about which the complete statistical information in form of correlation functions is known. It was shown that compact forms of macroscopic moduli can be derived for the class of materials known as "perfectly disordered." The work of Nomura and Chou [3] presented a method of evaluating the upper and lower bounds of the elastic moduli of short-fiber composites. They adopted a perturbation expansion of the local strain field by using a Green's function tensor. The correlations of elastic constants are evaluated based upon the characteristics of the geometry and distribution of the component phases. The bounds are then determined from a variational treatment. This approach is now followed to derive the effective transport constants of short-fiber composites.

DEFINITIONS AND BASIC EQUATIONS

The linear relation between the head flux q and gradient of temperature and the governing equation for a steady state are given by

$$\underset{\sim}{q} = \underset{\sim}{k} \, (- \underset{\sim}{\nabla} T)$$

$$\underset{\sim}{\nabla} \cdot \underset{\sim}{q} = 0 \tag{1}$$

where $\underset{\sim}{k}$ denotes thermal conductivity and is assumed to be a function of position only. It is understood that $\underset{\sim}{k}$ is symmetric due to the Onsager's principle [20].

We assume each component phase in a heterogeneous system is isotropic. By defining $-\partial_i T \equiv D_i$ and noting that $k_{ij} = k$ $(\underset{\sim}{r})$ δ_{ij}, the effective thermal conductivity of the composite, k_{ij}^* is defined by

$$<q_i> = k_{ij}^* <D_j> \tag{2}$$

where $< \cdot >$ denotes the spatial average. Decomposing k_{ij}, D_j, and q_i into a constant part and a fluctuating part and substituting them into (1), we have

$$\underset{\sim}{D}' = \underset{\sim\sim}{\Gamma k}' <\underset{\sim}{D}> + \underset{\sim}{\Gamma} (\underset{\sim}{k}' \underset{\sim}{D}')' \tag{3}$$

where ( )' denotes the fluctuating part, and Γ is an integral operator defined as

$$\Gamma_{ij} (\cdot) = \int_V g_{ij} (\underset{\sim}{r} \underset{\sim}{-r}') (\cdot) d\underset{\sim}{r}' \tag{4}$$

Here, g_{ij} (r) is constructed from the Green's function g(r) for an infinite medium by

$$g_{ij} (\underset{\sim}{r} \underset{\sim}{-r}') = - \frac{\partial}{\partial x_i} \frac{\partial}{\partial x_j'} g(\underset{\sim}{r} \underset{\sim}{-r}')$$

Equation (3) is of the Fredholm's integral equation of the second type. An expression of the effective thermal conductivity can be obtained by successive substitutions of equation (3) into itself and subsequently combining the result with equation (2)

$$\underset{\sim}{k} * = \underset{\sim}{<k>} + \underset{\sim}{<k}' \underset{\sim\sim}{\Gamma k}'> + \underset{\sim}{<k}' \underset{\sim\sim}{\Gamma k}' \underset{\sim\sim}{\Gamma k}'> + \cdots \tag{5}$$

It has been demonstrated in [3] that for short-fiber composites with ellipsoidal symmetry, equation (5) can be evaluated up to the third order term. The inverse of $\underset{\sim}{k}*$ is denoted by $\underset{\sim}{S}*$ which, following [3], is given by

$$\underset{\sim}{S}* = \underset{\sim}{<S>} + \underset{\sim}{<S}' \underset{\sim}{\Delta S}'> + \underset{\sim}{<S}' \underset{\sim\sim}{\Delta S}' \underset{\sim\sim}{\Delta S}'> + \cdots \tag{6}$$

where

$$\underset{\sim}{\Delta} = - \left(\underset{\sim}{<S>}^{-1} + \underset{\sim}{<S>}^{-1} \underset{\sim}{\Gamma}^S \underset{\sim}{<S>}^{-1} \right)$$

and Γ^s is constructed from $<S>^{-1}$ rather than $<\underset{\sim}{k}>$. It is also understood that $k = S^{-1}$ but $<k> \neq <S>^{-1}$.

EVALUATION OF THE SECOND AND THIRD ORDER TERMS

The derivations of equations (5) and (6) given in the previous section are valid for all locally inhomogeneous materials. The evaluation of these equations calls for multiple integration of higher order terms. In this section we focus our attention to short-fiber composites. The fibers are approximated by ellipsoidal inclusions of the same length embedded in a matrix. Both the fiber and matrix phases are isotropic and homogeneous. For unidirectionally aligned fibers the composite can be regarded as transversely isotropic. For composites possessing ellipsoidal symmetry, k* in equation (5) can be evaluated up to the third order term. Under the assumptions of the ellipsoidal shape of the inclusions, and their homogeneous distribution in the matrix, and taking the entire domain as having the same shape as the inclusion, the evaluation of equation (5) can be reduced simply to

$$\underset{\sim}{k}{}^* = \underset{\sim}{<k>} + \underset{\sim\sim}{<k'E>} + \underset{\sim\sim\sim}{<k'EE>} + \cdots \cdots \tag{7}$$

where

$$E_{11} = -\frac{k'}{<k>} (1-h(t))$$

$$E_{22} = E_{33} = \frac{-k'}{2 <k>} h(t)$$

$$E_{ij} = 0 \ (i{\neq}j)$$

and

$$h(t) = \frac{t^2}{t^2-1} \left[1 - \frac{1}{2} \left\{ \left(\frac{t^2}{t^2-1}\right)^{1/2} - \left(\frac{t^2-1}{t^2}\right)^{1/2} \ln\left(\frac{t+\sqrt{t^2-1}}{t-\sqrt{t^2-1}}\right) \right\} \right] \tag{8}$$

Here the x_1 axis is parallel to the major axis of the ellipsoidal inclusion and t denotes the inclusion aspect ratio (length of major axis/length of minor axis). The limiting values of h(t) are 2/3 and 1 for t = 1 and ∞, respectively.

Analogous to the treatment of inclusion problems by Eshelby [21], we call, for the problem of anisotropic thermal conductivity, E the modified Eshelby's tensor. Just as in the case of elastic moduli of short-fiber composites the modified Eshelby's tensor in this case depends only on the shape of the inclusion and is constant throughout that domain.

APPROXIMATION OF THE INFINITE SERIES SOLUTION

In this section, we derive, in closed form, an approximate expression of the infinite series of equation (5). After some lengthy but elementary calculations, we obtain

$$
\begin{aligned}
\underset{\sim}{k}^* ={}& \langle k\rangle + \langle k'\,\Gamma k'\rangle + \langle k'\,\Gamma k'\,\Gamma k'\rangle \\[4pt]
& + (\langle k'\,\Gamma k'\,\Gamma k'\,\Gamma k'\rangle - \langle k'\,\Gamma k'\rangle\,\Gamma\,\langle k'\,\Gamma k'\rangle) \\[4pt]
& + (\langle k'\,\Gamma k'\,\Gamma k'\,\Gamma k'\,\Gamma k'\rangle - \langle k'\,\Gamma k'\,\Gamma k'\rangle\,\Gamma\,\langle k'\,\Gamma k'\rangle \\[4pt]
& - \langle k'\,\Gamma k'\rangle\,\Gamma\,\langle k'\,\Gamma k'\,\Gamma k'\rangle) + (\langle k'\,\Gamma k'\,\Gamma k'\,\Gamma k'\,\Gamma k'\,\Gamma k'\rangle \\[4pt]
& - \langle k'\,\Gamma k'\rangle\,\Gamma\,\langle k'\,\Gamma k'\,\Gamma k'\,\Gamma k'\rangle - \langle k'\,\Gamma k'\,\Gamma k'\rangle\,\Gamma\,\langle k'\,\Gamma k'\,\Gamma k'\rangle \\[4pt]
& - \langle k'\,\Gamma k'\,\Gamma k'\,\Gamma k'\rangle\,\Gamma\,\langle k'\,\Gamma k'\rangle + \langle k'\,\Gamma k'\rangle\,\Gamma\,\langle k'\,\Gamma k'\rangle\,\Gamma\,\langle k'\,\Gamma k'\rangle) \\[4pt]
& + \cdots
\end{aligned}
\tag{9}
$$

Here we adopt the assumptions, following the previous section, that each $\Gamma k'$ in equation (9) can be replaced by the modified Eshelby's tensor E defined by (8), and that the series (9) is convergent. Thus, we have, by using the modified Eshelby's tensor

$$
\begin{aligned}
\underset{\sim}{k}^* \cong{}& \langle k\rangle + \langle k'\,E\rangle + \langle k'EE\rangle \\[4pt]
& + (\langle k'\,EEE\rangle - \langle k'\,E\rangle\,\langle EE\rangle) \\[4pt]
& + (\langle k'EEEE\rangle - \langle k'EE\rangle\,\langle EE\rangle - \langle k'\,E\rangle\,\langle EEE\rangle) \\[4pt]
& + (\langle k'\,EEEEE\rangle - \langle k'\,E\rangle\,\langle EEEE\rangle - \langle k'\,EE\rangle\,\langle EEE\rangle - \langle k'\,EEE\rangle\,\langle EE\rangle \\[4pt]
& + \langle k'\,E\rangle\,\langle EE\rangle\,\langle EE\rangle) \\[4pt]
& + \cdots \cdots \\[6pt]
={}& \langle k\rangle + \langle k'\,E(I-E)^{-1}\rangle\,(I + \langle EE(I-E)^{-1}\rangle)^{-1}
\end{aligned}
\tag{10}
$$

It should be noted that Voigt's rule of mixtures can be obtained by retaining only the first term of the right hand side of (10) and Kroner's approximation [8] can also be obtained by neglecting the product of average quantitites in equation (10). Zeller and Dederichs [18] derived an expression similar to equation (10) by using the T-matrix approximation in scattering theory. It can be shown that for a binary system, equation (10) is identical with an upper bound of k* which will be discussed in the next section.

Parallel discussions also hold for the inverse of $\underset{\sim}{k}^*$ (denoted by $\underset{\sim}{S}^*$ in the previous section). The result is

$$\underset{\sim}{S}^* \cong <\underset{\sim}{S}> + <\underset{\sim}{S}'\underset{\sim}{F}(\underset{\sim}{I}-\underset{\sim}{F})^{-1}> (\underset{\sim}{I} + <\underset{\sim}{F}\underset{\sim}{F}(\underset{\sim}{I}-\underset{\sim}{F})^{-1}>)^{-1} \tag{11}$$

where $F = \underset{\sim}{\Delta}\underset{\sim}{S}'$ and $\underset{\sim}{\Delta}$ is given in equation (6). It also can be shown that equation (11) is identical with a lower bound of $\underset{\sim}{k}^*$ to be derived in the following section for a binary system.

UPPER AND LOWER BOUNDS

Following a development parallel to that given in [3], explicit bounds of $\underset{\sim}{k}^*$ have been derived based upon the variational principle.

$$\left(<S> \delta_{ij} - \frac{<S'^2>^2 \Delta_{ij}^2}{<S'^3 \Delta_{ik}\Delta_{kj} - S'^2 \Delta_{ij}>} \right)^{-1} \leqq k_{ij}^* \leqq <k> \delta_{ij} - \frac{<k'E_{ij}>^2}{<k' E_{im}E_{mj}> - <k'E_{ij}>} \tag{12}$$

where E_{ij} is given in equation (8), and

$$\Delta_{11} = -\left(\frac{1}{<S>} + \frac{1}{<S>^2} \Gamma_{11}^S \right)$$

$$\Delta_{22} = \Delta_{33} = -\left(\frac{1}{<S>} + \frac{1}{<S>^2} \Gamma_{22}^S \right)$$

$$\Gamma_{11}^S = -<S> (1-h(t))$$

$$\Gamma_{22}^S = -<S> h(t)/2$$

In summary, the bounds of thermal conductivities can be expressed in the following explicit forms

$$\left\{ \frac{1}{<\frac{1}{k}>} - \frac{<\left(\frac{1}{k}\right)'^2>^2 h(t)}{<\left(\frac{1}{k}\right)'^3> h(t) + <\frac{1}{k}><\left(\frac{1}{k}\right)'^2>} \right\}^{-1} \leqq k_{11}^*$$

$$\leqq <k> - \frac{<k'^2>^2 (1-h(t))}{<k'^3> (1-h(t)) + <k> <k'^2>}$$

$$\left\{ \frac{1}{<\frac{1}{k}>} - \frac{<\left(\frac{1}{k}\right)'^2>^2 (1-h(t)/2)}{<\left(\frac{1}{k}\right)'^3> (1-h(t)/2) + <\frac{1}{k}> <\left(\frac{1}{k}\right)'^2>} \right\}^{-1} \leqq k_{22}^* = k_{33}^*$$

$$\leqq <k> - \frac{<k'^2>^2 h(t)}{<k'^3> h(t) + 2 <k> <k'^2>} \tag{13}$$

For the special case of a binary system and denoting the thermal conductivity and volume fraction of the fiber and matrix phases by k_f, V_f and k_m, V_m, respectively, equation (13) becomes

$$\left\{ \frac{V_f}{k_f} + \frac{V_m}{k_m} - \frac{V_f V_m \left(\frac{1}{k_f} - \frac{1}{k_m}\right)^2 h(t)}{\left(V_m - V_f\right)\left(\frac{1}{k_f} - \frac{1}{k_m}\right) h(t) + \frac{V_f}{k_f} + \frac{V_m}{k_m}} \right\}^{-1} \leqq k_{11}^*$$

$$\leqq V_f k_f + V_m k_m - \frac{V_f V_m \left(k_f - k_m\right)^2 (1-h(t))}{\left(V_m - V_f\right)\left(k_f - k_m\right)(1-h(t)) + V_f k_f + V_m k_m}$$

$$\left\{ \frac{V_f}{k_f} + \frac{V_m}{k_m} - \frac{V_f V_m \left(\frac{1}{k_f} - \frac{1}{k_m}\right)^2 (1-h(t)/2)}{\left(V_m - V_f\right)\left(\frac{1}{k_f} - \frac{1}{k_m}\right)(1-h(t)/2) + \left(\frac{V_f}{k_f} + \frac{V_m}{k_m}\right)} \right\}^{-1} \leqq k_{22}^* = k_{33}^* \tag{14}$$

$$\leqq V_f k_f + V_m k_m - \frac{V_f V_m \left(k_f - k_m\right)^2 h(t)}{\left(V_m - V_f\right)\left(k_f - k_m\right) h(t) + 2 \left(V_f k_f + V_m k_m\right)}$$

For the special case of spherical inclusions, $(h(t) = 2/3)$, equation (14) is simplified as

$$\left(\frac{V_f}{k_f} + \frac{V_m}{k_m} - \frac{2 V_f V_m \left(\frac{1}{k_f} - \frac{1}{k_m}\right)^2}{2\left(V_m - V_f\right)\left(\frac{1}{k_f} - \frac{1}{k_m}\right) + 3\left(\frac{V_f}{k_f} + \frac{V_m}{k_m}\right)} \right)^{-1} \leqq k^* \leqq V_f k_f + V_m k_m$$

$$\frac{V_f V_m \left(k_f - k_m\right)^2}{\left(V_f - V_m\right)\left(k_f - k_m\right) + 3\left(V_f k_f + V_m k_m\right)} \tag{15}$$

In the case of long continuous fibers, $h(t) = 1$, equation (14) becomes

$$k_{11}^* = V_f k_f + V_m k_m$$

$$\frac{\left(k_m + k_f\right)k_m k_f}{\left(V_f k_m + V_m k_f\right)^2 + k_m k_f} \leq k_{22}^* = k_{33}^* \leq \frac{\left(V_m k_m + V_f k_f\right)^2 + k_m k_f}{k_m + k_f}$$

$$\tag{16}$$

CONCLUSIONS AND DISCUSSIONS

We have presented bounds of effective thermal conductivities for unidirectional short-fiber composites. Our approach has taken into account the correlations of thermal conductivity up to the third order term. Beran [5] also has considered third order correlations but his treatment, based upon the concept of turbulence theory, is restricted to the case of statistically isotropic material systems. For this case of isotropic thermal conductivity our results are identical to Beran's. Hashin and Shtrikman [13] also treated the isotropic case of spherical inclusions and their approach is equivalent to considering the correlation function up to the second order.

The effect of anisotropy was first examined by Willis [12], who applied the

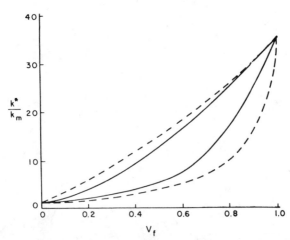

Figure 1. The variation of k_{11}^*/k_m with filler volume fraction for a graphite/epoxy system, using $k_f = 60$ BTU (hr-ft² - °F/in), $k_m = 1.7$ BTU (hr-ft² - °F/in), and $1/d = 1$. Solid lines: present theory, broken line: Hashin and Shtrikman [13].

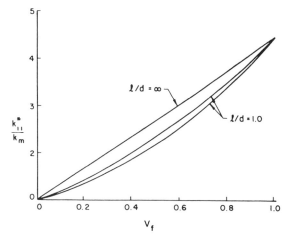

Figure 2. *The variation of k_{11}/k_m with filler volume fraction for an E-glass/epoxy system, using $k_f = 7.5$ BTU/hr-ft^2 - °F/in), $k_m = 1.7$ BTU (hr-ft^2 - °F/in), and $1/d = 1$ and ∞.*

variational approach of Hashin and Shtrikman [13] to consider the thermal conductivity of unidirectional short-fiber composites.

Figure 1 presents numerical results of a graphite/epoxy system where the fillers assume spherical shape. The present theory provides tighter bounds than those of [13]. Figure 2 illustrates the variation of k_{11}^*/k_m with fiber volume fraction of an E-glass/epoxy system for the limiting cases of $1/d \rightarrow \infty$ and $1/d = 1$. The bounds of k_{11}^* converge to a single straight line for continuous fibers as indicated by equation (16).

ACKNOWLEDGMENT

This research is supported by the Department of Energy under contract number DE-AC02-79ER10511.

REFERENCES

1. Chou, T. W. and Kelly, A., "Fiber Composites," *Mat. Sci. Eng., 25,* 35 (1976).
2. Chou, T. W., Nomura, S., and Taya, M., "A Self-Consistent Approach to the Elastic Stiffness of Short-Fiber Composites," *J. Comp. Mat.* (in press).
3. Nomura, S. and Chou, T. W., "Bounds for Elastic Moduli of Short-Fiber Composites," submitted for publication.
4. Nomura, S. and Chou, T. W., "Fiber Orientation Effects on the Elastic Moduli of Short-Fiber Composites," *Fibre Sci. Tech.* (in press).
5. Beran, M., *Nuovo Cimento Ser. X, 38,* 771 (1965).
6. Beran, M. and Molyneux, J., *Quart. Appl. Math., 24,* 107 (1966).
7. Hori, M. and Yonezawa, F., "Statistical Theory of Effective Electrical, Thermal, and Magnetic Properties of Random Heterogeneous Materials IV," *J. Math. Phys., 16,* 352 (1975).
8. Kröner, E., "Elastic Moduli of Perflectly Disordered Composite Materials," *J. Mech. Phy. Solids, 15,* 319 (1967).

9. Budiansky, B., "On the Elastic Moduli of Some Heterogeneous Materials," *J. Mech. Phys. Solids, 13*, 223 (1965).
10. Hill, R., "A Self-Consistent Mechanics of Composite Materials," *J. Mech. Phys. Solids, 13*, 213 (1965).
11. Hashin, Z., "Assessment of the Self-Consistent Scheme Approximation: Conductivity of Particulate Composites," *J. Comp. Mat., 2*, 284 (1968).
12. Willis, J. R., "Bounds and Self-Consistent Estimates for the Overall Properties of Anisotropic Composites," *J. Mech. Phys. Solids, 25*, 185–202 (1977).
13. Hashin, Z. and Shtrikman, S., *J. Appl. Phys., 30*, 3125 (1962).
14. Brown, W. F., *Trans. Rheology Soc.*, Pt. 9, 357 (1965).
15. Beran, M. *Statistical Continuum Theories.* Interscience (1968).
16. Hori, M., "Statistical Theory of Effective Electrical, Thermal, and Magnetic Properties of Random Heterogeneous Materials II," *J. Math. Phys., 14*, 1942 (1973).
17. Dederichs, P. H. and Zeller, R., "Variational Treatment of the Elastic Constants of Disordered Materials," *Z. Physik, 259*, 103 (1973).
18. Zeller, R. and Dederichs, P. H., "Elastic Constants of Polycrystals," *Phys. Stat. Sol., b55*, 831 (1973).
19. Kröner, E., "Bounds for Effective Elastic Moduli of Disordered Materials," *J. Mech. Phys. Solids, 25*, 137 (1977).
20. Onsager, L., *Phys. Rev. 37*, 405–26 (1931).
21. Eshelby, J. D., "The Determination of the Elastic Field of an Ellipsoidal Inclusion and Related Problems," *Proc. Roy. Soc., A241*, 376 (1957).

23

Thermal Conductivities of a Cracked Solid

A. HOENIG

INTRODUCTION

THE EFFECT OF CRACKS ON THE MACROSCOPIC PHYSICAL PROPERTIES OF bodies had received a great deal of attention in recent years, with the preponderance of work having been done in connection with the gross elastic properties of cracked bodies; see, e.g., Watt *et al.* [1] and Cleary *et al.* [2] for relevant bibliographies. Since many physical processes are known to cause microcracking and since the presence of such cracks can cause large changes in the macroscopic properties, this attention is well focussed.

Lately, some workers such as Hoenig [3], Cleary [2] and Kachanov [4] have begun to realize that the results of elastic studies can be generalized in a straightforward way to yield results appropriate to other physical phenomena. In this paper, a close connection between electrical and thermal conductivity will be exploited to generate expressions for the effective thermal conductivities of a cracked body. The cracks are hypothesized to be all elliptical, with identical planform aspect ratios but with varying major axes. The cracks are distributed homogeneously and isotropically throughout the body, and the cracks are sufficiently small so that macroscopically the cracked body appears homogeneous and isotropic. The cracks may be dry or filled with a material of conductivity different from that of the matrix phase. The results will generalize those of Hasselman [5].

The analytical methods to be discussed can be generalized to the case where the cracks are distributed in some anisotropic (but still homogeneous) manner, and these results will be presented in a subsequent paper.

ANALYSIS

An analogy exists between σ, the electric conductivity, J and E, the current and electric field vectors respectively, and between k, the thermal conductivity, and q and grad T, the vector heat flow and temperature. Notice that in the governing equations describing the two phenomena, corresponding quantities in the following lists play identical roles:

σ	k
J	q
E	grad T
$J = \sigma E$	$q = k(\text{grad } T)$
div $J = 0$	div $q = 0$

In the same way, one can show that this analogous behavior persists in the specification of boundary conditions [3]. These observations imply that conclusions about electric behavior apply to thermal problems when the substitutions

$$J \rightarrow q$$
$$E \rightarrow \text{grad } T \tag{1}$$
$$\sigma \rightarrow k$$

are carried out. Hoenig [3] has presented a study of the effects of cracks on the electrical conductivities of bodies. These results can be applied to thermal problems by means of the above substitutions. This analysis assumes that heat transfer across dry cracks by radiation or convection will not take place.

In what follows, let the suffix "F" (standing for 'fluid,' a common crack "filler") refer to the quantities associated with the crack. The suffix "o" refers to the surrounding matrix phase. Let f be the volume fraction of the crack phase. Then, from the equation immediately following Equation (5.1) in [3],

$$\frac{1}{k} = \frac{1}{k_o} + f\left\{1 - \frac{k_F}{k}\right\}\frac{<\text{grad } T>}{q^\infty} \tag{2}$$

where k is the effective thermal conductivity to be calculated, and the rightmost bracketed factor is to be interpreted as being the value of the ratio (grad $T)/q^\infty$ averaged over all possible crack orientations. The heat flow applied to the body is q^∞. If the cracks are considered to be the limit of ellipsoids whose thicknesses approach zero, then this average value will approach infinity and the volume fraction f will approach zero in such a way so that their product remains finite.

The problem of determining k in (2) has been reduced by the above comments to the problem of evaluating the expression (grad $T)/q\infty$ for a single crack and then averaging this expression over all orientations of the crack. The process of evaluation must somehow take into account the influence of the other cracks of the body; this is done by invoking a *self-consistent hypothesis,* initially articulated by Budiansky [6] and Hill [7]. According to their hypothesis, each crack in the matrix 'sees' itself as being embedded in an uncracked body, characterized however by the as-yet-unknown effective conductivity (that is, the macroscopic conductivity) of the cracked body. In this manner, the difficult problem of evaluating the right-most factor of (2) in the context of a crack subject to the complex influence of neighboring cracks is replaced by the substantially simpler one of evaluating this factor by considering a single crack in a specially defined but homogeneous matrix. An implicit assumption in this procedure is that the crack concentration is sufficiently small so that the effect of crack intersections can be neglected.

This self-consistent method (SCM) has been employed in [3],and we may here appropriate these results by means of the thermal-electric analogy (see above). Before doing so, write f as

$$f = 8/3\alpha E(t)\varepsilon \tag{3}$$

where $E(t)$ is the complete elliptic integral of the second kind whose argument is $t = (1 - \gamma^2)^{1/2}$, and γ is the planform aspect ratio, α is the thickness aspect ratio, and ε is the so-called *crack density parameter,* given for elliptic cracks by

$$\varepsilon = \frac{2Na^3}{\pi}\gamma \tag{4}$$

For this analysis, assume that both aspect ratios remain constant throughout the sample, and that the aspect ratio α is very close to zero. With these conventions, we proceed to examine some results.

DRY CRACKS

From Equation (5.7) of [3], we immediately infer the result

$$k/k_o = 1 - 8/9\varepsilon \tag{5}$$

The effective conductivity varies linearly with crack density, and the critical value of ε at which k vanishes (were the SCM valid for such a pronounced degree of cracking) is 9/8. Clearly, for a sufficiently high degree of cracking, the flawed body will cease conducting heat (except by radiation across the faces of the cracks), but it is beyond the power of the SCM to predict this point. Hasselman [5] derives the formula

$$k/k_o = (1 + 8/9\varepsilon)^{-1} \tag{6}$$

which agrees with (5) for small values of crack density. The two formulas would agree had Hasselman used the SCM (or, alternatively, had Hasselman's dilute concentration approximation been used here). Note that the effect of crack planform enters only via the crack density factor. It seems therefore likely that these results will hold with little error for all cracks of convex shape, provided that a general definition of crack density in terms of crack area A and crack perimeter P be used. This observation was first offered by Budiansky and O'Connell [8] who gave such a definition as

$$\varepsilon = \frac{2N}{\pi} <A^2/P>. \tag{7}$$

SATURATED AND FILLED CRACKS

These results will apply to the case of cracks saturated by a fluid or other material of conductivity k_f. Applying the analogy to Equation (5.9) of [3], the resulting formula for effective conductivity involves the effects of crack ellipticity and saturation in complicated manners:

$$k_o/k = 1 - (8/9)E\varepsilon\left\{\frac{1}{\dfrac{k}{k_o}\Omega + \dfrac{\gamma^2}{t^2}(K-E)} + \frac{1}{\dfrac{k}{k_o}\Omega + \dfrac{1}{t^2}(E - \gamma^2 K)}\right\} \tag{8}$$

Here, the saturation parameter is defined as

$$\Omega = \frac{k_o}{\alpha k_f}$$

and K and E are the complete elliptic integrals of the first and second kinds of the argument t. Two special cases of Equation (8) suggest themselves: circular cracks ($\gamma = 1$), and long, flat, ribbon-like cracks ($\gamma \to 0$).

$$k_o/k = 1 - \frac{8\pi\varepsilon/9}{k\Omega/k_o + \pi/4} \tag{8a}$$
$$\text{(circular cracks)}$$

$$k_o/k = 1 - \frac{8}{9}\left[\frac{1}{k\Omega/k_o} + \frac{1}{k\Omega/k_o + 1}\right]\varepsilon \tag{8b}$$
$$\text{(ribbon cracks)}$$

The smaller is the parameter Ω, the more conductive are the cracks with respect to the matrix. $\Omega \to 0$ corresponds to the infinitely conductive inclusions, and $\Omega \to \infty$ corresponds to the case of dry cracks.

DISCUSSION

Graphs showing the dependence of effective conductivity upon crack density are displayed in Figures 1 and 2 for dry and filled cracks, and for circular

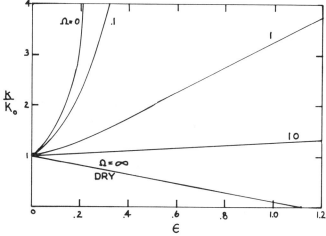

Figure 1. *Self-consistent effective thermal conductivities for dry and filled randomly distributed circular cracks.*

and ribbon-like cracks. As noted above, planform does not effect the relationship between conductivity and crack density for dry cracks. Otherwise, both planform and saturation parameter can have substantial effects, although for intermediate values of saturation Ω and for small values of ε (less than 0.5), the effect of crack planform may be less than 10 percent. The practical effect of this observation may be to mitigate concern about the exact crack shape for these ranges of Ω and ε.

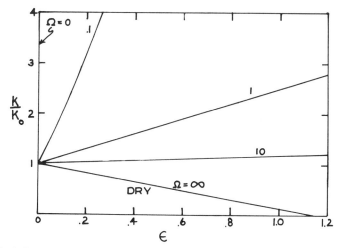

Figure 2. *Self-consistent effective conductivities for dry and filled, randomly distributed long, ribbon-like cracks.*

As Hasselman [5] notes, the dependence of conductivity upon crack density may be fortuitous, as crack density and size are probably experimentally determined far easier than are crack width and volume. There are at least two reasons contributing to this relative ease in measuring ε. Firstly, crack density may be accurately estimated by measuring changes in other physical properties (such as electric conductivity, elastic wave speeds, etc.) and relating these changes to levels of crack density. Secondly, crack density may also be obtained from measurements of the density of crack traces on a cross-section of a cracked specimen. These crack traces will be intersections of each crack with the planar cross-section, and so will be a set of randomly distributed straight lines of various lengths on the cross-section. The crack density can be related to the crack trace density per unit area and to the expected value of the square of the length of these traces. This method is discussed in detail by Budiansky and O'Connell [8].

The author gratefully acknowledges the support of the Mathematics Department of the John Jay College of Criminal Justice in the City University of New York.

NOMENCLATURE

σ electric conductivity
J electric current
E electric field
k_o thermal conductivity of uncracked body
q vector heat flow
T temperature
k effective thermal conductivity of cracked body
q^∞ heat flow applied to body
f volume fraction of crack phase
a,b,c major axes of ellipsoid; $a \geqslant b \geqslant c$
α thickness aspect ratio; $\alpha = c/a$
γ planform aspect ratio; $\gamma = b/a$
ε crack density parameter
N number of cracks/unit volume

REFERENCES

1. Watt, J. P., Davies, G. F. and O'Connell, R. J., "The Elastic Properties of Composite Materials," *Rev. Geophys. Space Phys.*, 14, (1976), pp. 541–63.
2. Cleary, M. P., Chen, I. W. and Lee, S. M., "Self-consistent techniques for heterogeneous media," *J. Eng. Mech. Div., ASCE*, 106, N5, (1980), pp. 861–87.
3. Hoenig, A., "Electric Conductivities of a Cracked Solid," *Pageoph.*, 117, (1978/79), pp. 690–710.
4. Kachanov, M., "Continuum Models of Media with Cracks," *J. Eng. Mechanics Div., ASCE*, 106, EM5, (1980), pp. 1039–51.

5. Hasselman, D. P. H., "Effect of Cracks on Thermal Conductivity," *J. Compos. Mater.,* 12, (1978), pp. 403–07.
6. Budiansky, B., "On the Elastic Moduli of Some Heterogeneous Materials," *J. Mech. Phys. Solids,* 13, (1965), pp. 223–27.
7. Hill, R., "A Self-Consistent Mechanics of Composite Materials," *J. Mech. Phys. Solids,* 13, (1965), 213–22 (1965).
8. Budiansky, B. and O'Connell, R. J., "Elastic Moduli of a Cracked Solids," *Int. J. Solids Structures,* 12, (1976), 81–97.

24

Moisture Diffusivity of Unidirectional Composites

K. KONDO AND T. TAKI

INTRODUCTION

ADVANCED FIBER-REINFORCED EPOXY COMPOSITE MATERIALS HAVE BEEN
gaining wide use in aerospace structure application due to their excellent
performance characteristics. However, it is well recognized that absorbed atmospheric moisture can have an undesirable effect on the mechanical properties of such materials [1]. In order to understand the response of the composite materials to moist environments, their moisture diffusivity must be investigated. This paper is concerned with the moisture diffusivity of unidirectional fiber-reinforced composite materials.

The moisture diffusivity of composite materials can be obtained by using
an analogy between the moisture diffusion and the thermal conduction or the
longitudinal shear loading. Shen and Springer [2] predicted the diffusivity of
unidirectional composites from the heat conductivity obtained by Springer
and Tsai [3]. Shirrel and Halpin [4] derived the moisture diffusivity from the
Halpin-Tsai equation [5] for the longitudinal shear modulus. The effective
moisture diffusivity can also be obtained by the numerical methods as used to
predict the longitudinal shear modulus by Chen and Cheng [6] and Adams
and Doner [7]. Although the fibers in actual unidirectional composites are
randomly distributed, most of the theoretical predictions of unidirectional
composite behavior have assumed a regular periodic array of fibers.

Hashin [8] obtained the upper and lower bounds of the longitudinal shear
modulus for actual unidirectional composites, and Nomura and Chou [9]
provided tighter bounds. Chou, Nomura, and Taya [10] predicted the
longitudinal shear modulus of actual unidirectional composites by the self-
consistent approach proposed by Hill [11]. Tsai [12] proposed a procedure to
predict the longitudinal shear modulus of actual composites by a linear com-

bination of the upper and lower bounds by Hashin [8]. The bounds were assigned a numerical value of $C = 1$ for the upper bound and $C = 0$ for the lower bound, the term C being called a contiguity factor. For a glass-fiber composite, $C = 0.2$ was found to be reasonable by comparison with experimental data and for a boron epoxy composite, $C = 0.05$.

In this paper, the transverse diffusivity of actual unidirectional composites is predicted in terms of the constituent diffusivities and the degree of fiber packing randomness. The theoretical results are compared with experimental data obtained by the authors and other researchers [2,4].

THEORY OF MOISTURE DIFFUSION

Relative moisture concentration in a material is defined as

$$r = C/C_\infty \qquad (1)$$

where C is the moisture concentration in the material and C_∞ is the equilibrium moisture concentration under given temperature and relative humidity. It is assumed that if different materials are in contact with each other, their relative moisture concentrations are the same at the interface. It should be noted that their moisture concentrations at the interface are not the same in general. Thus, the relative moisture concentration corresponds to the temperature in the thermal conduction problem.

A unidirectional fiber-reinforced composite material is referred to cartesian coordinates x, y, and z where x and y are in the transverse plane while z is in the fiber direction as shown in Figure 1. If the fibers and matrix are assumed to be transversely isotropic, then Fick's law gives

$$J_x = -D \frac{\partial C}{\partial x} = -DC_\infty \frac{\partial r}{\partial x}$$

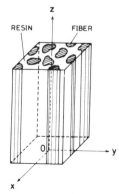

Figure 1. *Coordinate system for undirectional composite.*

$$J_y = -DC_\infty \frac{\partial r}{\partial y}$$

$$J_z = -D_z C_\infty \frac{\partial r}{\partial z} \qquad (2)$$

where J_x, J_y, and J_z are the x, y, and z components of moisture flux, and D and D_z are the transverse and longitudinal moisture diffusivities of the fibers or the matrix. Application of the law of mass conservation to equations (2) gives

$$D \left(\frac{\partial^2 r}{\partial x^2} + \frac{\partial^2 r}{\partial y^2} \right) + D_z \frac{\partial^2 r}{\partial z^2} = \frac{\partial r}{\partial t} \qquad (3)$$

If the fibers are all identical and arranged symmetrically with rotation about z axis through $\pi/3$ or $\pi/2$, or if the fibers are randomly dispersed, the composite is transversely isotropic. Then, the overall moisture diffusivities are defined by

$$\langle J_x \rangle = -D_t C_{c\infty} \langle \frac{\partial r}{\partial x} \rangle \qquad\qquad \langle J_y \rangle = -D_t C_{c\infty} \langle \frac{\partial r}{\partial y} \rangle$$

$$\langle J_z \rangle = -D_l C_{c\infty} \langle \frac{\partial r}{\partial z} \rangle \qquad (4)$$

where $\langle \quad \rangle$ denotes the volume average, and D_t and D_l are the effective transverse and longitudinal diffusivities, $C_{c\infty}$ being given by

$$C_{c\infty} = C_{f\infty} V_f + C_{m\infty}(1 - V_f) \qquad (5)$$

in which V_f is the volume fraction of the fibers, the subscripts f and m refer to the fibers and matrix respectively.

PREDICTION OF TRANSVERSE MOISTURE DIFFUSIVITY

The transverse moisture diffusivity of the unidirectional composite can be obtained by analyzing the steady-state two-dimensional diffusion problem in the x-y plane. The composite material is assumed to be subjected to the boundary relative moisture concentration

$$r(s) = y \qquad (6)$$

Then, the divergence theorem gives

$$< \frac{\partial r}{\partial y} > = 1 \tag{7}$$

and it follows from equations (2) that

$$<J_y> = -<DC_\infty \frac{\partial r}{\partial y} > \tag{8}$$

Substitution of equations (7) and (8) into equation (4) yields

$$D_t = <DC_\infty \frac{\partial r}{\partial y} >/C_{c\infty} \tag{9}$$

From equation (3), the steady-state relative moisture concentration is described by

$$\frac{\partial^2 r_f}{\partial x^2} + \frac{\partial^2 r_f}{\partial y^2} = 0 \qquad , \qquad \frac{\partial^2 r_m}{\partial x^2} + \frac{\partial^2 r_m}{\partial y^2} = 0 \tag{10}$$

At the interface between the fiber and the matrix, there are imposed the continuity conditions

$$r_f = r_m \qquad , \qquad D_f C_{f\infty} \frac{\partial r_f}{\partial n} = D_m C_{m\infty} \frac{\partial r_m}{\partial n} \tag{11}$$

when n is a coordinate normal to the interface. Obtaining the relative moisture concentration field r_f and r_m governed by equations (6),(10), and (11), the effective transverse diffusivity D_t can be determined from equation (9).

Firstly, the moisture diffusivity is predicted for a regular periodic array of fibers. It is assumed that the fibers are of identical circular cross section and form a hexagonal or square array in the transverse plane as shown in Figure 2. Because of the symmetry of array, it is sufficient to consider only a typical repeating unit of the composite as represented in Figure 3. Assuming that the fibers do not absorb the moisture, the diffusion problem is solved by the finite element method utilizing the eight-node isoparametric quadrilateral elements. The prescribed boundary relative moisture concentrations are

$$r = 0 \text{ at } y = 0 \qquad , \qquad r = 1 \text{ at } y = 1 \tag{12}$$

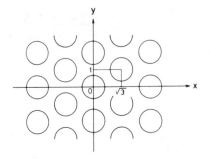

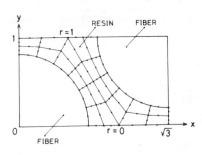

Figure 2. Transverse plane of hexagonal array composite.

Figure 3. Finite element idealization of typical repeating unit of hexagonal array composite.

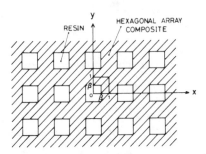

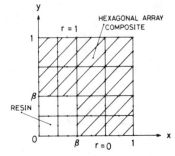

Figure 4. Transverse plane of random model.

Figure 5. Finite element idealization of typical repeating unit of random model.

Next, the diffusivity for a random array of fibers is predicted by establishing a random model which has a square array of cylinders of matrix embedded in the hexagonal packing composite as shown in Figure 4. The ratio of the side length of the square cylinder to the distance between adjacent square centers is β. Therefore the volume fraction of the pure matrix cylinders is β^2. Because of the geometrical symmetry, it is sufficient to consider a repeating unit as depicted in Figure 5. The relative moisture concentration field is obtained by the finite element method utilizing the eight-node isoparametric quadrilateral elements with the effective diffusivity of the hexagonal array composite determined beforehand.

EXPERIMENTS ON MOISTURE DIFFUSIVITY

Moisture absorption tests at environmental constant relative humidity and temperature have been conducted for neat epoxy resins and the corresponding graphite-fiber reinforced composites in order to determine their moisture diffusivities. The specimens with nominal thickness of 1 mm have been manufactured by the suppliers as indicated in Table 1. The coupons for the

Table 1. Materials of specimens for absorption test.

Supplier	Epoxy Resin	Graphite Fiber-Reinforced Composite (V_f: Fiber Volume Fraction)	
Toray	#3130	Torayca T300/#3130	($V_f = 0.60$)
Mitsubishi Rayon	410	Pyrofil A410-150	($V_f = 0.56$)

absorption tests are nominally 40 mm long by 20 mm wide. Initial dry state is achieved by storing the specimens in a desiccator chamber for several months. The environmental constant relative humidity is created by aqueous sulfuric-acid solution placed at the bottom of an airtight container of the specimens following the ASTM standard [13], the container being placed in an oven at constant temperature. Each specimen is weighed periodically and the percent-age weight gain is plotted versus the square root of time. The diffusivity can be determined from the initial slope of the obtained curve following the method proposed by Shen and Springer [2].

RESULTS AND DISCUSSIONS

The theoretical prediction of the transverse diffusivity D_t for the hexagonal or square array composites is shown in Figure 6 as a function of the fiber volume fraction V_f. It can be seen that the diffusivity for the hexagonal array is higher than that for the square array. The difference between the two results is larger at higher values of V_f while it is small at the lower values of V_f.

The effective transverse moisture diffusivity for the random array com-

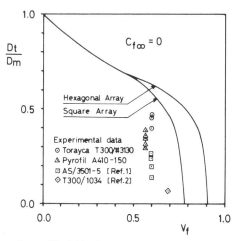

Figure 6. *Transverse moisture diffusivities for regular periodic array composites and experimental data.*

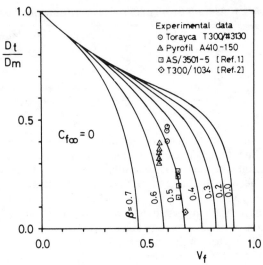

Figure 7. Transverse moisture diffusivities for random array composites and experimental data.

posites is plotted in Figure 7 as a function of the fiber volume fraction for several values of β. It is seen that if β increases, that is, if the degree of the randomness increases, the transverse diffusivity decreases remarkably. The influence of β on the diffusivity is small at lower values of V_f while it is large at higher values of V_f. However, much higher values of V_f can be attained only by composites with low degree of randomness.

The obtained experimental results of the diffusivities of the resins and composites are represented in Table 2 for different hygrothermal environments. Each effective transverse diffusivity divided by that of the corresponding neat

Table 2. Diffusivities of epoxy resins and graphite fiber-reinforced composites obtained by absorption tests (mm²s⁻¹).

Material	Environment; Temperature, °C/Relative Humidity, %				
	30°C		75°C		
	65%	90%	40%	65%	90%
#3130	6.99×10^{-8}	4.64×10^{-8}	7.01×10^{-7}	7.86×10^{-7}	8.88×10^{-7}
Torayca T300/#3130	—	—	3.12×10^{-7}	3.12×10^{-7}	4.12×10^{-7}
410	1.23×10^{-7}	1.29×10^{-7}	1.50×10^{-6}	1.63×10^{-6}	1.58×10^{-6}
Pyrofil A410-150	2.88×10^{-8}	4.64×10^{-8}	4.41×10^{-7}	6.30×10^{-7}	5.41×10^{-7}

resin D_t/D_m, is shown in Figures 6 and 7 along with the experimental data obtained by DeIasi and Whiteside [1] and Shen and Springer [2] for graphite composites. While the theoretical predictions for the hexagonal and square array are much higher than the experimental data, the prediction based on the random model with the value $\beta \cong 0.5$ provides good agreement with the experimental results. The scattering of the test data may be attributed to the fiber packing randomness of the actual unidirectional composites.

In the analogy between the moisture diffusion and the transverse thermal conduction or the longitudinal shear loading, the relative moisture concentration r and the moisture diffusivity multiplied by the equilibrium moisture concentration $D_t C_\infty$ correspond to the temperature T and the transverse heat conductivity K_t, respectively, or they correspond to the axial displacement w and the longitudinal shear modulus G, respectively. However, based on an erroneous assumption that the moisture concentration C and the diffusivity D_t correspond to T and K_t, or w and G, the transverse diffusivities were predicted by Shen and Springer [2] and Shirrell and Halpin [4]. Therefore their predictions for the composites with fibers which do not absorb the moisture $(C_{f\infty} = 0)$ should be divided by $(1-V_f)$. Similarly, the longitudinal diffusivity for the composites with $C_{f\infty} = 0$ should be

$$D_l = D_{mz} \tag{13}$$

instead of $D_l = (1-V_f)D_{mz}$ as suggested in References [2,4].

Using the correct analogy, the transverse diffusivities are shown in Figure 8 based on the results obtained by Springer and Tsai [3] for the heat conductivity and by Halpin and Tsai [5] for the longitudinal shear modulus. It can be seen that their predictions do not agree with the experimental data. The upper and

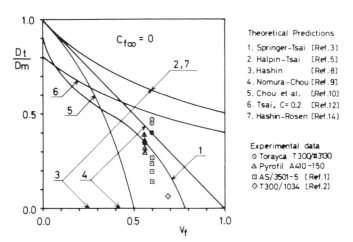

Figure 8. *Predictions of transverse diffusivity based on analogy and experimental data.*

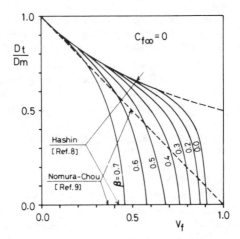

Figure 9. Comparison of present results with upper and lower bounds of transverse diffusivity for random array composites.

lower bounds of diffusivity for the actual composites obtained by Hashin [8] and the bounds improved by Nomura and Chou [9] are also shown in Figure 8 along with the prediction of Chou, Nomura, and Taya [10] by the self-consistent method [11] and Tsai's prediction [12] with the contiguity factor $C = 0.2$. The result of Hashin and Rosen [14] for completely random array composites coincides with the Halpin-Tsai equation [5] and also with Hashin's upper bound [8]. While all the experimental data fall between Hashin's bounds [8], some data fall outside Nomura and Chou's bounds [9]. And it is difficult to find a numerical value of the contiguity factor C which gives a good agreement with the experimental data.

Comparison of the present results with the upper and lower bounds for actual composites is shown in Figure 9. It should be noted that all the present results are inside Hashin's bounds [8], but are not inside Nomura and Chou's bounds [9].

Adams and Tsai [15] investigated the influence of random fiber packing on the transverse stiffness of the unidirectional composites by using realistic random models. However the models are not applicable directly to the moisture diffusivity because fibers with $D_f C_{f\infty} = 0$ correspond to those with $G = 0$, that is, the holes for the longitudinal shear loading.

CONCLUSIONS

Formulating the moisture diffusion problem in terms of the relative moisture concentration, the transverse diffusivity of the unidirectional fiber-reinforced composite materials has been predicted for the regular periodic array of fibers and the random array of fibers. The theoretical prediction for the regular array packing of fibers has been found to be much higher than the experimental data obtained by the authors and other researches. And the prediction based on the random model with the degree of randomness $\beta = 0.5$ has provided good agreement with the experimental data. The analogy between the moisture diffusion and the transverse thermal conduction or the longitudinal shear loading has been discussed, and the comparison of the various predictions based on the analogy available in the literature has been made.

NOMENCLATURE

C	moisture concentration (g mm^{-3})
C_∞	equilibrium moisture concentration (g mm^{-3})
D, D_z	transverse and longitudinal moisture diffusivities of material (mm^{2}s^{-1})
D_t, D_l	effective transverse and longitudinal diffusivities of composite (mm^{2}s^{-1})
J_x, J_y, J_z	x, y, and z components of moisture flux (g mm^{-2}s^{-1})
r	relative moisture concentration (dimensionless) = C/C_∞
s	boundary of composite
t	time (s)
V_f	volume fraction of fibers in composite
x, y, z	coordinates as shown in Figure 1
β	randomness of fiber packing defined in Figure 4
$<\ >$	volume average

Subscript

c	pertaining to composite
f	pertaining to fiber
m	pertaining to matrix

REFERENCES

1. Delasi, R. and Whiteside, J. B., "Effect of Moisture on Epoxy Resins and Composites," in *Advanced Composite Materials—Environmental Effects*, J. R. Vinson, ed. ASTM STP 658, American Society for Testing and Materials, 2-20 (1978).
2. Shen, C. H. and Springer, G. S., "Moisture Absorption and Desorption of Composite Materials," *J. Composite Materials, 10*, 2-20 (1976).
3. Springer, G. S. and Tsai, S. W., "Thermal Conductivities of Unidirectional Composites," *J. Composite Materials, 1*, 166-173 (1967).
4. Shirrell, C. D. and Halpin, J., "Moisture Absorption and Desorption in Epoxy Composite Laminates," Composite Materials: Testing and Design (Fourth Conference), ASTM STP 617, American Society for Testing and Materials, 514-528 (1977).
5. Halpin, J. and Tsai, S. W., "Effects of Environmental Factors of Composite Materials," AFML-TR 67-423 (1969).
6. Chen, C. H. and Cheng, S., "Mechanical Properties of Fiber Reinforced Composites," *J. Composite Materials, 1*, 30-41 (1967).
7. Adams, D. F. and Doner, D. R., "Longitudinal Shear Loading of a Unidirectional Composite," *J. Composite Materials, 1*, 4-17 (1967).
8. Hashin, Z., "Theory of Fiber Reinforced Materials," NASA CR-1974 (1972).

9. Nomura, S. and Chou, T. W., "Bounds of Effective Thermal Conductivity of Short-Fiber Composites," *J. Composite Materials, 14,* 120–129 (1980).
10. Chou, T. W., Nomura, S., and Taya, M., "A Self-Consistent Approach to the Elastic Stiffness of Short-Fiber Composites," *J. Composite Materials, 14,* 178–188 (1980).
11. Hill, R., "A Self-Consistent Mechanics of Composite Materials," *J. Mech. Phys. Solids, 13,* 213–222 (1965).
12. Tsai, S. W., "Structural Behavior of Composite Materials," NASA CR-71 (1971).
13. "Standard Recommended Practice for Maintaining Constant Relative Humidity by Means of Aqueous Solutions," ASTM Designation: E104-51 (1971).
14. Hashin, Z. and Rosen, B. W., "The Effective Moduli of Fiber-Reinforced Materials," *J. Appl. Mech., 31,* 223–232 (1964).
15. Adams, D. F. and Tsai, S. W., "The Influence of Random Filament Packing on the Transverse Stiffness of Unidirectional Composites," *J. Composite Materials, 3,* 368–381 (1969).
16. Kondo, K. and Taki, T., "Transverse Moisture Diffusivity of Unidirectionally Fiber-Reinforced Composite," Proc. Japan-U.S. Conference on Composite Materials: Mechanics, Mechanical Properties and Fabrication, Tokyo, 308–317 (1981).

MATHEMATICAL FORMULATION

(1) Derivation of equation (7) from equation (6) by using the divergence theorem is as follows:

$$< \frac{\partial r}{\partial y} > = \frac{1}{V} \int_v \frac{\partial r}{\partial y} \, dV = \frac{1}{V} \int_s \gamma n_y ds$$

$$= \frac{1}{V} \int_s y n_y ds = \frac{1}{V} \int_v \frac{\partial y}{\partial y} \, dv$$

$$= \frac{1}{V} \int_v dV = 1$$

(2) Derivation of equation (13) based on the analogy between the longitudinal thermal conduction and the longitudinal moisture diffusion is as follows: The longitudinal thermal conductivity was given by Springer and Tsai [3] as

$$K_l = K_{fz} V_f + K_{mz}(1 - V_f)$$

Therefore using the correct similarity, we have

$$D_l C_{c\infty} = D_{fz} C_{f\infty} V_f + D_{mz} C_{m\infty}(1 - V_f)$$

Introducing equation (5) into this equation, we obtain

$$D_l = \frac{D_{fz}C_{f\infty}V_f + D_{mz}C_{m\infty}(1-V_f)}{C_{f\infty}V_f + C_{m\infty}(1-V_f)}$$

which, setting $C_{f\infty} = 0$, gives equation (13). It should be noted that the result of Shen and Springer [2] is written as

$$D_l = D_{fz}V_f + D_{mz}(1-V_f)$$

which, setting $D_{fz} = 0$, gives

$$D_l = D_{mz}(1-V_f)$$

25

Moisture and Thermal Expansion Properties of Unidirectional Composite Materials and the Epoxy Matrix

D. S. CAIRNS AND D. F. ADAMS

INTRODUCTION

STRUCTURAL COMPOSITE MATERIALS TYPICALLY CONSIST OF A PRIMARY load-carrying material phase, such as fibers, held together by a binder of matrix material, often an organic polymer. The fibers are conveniently oriented unidirectionally in layers (plies), a complex structural laminate being built up of individual plies oriented as required to provide the required stiffness and strength in any given direction. That is, the unidirectional ply is the basic element of construction.

Since polymer matrix materials typically exhibit thermal expansion coefficients which are much higher than those of fibers, and the fibers may be thermally as well as mechanically anisotropic, complex stress states are induced in the composite due to temperature changes. Since high performance composite materials are normally cured at elevated temperatures, pre-existing thermal stresses are assured at essentially any use temperature. In addition, polymers are hygroscopic, and have high moisture expansion coefficients. Most fibers do not absorb moisture. Thus, in the presence of moisture, additional expansion mismatch stresses are induced.

The longitudinal and transverse coefficients of thermal expansion and moisture expansion of the orthotropic unidirectional ply must be known for design purposes. These composite properties can be experimentally measured, or they can be predicted, based upon the constituent material properties. If the latter approach is taken, then the fiber and matrix properties must be measured.

Methods are presented here for measuring the properties of both the matrix and the resulting unidirectional composite, and for using the matrix properties in a micromechanics analysis to predict the unidirectional composite

300

properties. Analytical predictions are then correlated with experimental data. Full details are included in Reference [1].

EXPERIMENTAL PROCEDURES

Since little has been done to date to determine the moisture expansion coefficients of unidirectional composite materials, the main thrust of the present study [1] was to characterize a typical matrix material and two composites with very different fiber materials, but with this same matrix. Since the Hercules AS/3501-6 graphite/epoxy composite system is widely used [2], as is the Hercules 3501-6 epoxy resin system, these materials were chosen for the study. Owens-Corning S2 glass fiber [3] in the same 3501-6 epoxy matrix was chosen as the third material system. These three materials were also used for the thermal expansion tests.

Moisture Expansion Tests

Three Blue M Stabiltherm Model OV-U60A Gravity Ovens were used to provide heat for the accelerated moisture conditioning. Each oven contains a semi-vaporproof insert made of Plexiglas to provide a chamber for the moisture conditioning. The chambers are not totally vaporproof since holes for instrumentation allow some water vapor to escape. Housed inside the chambers are the quartz glass dilatation-measuring assemblies and the specimens used for determining weight gain and diffusivity constants. On the outside of the ovens are Daytronic Model DS200 Linear Variable Differential Transformers (LVDTs) with calibration assemblies, used to monitor the dilatation of the specimens via a quartz glass pushrod. A Mettler Model HL 32 analytical balance was used for the constant monitoring of the specimen weights.

A unique data acquisition system for the moisture experiments was developed by the Composite Materials Research Group at the University of Wyoming. It is based on a Zilog Z-80 microprocessor which, upon starting a test, requires no additional operator assistance. Specimen names are input to a floppy disc. Weight, dilatation, and time are recorded. The system provides several advantages over other moisture dilatation systems. For example, it constantly monitors times, displacements, and weights. It records power failures and is capable of restoring the balance after a power failure. This is particularly useful since minimal specific history of a test needs to be known.

Thermal Expansion Tests

The thermal expansion data acquisition was much less automated than the moisture system. The specimen was placed in a quartz tube and a quartz pushrod was mounted against it. The other end of the pushrod was in contact with an LVDT interfaced to a Daytronic Model 201C amplifier. Temperature was measured by an iron-constantan thermocouple. Nonlinearity of the thermocouple was one percent or less in the present tests. An electric cold-compensation junction was used to provide a stable reference point for the thermocouple. Displacement and temperature values were recorded on an X-Y plotter. These data were reduced manually and best-fit curves were generated numerically.

EXPERIMENTAL DATA

Moisture Expansion

The following data represent typical results of the experimental portion of the present study. Full details are included in Reference [1]. Figures 1 and 2 represent strain (transverse strain in the case of the unidirectional composites) versus moisture absorption plots for the three types of specimens. The highly nonlinear moisture expansion curves for the S2 glass/3501-6 specimens will be noted. The 3501-6 epoxy resin curves appear to be the most linear. The AS/3501-6 and S2/3501-6 composite specimen data were fit to cubic equations of the form:

$$\varepsilon = a_o + a_1 M + a_2 M^2 + a_3 M^3 \tag{1}$$

where

ε = linear strain
M = moisture absorption (weight percent)
a_o, a_1, a_2, a_3 = curve-fit coefficients based on experimental data

This equation was used since it most closely modeled the behavior of the moisture expansion of the composites. The epoxy resin showed only slight nonlinearity during the initial moisture expansion, so a linear moisture expansion behavior was assumed. The coefficients vary as shown in Table 1.

Long periods of time are required for moisture expansion testing, the data shown in Figures 1 and 2 representing 30 days of exposure. Moisture diffu-

Table 1. Transverse moisture expansion curve-fit coefficients.[a]

	Coefficients			
Material	a_o (10⁻⁵)	a_1 (10⁻³)	a_2 (10⁻⁵)	a_3 (10⁻³)
3501-6 Neat Epoxy	0	3.2	0	0
AS/3501-6 Graphite/Epoxy	10.42900	3.65039	1.30224	−0.37836
S2/3501-6 Glass/Epoxy	8.05688	6.02900	3.90029	5.42363

[a]For use in Equation (1).

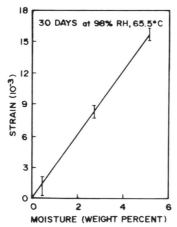

Figure 1. *Hercules 3501-6 neat epoxy moisture expansion data with curve-fit and experimental scatter indicated.*

sion for the systems under consideration appears to follow Fick's Second Law for anisotropic media, i.e.

$$\frac{\partial c}{\partial t} = \frac{\partial}{\partial x_i}\left[D_{ij}\,\frac{\partial c}{\partial x_j}\right]$$ (2)

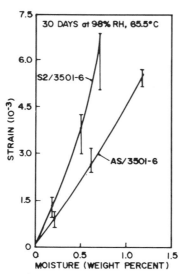

Figure 2. *Unidirectional composite transverse moisture expansion data with curve-fits and experimental scatter indicated.*

where

c = concentration of moisture
D_{ij} = diffusivity tensor
t = time
x_i, x_j = orthogonal coordinates
i, j = 1,2,3

In-depth studies of moisture diffusion may be found in References [4–7].

Thermal Expansion

From experiment to experiment, the transverse thermal expansion data showed far less scatter than the moisture expansion data. Representative examples are shown in Figures 3 and 4. The curve-fits are of the following form:

$$\varepsilon = b_o + b_1\, T + b_2\, T^2 \tag{3}$$

where

ε = linear strain
T = temperature (°C)
b_o, b_1, b_2 = curve-fit coefficients based on experimental data

This response has been noted previously [8,9]. That is, a parabolic thermal expansion curve is seen for the 3501-6 epoxy resin, and also for composites incorporating this epoxy. Values of the curve-fit parameters may be found in Table 2.

It will be noted that the 3501-6 epoxy shows the highest thermal expansion,

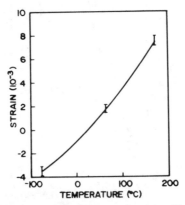

Figure 3. *Hercules 3501-6 neat epoxy thermal expansion data with curve-fit and experimental scatter indicated (no moisture conditioning).*

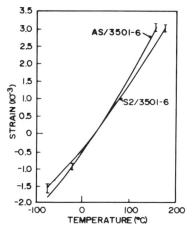

Figure 4. *Unidirectional composite transverse thermal expansion data with curve-fits and experimental scatter indicated (no moisture conditioning).*

followed by the AS/3501 and S2/3501-6, respectively. This is as expected since the transverse modulus of an AS graphite fiber is lower and the transverse thermal expansion is higher than for an S2 glass fiber. The fiber volume of the S2/3501-6 composite was also slightly higher than that of the AS/3501-6 composite, further reducing the thermal expansion of the glass/epoxy composite.

ANALYTICAL/EXPERIMENTAL CORRELATIONS

The Composite Materials Research Group at the University of Wyoming has devoted much time and effort to the development of micromechanics analyses of composite materials [10–13]. Micromechanics refers to the analysis of an individual composite lamina on a local scale, in an effort to

Table 2. *Transverse thermal expansion curve-fit coefficients.[a]*

Material	Average Moisture Content (Wt. %)	Coefficients		
		b_0 (10^{-4})	b_1 (10^{-5})	b_2 (10^{-8})
3501-6 Neat Epoxy	0	−9.72023	3.83445	6.12029
S2/3501-6	0	−3.52604	1.46491	1.50351
AS/3501-6	0	−3.08472	2.00603	2.50517
3501-6 Neat Epoxy	6.23	−11.25670	4.48252	5.98654
S2/3501-6	1.16	3.33400	1.38309	3.97577
AS/3501-6	1.87	−6.74528	2.04744	4.54569

[a]For use in Equation (3), based on a nominal zero value at 23°C.

predict stresses, strains, lamina stiffness properties, and failure under mechanical, moisture, and thermal loadings.

The micromechanics analysis employs a finite element scheme, incorporating a generalized plane strain assumption to solve the displacement boundary value problem, defined by a regular square packing array of fibers. It evaluates stresses, strains, and material properties after each incremental load is applied, permitting nonlinear moisture and thermal response of the constituent materials. Complete details of the analysis and related computer program are presented in References [10–13].

In general, constituent material experimental data can be input directly to the analysis. The mechanical response of the matrix material has been characterized at the University of Wyoming [8,14]. Fiber properties are available in References [2,3]. These properties are listed in Table 3. Transverse fiber properties are best estimates. The stress-strain response of the matrix is evaluated at each increment using the Richard-Blacklock relation [15].

The micromechanics analysis was first run to simulate the composite curing cycle cooldown to room temperature. This cooldown induces stresses in the composite material due to the mismatch of thermal expansion coefficients of the polymer matrix and fiber. For the dry thermal expansion tests, the composite material was cycled down to −73 °C, then up to the cure temperature of the matrix material (175 °C). For the moisture correlations, the composite was first cooled to room temperature. Then, it was cycled up to the moisture-conditioning temperature of 66 °C. Moisture was added in 0.5% increments to the matrix. This provided the moisture expansion correlations.

Moisture Expansion

There is some discrepancy in the measured properties for the matrix material [12]. The matrix is assumed to be isotropic. Unfortunately, data generated from shear tests suggest a different stress-strain response than data generated from tensile tests. The micromechanics analysis was run using both sets of data, with the results for the AS/3501-6 being presented in Figure 5. While the differences in matrix properties may be significant in predicting other composite material properties, little difference was noted in the prediction of moisture expansion.

It was also desired to study the effect of curing stresses on the moisture expansion data. Therefore, curves for which curing stresses were neglected are presented along with the tensile and shear data curves. Tensile data were used for this analysis. Interestingly, all three micromechanics-predicted moisture expansion curves vary by only a few percent from each other. Figure 6 shows the micromechanics predictions of moisture expansion for the S2/3501-6 composite. Correlation of experimental data with predicted values is not nearly as close as was shown for the AS/3501-6 composite (Figure 5). However, the differences between predictions obtained neglecting the curing cycle (using tensile data) and including the curing cycle are again small. The reason for the predicted values being lower may be due to a degradation of

Table 3. Constituent material properties for AS Graphite Fiber, S2 Glass Fiber, and 3501-6 epoxy resin.

				Hercules 3501-6 Epoxy Matrix (Room Temperature, Dry) [11]	
Property		Hercules AS Graphite Fiber [1]	Owens-Corning S2 Glass Fiber [2]	(From Tensile Test Data)	(From Shear Test Data)
Longitudinal Modulus, E_L	GPa (Msi)	220 (32)	86.2 (12.5)	4.27 (0.62)	5.79 (0.84)
Transverse Modulus, E_T	GPa (Msi)	14 (2.0)	86.2 (12.5)	4.27 (0.62)	5.79 (0.84)
Longitudinal Shear Modulus, G_{LT}	GPa (Msi)	34.5[a] (5.0)	35.3 (5.1)	1.59 (0.23)	2.14 (0.31)
Transverse Shear Modulus, G_{TT}	GPa (Msi)	5.5 (0.8)	35.3 (5.1)	1.59 (0.23)	2.14 (0.31)
Major Poisson's Ratio, ν_{LT}		0.20	0.22	0.34	0.34[a]
In-Plane Poisson's Ratio, ν_{TT}		0.25	0.22	0.34	0.34[a]
Longitudinal Tensile Strength σ_L^u	MPa (ksi)	3100 (450)	4825 (700)	82.7 (12.0)	170 (24.6)
Transverse Tensile Strength, σ_L^u	MPa (ksi)	345[a] (50)	4825 (700)	82.7 (12.0)	170 (24.6)
Longitudinal Shear Strength, τ_{LT}^u	MPa (ksi)	1550 (225)	2410[a] (350)	41.4 (6.0)	84.8 (12.3)
Transverse Shear Strength, τ_{TT}^u	MPa (ksi)	172 (25)	2410[a] (350)	41.4 (6.0)	84.8 (12.3)
Longitudinal Coefficient of Thermal Expansion, α_L ($10^{-6}/\,^{\circ}C$)		−0.36	5.0	40.9	40.9
Transverse Coefficient of Thermal Expansion, α_T ($10^{-6}/\,^{\circ}C$)		9.0[a]	5.0	40.9	40.9
Coefficient of Moisture Expansion, β ($10^{-3}/\%\,M$)		0	0	3.2	3.2

[a]Estimated.

307

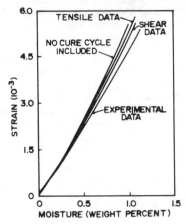

Figure 5. Micromechanics predictions of transverse moisture expansion in a unidirectional AS/3501-6 graphite/epoxy composite.

the fiber-matrix interface bond during moisture absorption [16,17]. The micromechanics analysis did not take this into account; a perfect fiber-matrix bond was assumed.

The matrix material appeared to have an essentially linear coefficient of moisture expansion, while the S2/3501-6 and AS/3501-6 were seen to be nonlinear. Table 4 presents predictions using the micromechanics analysis. The values shown are incremental strain values. Note that there is a slight decrease in incremental strain for each increment of moisture until the foot-

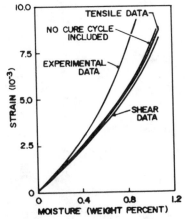

Figure 6. Micromechanics predictions of transverse moisture expansion in a unidirectional S2/3501-6 glass/epoxy composite.

***Table 4.** Micromechanics predictions of moisture-induced transverse strain at 66°C.*

Moisture in Matrix (%)	Moisture in Composite (%)	Incremental Transverse Strain (10^{-4})
S2/3501-6 Glass/Epoxy Composite		
0.5	0.107	5.647
1.0	0.214	5.646
1.5	0.321	5.644
2.0	0.429	5.642
2.5	0.536	5.639
3.0	0.643	5.635
3.5	0.750	5.632[a]
4.0	0.857	6.587[a]
4.5	0.964	6.693
5.0	1.072	6.923
5.5	1.179	7.125
6.0	1.286	7.268
AS/3501-6 Graphite/Epoxy Composite		
0.5	0.1401	7.502
1.0	0.2802	7.497
1.5	0.4202	7.490
2.0	0.5605	7.480
2.5	0.7006	7.467
3.0	0.8407	7.452[b]
3.5	0.9808	7.541[b]
4.0	1.1210	7.651
4.5	1.2611	7.806
5.0	1.4012	7.954
5.5	1.5413	8.107
6.0	1.6814	8.219

[a]Large incremental strain increase.
[b]Small incremental strain increase.

noted level is reached. Obviously, in the S2 glass/epoxy, something happens in the increment between 3.5 and 4.0 percent moisture by weight in the matrix material. At this point, the incremental strains take a jump upward and continue to increase. Crossman, et. al [18] also noted these nonlinearities. There is particular interest in determining coefficients of moisture expansion of graphite/epoxy composites since these materials are used extensively in spacecraft structures, where very small dimensional changes due to moisture egress are an important consideration. The nonlinearity is much more subtle than for the S2 glass/epoxy composites, as can be seen by the comparisons in Table 4, due to less mismatch of properties of the fiber and matrix in the transverse direction (see Table 3).

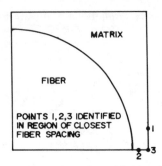

Figure 7. Quadrant of typical repeating unit modeled by micromechanics analysis.

During the moisture absorption process, the stress state in the matrix becomes very complicated. It is nonhomogeneous, triaxial, and nonlinear with moisture absorption. The stress state is further complicated with increasing moisture since an increase in moisture changes the matrix constitutive properties in a nonlinear fashion also. The region of closest fiber spacing appears to have the greatest influence on composite strain. Any strain of the matrix in this region has a greater effect due to being bounded by fibers which are much stiffer (especially the S2 glass) in the transverse direction. Figure 7 identifies three specific points in this region of closest fiber spacing.

Figure 8 is a plot of the in-plane minimum principal stresses present in the matrix material at Points 1, 2 and 3 versus weight percentage moisture in the matrix for the S2/3501-6 glass/epoxy composite during moisture absorption. For Point 1, up to 1.5 percent moisture in the matrix material, added

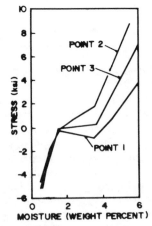

Figure 8. Minimum principal stresses in matrix of S2/3501-6 glass/epoxy unidirectional composite at points indicated in Figure 7, as a function of matrix moisture content.

moisture tends to decrease the thermally-induced compressive stresses present from cooldown after the curing process. But additional moisture tends to put the matrix material into compression again, indicating a limitation of expansion transverse to the fiber in this region. At the 3.5 percent moisture content, added moisture changes the slope of the minimum principal stress plot from negative to positive, indicating an increase in expansion due to moisture loading. In fact, the matrix material quickly goes into tension. That is, after a value of 3.5 percent moisture in the matrix, added moisture has a net positive effect because the added moisture tends to relieve any compressive forces, rather than contribute to them. Again, the plot for Point 2 shows the same general effect, i.e., the large tensile stress increase indicates a proportionally larger increase in composite strain. Points 1 and 3 would be expected to be somewhat similar since both lie on the midline of symmetry between adjacent fibers, while Point 2 does not.

Therefore, it can be seen that an inordinate increase in minimum principal stresses in the region of closest fiber packing causes a correspondingly large increase in composite transverse strain. It should be noted that this is a local effect. The bulk of the matrix material is in tension after cooldown from the curing temperature. After 1.5 percent moisture absorption of the matrix, the bulk of the matrix material then goes into compression.

The graphite/epoxy composite does not exhibit any of the above phenomena. The matrix material is in a much more uniform stress state. No inordinate changes in stress are seen with increasing moisture, which probably accounts for the far more linear transverse moisture expansion prediction compared to the S2 glass/epoxy prediction (Figure 5 versus Figure 6).

The physical behavior of S2 glass/epoxy discussed above does not provide a complete explanation of the nonlinear transverse moisture expansion response observed. Much more analysis needs to be done in this area. It is significant, however, that both theory and experiment show the nonlinear moisture expansion behavior.

Thermal Expansion

In general, the predicted composite response is nonlinear over a wide temperature range, following the same physical behavior as the 3501-6 epoxy resin. Figure 9 compares experimental data for thermal expansion of the unconditioned (dry) AS/3501-6 composite in the transverse direction. As with moisture expansion, thermal expansion is affected little by the noted differences between shear and tensile matrix data. Figure 10 presents the micromechanics predictions of unconditioned S2/3501-6 composite transverse thermal strain. Results using both tensile and shear data for the matrix are shown. It will be noted that these predicted results correlate better with the experimental data than did the AS/3501-6 transverse thermal expansion results (Figure 9).

It is interesting that the micromechanics analysis over-predicts the transverse thermal expansion of both the AS/3501-6 and the S2/3501-6 com-

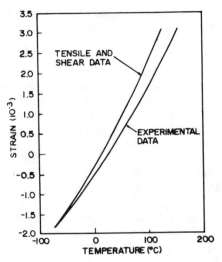

Figure 9. *Micromechanics predictions of transverse thermal expansion in a unidirectional AS/3501-6 graphite/epoxy composite (no moisture conditioning).*

posites. The transverse coefficient of thermal expansion of an individual graphite fiber is a best estimate since the fibers themselves are anisotropic [2]. A lower value does not change the coefficient of moisture expansion significantly, but it has a substantial influence on the coefficient of thermal expansion. It is, therefore, felt that more confidence may be placed in the ther-

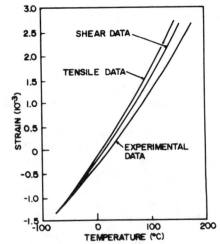

Figure 10. *Micromechanics predictions of transverse thermal expansion in a unidirectional S2/3501-6 glass/epoxy composite (no moisture conditioning).*

mal expansion predictions of the S2/3501-6 composite since the fiber properties are better known than for the AS graphite fiber.

DISCUSSION

Moisture Expansion Experiments

Using a carefully controlled testing method, moisture expansion data for both neat epoxy and unidirectional composites have been determined. The technique developed represents state-of-the-art data acquisition, with a flexibility to accommodate new and more accurate equipment as it becomes available. The data should serve as a base from which designers may account for moisture-induced stresses and strains in composites.

It is appropriate to review some of the major points already discussed. First, although using a matrix material with an essentially linear coefficient of moisture expansion, and fibers with assumed zero moisture absorption, a nonlinear moisture expansion curve was observed, both experimentally and in the micromechanics predictions. This is related to a nonlinear increase of the minimum principal stress in the matrix material in the region of closest fiber packing. The AS/3501-6 graphite/epoxy composite showed far less nonlinearity than did the S2/3501-6 glass/epoxy composite. Agreement between theory and experiment was quite good, being strongly influenced by fiber material properties such as transverse modulus and the transverse coefficient of thermal expansion. Unfortunately, these are best estimates based on literature data. Therefore, as an approximation, a linear moisture expansion response could probably be assumed for the AS/3501-6, depending on the required accuracy of the application. The slight nonlinearity of the initial portion of the 3501-6 epoxy resin expansion curves may have been due to moisture filling interstitial voids [18]. This causes the material to have less expansion for a given apparent moisture absorption during this process, since less strain is induced in the material. However, this deviation from linearity was observed to be only a few percent of full scale. This deviation can in no way account for the large nonlinearity of the composites, especially the S2/3501-6 system. The predicted values would not be significantly influenced since the nonlinearity is small, and occurs prior to the moisture level at which the predicted S2/3501-6 composite moisture expansion exhibits a large incremental increase.

There seems to be a relatively large discrepancy between the micromechanics predictions of moisture expansion and experimental data for the S2/3501-6 composite. While of the same order of magnitude, the predicted values are lower than the experimentally determined values. It has been postulated that moisture severely degrades mechanical properties at the fiber-matrix interface in glass/epoxy [16,17]. Since a perfect fiber-matrix bond was assumed here in the analysis, this effect would be totally ignored. A degradation of the fiber-matrix interface would cause a severe decrease in stiffness of the composite, thereby increasing moisture expansion beyond the value predicted assuming a perfect interface bond.

Fiber volumes for the S2/3501-6 composites did vary somewhat from one specimen to another, a 68 percent average fiber volume being measured and used in the micromechanics analysis. A sensitivity analysis showed the composite response to be very sensitive to fiber volume, which may account for some of the data scatter. While the data obtained represent a good beginning, further tests at different humidity levels and different temperatures need to be conducted to provide a greater degree of confidence in the data. This would also provide a family of curves to determine how these parameters affect moisture expansion. Specifically, strain versus percentage of equilibrium moisture content tests should be run, to verify that the matrix moisture expansion is indeed linear. Also, it would be interesting to numerically model an S2/3501-6 composite with a degraded fiber-matrix interface, to study the effects of the interface degradation hypothesis.

Thermal Expansion Experiments

The transverse thermal expansions of the AS/3501-6 and S2/3501-6 unidirectional composites are nonlinear over a wide temperature range. This comes as no surprise since the 3501-6 epoxy matrix material exhibits this behavior and the composites can be expected to behave in a similar manner.

The AS/3501-6 composite specimens did not correlate as well with the theory as did the S2/3501-6 specimens. In both cases, however, the agreement was acceptable. Since the transverse thermal expansion coefficient of the graphite fiber used in the analysis was an estimate, this could be a source of error. A value of $9 \times 10^{-6}/°C$ for the transverse coefficient of thermal expansion of the graphite fiber was assumed as a base value for the present analysis (Table 3). Fischbach [19] has stated that a maximum upper limit of $14 \times 10^{-6}/°C$ has been predicted for the thermal expansion of a perfect graphite fiber. "Perfect" means that all of the graphitic planes are perfectly aligned and no voids are present in the fiber. In medium modulus fibers, such as the Hercules AS graphite fiber used in the present study, the lower degree of order of the graphite planes causes a lower transverse coefficient of thermal expansion. Also, microvoids in the fiber tend to accommodate thermal expansion, decreasing the transverse thermal expansion coefficient. Therefore, strong arguments can be made for the $9 \times 10^{-6}/°C$ transverse coefficient of thermal expansion of the graphite fiber used as being a realistic value. Also, as stated previously, fiber volume has a strong influence and may be a source of error.

CONCLUSION

Since composites are gaining acceptance for use in primary structural applications, it is necessary to provide good thermal and moisture expansion data for designers. These data are not only necessary for calculating dimensional changes, but also for predicting thermal- and moisture-induced stresses in composite laminates. The data generated here are considered reliable for use as a design tool. Linear interpolation should be valid over small temperature ranges. More complete data are included in Reference [1].

ACKNOWLEDGMENTS

This study was performed in part under sponsorship of the United States Army Research Office, Durham, North Carolina, under Grant DAAG 29-79-C-0189, Dr. John C. Hurt, Project Monitor.

REFERENCES

1. Cairns, D. S. and Adams, D. F., "Moisture and Thermal Expansion of Composite Materials," Report UWME-DR-101-104-1, Department of Mechanical Engineering, University of Wyoming, Laramie, Wyoming (November 1981).
2. "Hercules Magnamite Graphite Fibers," Hercules, Incorporated, Magna, Utah (1978).
3. "Textile Fibers for Industry," Owens-Corning Fiberglas Corp., Toledo, Ohio, 8–30 (1971).
4. Crank, J. *The Mathematics of Diffusion*. New Rochelle, New York: Cambridge University Press, 3–27 (1975).
5. Springer, G. S. and Shen, C. H., "Moisture Absorption and Desorption of Composite Materials," *Journal of Composite Materials, 10* (1), 2–19 (January 1976).
6. Springer, G. S., "Moisture Content of Composites Under Transient Conditions," *Journal of Composite Materials, 11* (1), 107–122 (January 1977).
7. Shirrell, C. D., "Diffusion of Water Vapor in Graphite/Epoxy Composites," *Advanced Composite Materials—Environmental Effects*, ASTM STP 658, American Society for Testing and Materials, Philadelphia, Pennsylvania, 21–42 (1977).
8. Unpublished Experimental Data for Hercules 3501-6 Epoxy Resin, Composite Materials Research Group, Mechanical Engineering Department, University of Wyoming, Laramie, Wyoming (1978).
9. Freeman, W. T. and Campbell, M. D., "Thermal Expansion Characteristics of Graphite Reinforced Composite Materials," *Composite Materials: Testing and Design (Second Conference)*, ASTM STP 497, American Society for Testing and Materials, Philadelphia, Pennsylvania, 121–142 (1972).
10. Adams, D. F. and Miller, A. K., "Hygrothermal Microstresses in a Unidirectional Composite Exhibiting Inelastic Material Behavior," *Journal of Composite Materials, 11* (3), 285–299 (July 1977).
11. Miller, A. K. and Adams, D. F., "Micromechanical Aspects of the Environmental Behavior of Composite Materials," Report UWME-DR-701-111-1, Department of Mechanical Engineering, University of Wyoming, Laramie, Wyoming (January 1977).
12. Crane, D. A. and Adams, D. F., "Finite Element Micromechanical Analysis of a Unidirectional Composite Including Longitudinal Shear Loading," Department Report UWME-DR-001-101-1, Department of Mechanical Engineering, University of Wyoming, Laramie, Wyoming (February 1981).
13. Schaffer, B. G. and Adams, D. F., "Nonlinear Viscoelastic Analysis of a Unidirectional Composite Material," *Journal of Applied Mechanics, 48* (4), 859–865 (December 1981).
14. Walrath, D. E. and Adams, D. F., "Fatigue Behavior of Hercules 3501-6 Epoxy Resin," Report No. NADC-78139-60, Naval Air Development Center, Warminster, Pennsylvania (January 1980).
15. Richard, R. M. and Blacklock, J. M., "Finite-Element Analysis of Inelastic Structures," *AIAA Journal, 7* (3), 432–438 (March 1969).
16. Vaughan, D. J. and McPherson, E. L., "The Effects of Adverse Environmental Conditions on the Resin-Glass Interface of Epoxy Composites," *Proceedings of the 27th Annual Conference, Reinforced Plastics/Composites Institute*, The Society of the Plastics Industry, Inc., New York, New York, Section 21-C, 1–7 (1972).
17. Eakens, W. J., "Effect of Water on Glass Fiber Resin Bonds," *Interfaces in Composites*,

ASTM STP 452, American Society for Testing and Materials, Philadelphia, Pennsylvania, 137–148 (1969).
18. Crossman, F. W., Mauri, R. E., and Warren, W. J., "Moisture-Altered Viscoelastic Response of Graphite/Epoxy Composites," *Advanced Composite Materials—Environmental Effects,* ASTM STP 658, American Society for Testing and Materials, Philadelphia, Pennsylvania, 205–220 (1977).
19. Fischbach, D. B., Personal Correspondence, Professor of Ceramic Engineering, University of Washington, Seattle, Washington (August 1981).

26

In-Plane Thermal Expansion and Thermal Bending Coefficients of Fabric Composites

T. ISHIKAWA AND T. W. CHOU

1. INTRODUCTION

ALTHOUGH APPLICATIONS OF WOVEN FABRIC COMPOSITES HAVE BEEN made rather extensively, there is a lack of understanding of their thermo-mechanical behavior. Recently, Ishikawa [1], and Ishikawa and Chou [2–5] have examined the elastic properties of both hybrid and non-hybrid fabric composites. In their analyses, Ishikawa and Chou adopted three models for simulating the behavior of fabric composites; they are the "mosaic model," "fiber undulation model" and "bridging model."

In the mosaic model [1,2], a fabric composite is simplified as an assemblage of pieces of cross-ply laminates. Estimates of the upper and lower bounds of elastic constants of fabric composites can be achieved by using the mosaic model. The effect of fiber continuity and the resulting "softening effects" on composite elastic properties have been assessed by the fiber undulation model [3]. Both the mosaic model and fiber undulation model are essentially one dimensional and the latter is particularly suitable for modelling composites composed of plain weave fabrics. In the case of satin fabric composites, each region of interlaced threads is surrounded by straight threads. The characteristics of load distribution and transferring are best simulated by the bridging model [4,5]. Comparisons of the predictions based upon these three models with finite element analyses and experimental measurements are very favorable.

In this paper, the in-plane thermal expansion coefficients and thermal bending coefficients of fabric composites are examined. The three analytical models described above have been extended to treat the present problems. Results of the analysis are compared with experiments.

2. FABRIC STRUCTURE AND MOSAIC MODEL

A two-dimensional fabric is woven by two sets of interlaced threads or yarns known as warp and fill threads [6]. The types of fabrics can be categorized by both the geometrical and material patterns of repeat of the interlaced regions as shown in Figure 1. For non-hybrid composites, the geometrical pattern of repeat is identified by the parameter n_g as explained in Ref. [3].

The mosaic model treats a fabric composite as an assemblage of asymmetrical cross-ply laminates (Fig. 2). Based upon this model and the iso-strain and iso-stress assumptions, simple bounds of elastic constants can be derived in closed form. This approach has been used for the prediction of hybrid fabric composite elastic properties [2], which are in good agreement with experiments [7].

To develop the mosaic model for thermo-mechanical property analysis, it is necessary to recapitulate the constitutive equation for a laminated plate

$$\begin{Bmatrix} N \\ M \end{Bmatrix} = \begin{bmatrix} A & B \\ B & D \end{bmatrix} \begin{Bmatrix} \varepsilon^0 \\ \varkappa \end{Bmatrix} - \Delta T \begin{Bmatrix} \tilde{A} \\ \tilde{B} \end{Bmatrix} \tag{1.1}$$

Where

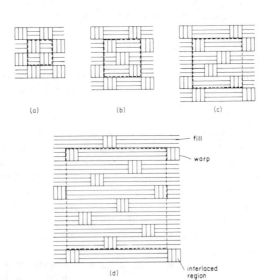

Figure 1. Examples of interlacing patterns in woven fabric (a) plain weave (n_g = 2); (b) twill weave (n_g = 3); (c) 4 harness satin (n_g = 4); and (d) 8 harness satin (n_g = 8). Dotted line indicates a basic two-dimensional repeating unit.

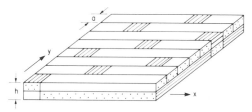

Figure 2. *Mosaic model for an 8 harness satin where a is thread width after resin impregnation and h is lamina thickness.*

$$(\tilde{A}_i, \tilde{B}_i) = \sum_{m=1}^{N} \int_{h_{m-1}}^{h_m} (1, z) \, q_i^{(m)} \, dz \quad (i = 1,2,6) \tag{1.2}$$

$$q_i^{(m)} = Q_{ij}^{(m)} \, \alpha_j^{(m)} \tag{1.3}$$

Here, ΔT indicates a small uniform temperature change, α_j denotes the thermal expansion coefficient, and the superscript (m) indicates a constituent lamina of the laminate. Subscripts of the quantities in Eq. (1.1) have been omitted. If inverted, Eq. (1.1) can be written as

$$\begin{Bmatrix} \varepsilon^0 \\ x \end{Bmatrix} = \begin{bmatrix} a^* & b^* \\ b^* & d^* \end{bmatrix} \begin{Bmatrix} N \\ M \end{Bmatrix} + \Delta T \begin{Bmatrix} \tilde{a}^* \\ \tilde{b}^* \end{Bmatrix} \tag{2.1}$$

where

$$\begin{Bmatrix} \tilde{a}^* \\ \tilde{b}^* \end{Bmatrix} = \begin{bmatrix} a^* & b^* \\ b^* & d^* \end{bmatrix} \begin{Bmatrix} \tilde{A} \\ \tilde{B} \end{Bmatrix} \tag{2.2}$$

The definitions of N, M, A, B, D, a^*, b^*, d^*, ε^0, x, Q, h, and z follow those of the classical laminated plate theory [8–10] and they are not repeated here. Also, $\tilde{a}^*$ and $\tilde{b}^*$ in Equation (2) denote, respectively, the in-plane thermal expansion and thermal bending coefficients.

We adopt the mosaic model and consider a long composite strip free of externally applied loading. The averaged strains and curvatures of a one-dimensional strip of width a in Figure 2 along the fill and warp directions due to a uniform temperature change ΔT can be expressed in the following general forms.

$$\bar{\varepsilon}_i^0 = \frac{1}{n_g a} \int_0^{n_g a} \Delta T \, \tilde{a}_i^* (\xi) \, d\xi$$

$$= \Delta T \, a_i^* \quad (i = 1,2) \tag{3}$$

$$\bar{\kappa}_i = \frac{1}{n_g a} \int_0^{n_g a} \Delta T \, \tilde{b}_i^* \, (\xi) \, d\xi$$

$$= \Delta T \, \frac{n_g - 2}{n_g} \, \tilde{b}_i^* \quad (i = 1,2)$$

(4)

Here, ξ stands for x or y and the bar denotes average of a quantity. Because of the nature of the cross-ply laminates $\tilde{a}_6^*$ and $\tilde{b}_6^*$ vanish. From Eqs. (3) and (4), the average thermal expansion and thermal bending coefficients for the mosaic model are given by

$$\bar{a}_i^* = \tilde{a}_i^*$$

$$\bar{b}_i^* = \left(1 - \frac{2}{n_g}\right) \tilde{b}_i^*$$

(5)

Equation (5) was first derived in Ref. [1].

3. FIBER UNDULATION MODEL

The fiber continuity omitted in the mosaic model has been taken into account in the fiber undulation model. An idealized one-dimensional strip along a yarn in a fabric composite is depicted in Figure 3 where ξ is now taken to be the x direction. The undulation shape is defined by the parameters $h_1(x)$, $h_2(x)$, a_u, a_0 and a_2 where $a_0 = (a - a_u)/2$ and $a_2 = (a + a_u)/2$. The parameter a_u is the length of undulation and it can assume value between 0 and a. Figure 3 indicates the existence of regions of pure matrix material, which is consistent with experimental observations [11,12].

To simulate the observed configuration of fiber undulation, the following undulation shapes are assumed for the fill and warp threads, respectively

$$h_1(\xi) = \begin{cases} 0 & (0 \leqslant \xi \leqslant a_0) \\ \left[1 + \sin\left\{\left(\xi - \frac{a}{2}\right)\frac{\pi}{a_u}\right\}\right]h_t/4 & (a_0 \leqslant \xi \leqslant a_2) \\ h_t/2 & (a_2 \leqslant \xi \leqslant n_g a/2) \end{cases}$$

(6)

$$h_2(\xi) = \begin{cases} h_t/2 & (0 \leqslant \xi \leqslant a_0) \\ \left[1 - \sin\left\{\left(\xi - \frac{a}{2}\right)\frac{\pi}{a_u}\right\}\right]h_t/4 & (a_0 \leqslant \xi \leqslant a/2) \\ -\left[1 + \sin\left\{\left(\xi - \frac{a}{2}\right)\frac{\pi}{a_u}\right\}\right]h_t/4 & (a/2 \leqslant \xi \leqslant a_2) \\ -h_t/2 & (a_2 \leqslant \xi \leqslant n_g a/2) \end{cases}$$

(7)

Here, ξ again represents x or y.

By assuming no in-plane force and moment and following the derivations of Eqs. (3) and (4), we obtain for the fiber undulation model

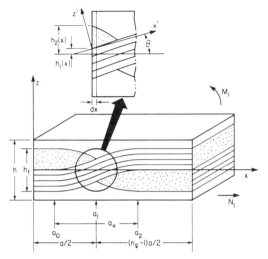

Figure 3. *Fiber undulation model.*

$$\bar{\tilde{a}}_i^{*U} = \left(1 - \frac{2a_u}{n_g a}\right) \tilde{a}_i^* + \frac{2}{n_g a} \int_{a_0}^{a_2} \tilde{a}_i^* (\xi)\, d\xi \tag{8}$$

$$\bar{\tilde{b}}_i^{*U} = \left(1 - \frac{2}{n_g}\right) \tilde{b}_i^* + \frac{2}{n_g a} \int_{a_0}^{a_2} \tilde{b}_i^* (\xi)\, d\xi \tag{9}$$

Here, the superscript U signifies the fiber undulation model. It is understood that $\tilde{a}_6^{*U}$ and $\tilde{b}_6^{*U}$ vanish for cross-ply constructions. Since $\tilde{b}_i(\xi)$ is an odd function of ξ with respect to $\xi = a_1$ due to the assumed form of $h_1(\xi)$, the integration in Eq. (9) vanishes and

$$\bar{\tilde{b}}_i^{*U} = \left(1 - \frac{2}{n_g}\right) \tilde{b}_i^* \tag{10}$$

Equation (10) is identical to Eq. (5) obtained from the mosaic model and this indicates that fiber undulation has no effect on the thermal bending coefficients. Identical conclusions were obtained for the bending-stretching coupling constant in Ref. [3].

For the in-plane thermal expansion coefficient, it is necessary to evaluate the integration in Eq. (8). This is done based upon the assumption that the classical laminated plate theory is applicable to each infinitesimal piece of width dx of the one-dimensional strip shown in Figure 3. The following steps are taken to obtain $\tilde{a}_i^* (\xi)$. First, $\tilde{A}_i (\xi)$ and $\tilde{B}_i (\xi)$ are evaluated from Eqs. (1.2) and (1.3) for $0 \leqslant \xi \leqslant a/2$ and the results are

$$\tilde{A}_i(\xi) = q_i^M \left(h_1(\xi) - h_2(\xi) + h - h_t/2 \right)$$
$$+ q_i^F (\theta)\, h_t/2 + q_i^W \left(h_2(\xi) - h_1(\xi) \right) \tag{11}$$

$$\tilde{B}_i(\xi) = \frac{1}{2}\, q_i^F (\theta) \left(h_1(\xi) - h_t/4 \right) h_t$$
$$+ \frac{1}{4}\, q_i^W \left(h_2(\xi) - h_1(\xi) \right) h_t \tag{12}$$

Where the superscripts F, W and M signify the fill thread, warp thread and matrix regions, respectively. The quantity of q_i is determined from Eq. (1.3); $q_i^F(\theta)$, in particular, is determined from the local stiffness matrix $Q_{ij}^F(\theta)$, following the procedures outlined in Refs. [3,4]. Furthermore, the off-axis thermal expansion coefficients $\alpha_i^F(\theta)$ of Eq. (1.3) are given by

$$\alpha_1^F(\theta) = \cos^2\theta\; \alpha_L^F + \sin^2\theta\; \alpha_T^F$$
$$\alpha_2^F(\theta) = \alpha_T^F \tag{13}$$
$$\alpha_6^F(0) = 0$$

where α_L and α_T denote, respectively, thermal expansion coefficients parallel and transverse to the fiber direction in a unidirectional fiber composite. Thus, $\tilde{A}_i(\xi)$ and $\tilde{B}_i(\xi)$ can be determined from Eqs. (11) and (12). Also, $a_{ij}^*(\xi)$, $b_{ij}^*(\xi)$, and $d_{ij}^*(\xi)$ in Eq. (2.1) are obtained by inversion of $A_{ij}(\xi)$, $B_{ij}(\xi)$ and $D_{ij}(\xi)$ [3,4]. Finally $\tilde{\alpha}_i(\xi)$ can be derived from Eq. (2.2). Numerical integration of Eq. (8) has been conducted and the results for $\bar{\alpha}_1^{*U}$ and $\bar{b}_1^{*U}$ as functions of $1/n_g$ are given in Figure 5. It should be noted that the balanced thermal property such as $\bar{\alpha}_1^* = \bar{\alpha}_2^*$ for a fabric composite can be realized if the above procedure of calculation is conducted for one-directional strips along both the fill and warp directions.

4. BRIDGING MODEL

The models presented in Sections II and III are essentially one-dimensional and they are inadequate for simulating the behavior of satin composites where the interlaced regions are not connected. To overcome this deficiency, the bridging model has been proposed [4] to better take into account the load transfer mechanism. Consider, for example, an 8 harness satin. The repeating unit in Figure 4a is first modified to a square shape (Fig. 4b) of the same area for simplicity of calculations. A schematic view of the bridging model is shown in Figure 4c which contains an interlaced region and its surrounding area. The four regions denoted by A, B, C, and E consist of straight threads, and hence, can be regarded as pieces of cross-ply laminates. Region C has an interlaced structure where only the fill thread is assumed to be undulated and $n_g = 2$. The undulation of the warp thread has been neglected. Thus, it is

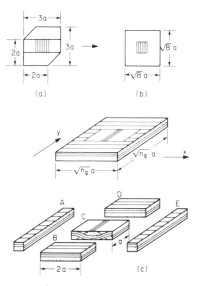

Figure 4. Concept of the bridging model: (a) shape of the repeating unit of 8 harness satin; (b) modified shape for the repeating unit; (c) idealization for the bridging model.

necessary to consider a bridging with an undulated warp thread for evaluating $\bar{a}_2^*$ and $\bar{b}_2^*$.

The regions B and D are stiffer than the undulated region C and hence, they carry more load when an external load is applied in the x direction. Regions B, C, and D are termed bridging regions. For the present problem, assuming no external loading, the equilibrium of the bridging regions requires

$$a \begin{Bmatrix} N^U \\ M^U \end{Bmatrix} + (\sqrt{n_g} - 1)a \begin{Bmatrix} N \\ M \end{Bmatrix} = 0 \qquad (14)$$

where the subscript U again denotes the undulated region, and N and M without superscript are for the cross-ply laminate. Furthermore, under the assumption of uniform strain and curvature in the bridging regions B, C, and D, we define

$$\varepsilon_i^{0U} = \bar{\varepsilon}_i^0$$
$$\kappa_i^U = \bar{\kappa}_i \qquad (15)$$

where the bar denotes average of the bridging regions.

Substituting Eq. (1) into Eq. (14) and taking into account Eq. (15), we have

$$\left(\begin{bmatrix} A^U & B^U \\ B^U & D^U \end{bmatrix} + (\sqrt{n_g} - 1) \begin{bmatrix} A & B \\ B & D \end{bmatrix} \right) \begin{Bmatrix} \bar{\varepsilon}^0 \\ \bar{\kappa} \end{Bmatrix}$$

$$= \Delta T \left[\begin{Bmatrix} \tilde{A}^U \\ \tilde{B}^U \end{Bmatrix} + (\sqrt{n_g} - 1) \begin{Bmatrix} \tilde{A} \\ \tilde{B} \end{Bmatrix} \right] \qquad (16)$$

The quantities on the left hand side of Eq. (16) can be related to the average elastic stiffness in the bridging regions as [4]

$$\begin{bmatrix} A^U & B^U \\ B^U & D^U \end{bmatrix} + (\sqrt{n_g} - 1) \begin{bmatrix} A & B \\ B & D \end{bmatrix} = \sqrt{n_g} \begin{bmatrix} \bar{A} & \bar{B} \\ \bar{B} & \bar{D} \end{bmatrix} \qquad (17)$$

Hence, Eq. (16) can be rewritten as follows:

$$\begin{Bmatrix} \bar{\varepsilon}^0 \\ \bar{\kappa} \end{Bmatrix} = \Delta T \begin{bmatrix} \bar{a}^* & \bar{b}^* \\ \bar{b}^* & \bar{d}^* \end{bmatrix} \left(\frac{1}{\sqrt{n_g}} \begin{Bmatrix} \tilde{A}^U \\ \tilde{B}^U \end{Bmatrix} + 1 - \frac{1}{\sqrt{n_g}} \begin{Bmatrix} \tilde{A} \\ \tilde{B} \end{Bmatrix} \right) \qquad (18)$$

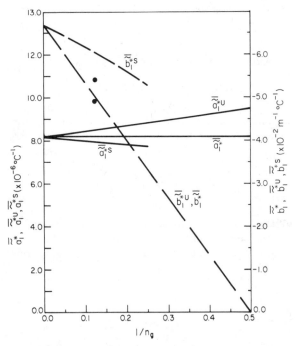

Figure 5. The variation of the thermal deformation coefficients with $1/n_g$ for a graphite/epoxy composite, $V_f = 60\%$ and $a_u/a = 1.0$. $\tilde{a}_1^*$: ———, $\tilde{b}_1^*$: ——— and ● : experimental results of b_1 at 300°K.

Here, a^*, b^* and d^* are obviously obtained by inverting A, B, and D. In comparing to Eq. (2.1) the quantities in the parenthesis on the right hand side of Eq. (18) can be regarded as the average values of A_i and B_i for the bridging regions and hence, they are denoted by $\bar{A}_i$ and $\bar{B}_i$. Thus we obtain

$$\left\{ \begin{matrix} \tilde{\bar{a}}^* \\ \tilde{\bar{b}}^* \end{matrix} \right\} = \begin{bmatrix} \bar{a}^* & \bar{b}^* \\ \bar{b}^* & \bar{d}^* \end{bmatrix} \begin{bmatrix} \tilde{\bar{A}} \\ \tilde{\bar{B}} \end{bmatrix} \tag{19}$$

Finally, the whole satin composite of Figure 4(b) can be regarded as a linkage of the regions A, B-C-D, and E in series. The average strains and curvature for the entire model are given by

$$\left\{ \begin{matrix} \bar{\varepsilon}^{0S} \\ \bar{\varkappa}^{S} \end{matrix} \right\} = \frac{1}{\sqrt{n_g}} \left(2 \left\{ \begin{matrix} \bar{\varepsilon}^0 \\ \bar{\varkappa} \end{matrix} \right\} + (\sqrt{n_g} - 2) \left\{ \begin{matrix} \varepsilon^0 \\ \varkappa \end{matrix} \right\} \right)$$

$$= \Delta T \left(\frac{2}{\sqrt{n_g}} \left\{ \begin{matrix} \tilde{\bar{a}}^* \\ \tilde{\bar{b}}^* \end{matrix} \right\} + \left(1 - \frac{2}{\sqrt{n_g}} \right) \left\{ \begin{matrix} \tilde{a}^* \\ \tilde{b}^* \end{matrix} \right\} \right) \tag{20}$$

where the superscript S signifies the properties of the satin composite and ε^0 and $\varkappa$ denote mid-plane strain and curvature for the cross-plies in regions A and E of Figure 4. Equation (20) implies

$$\left\{ \begin{matrix} \tilde{\bar{a}}^{*S} \\ \tilde{\bar{b}}^{*S} \end{matrix} \right\} = \frac{2}{\sqrt{n_g}} \left\{ \begin{matrix} \tilde{\bar{a}}^* \\ \tilde{\bar{b}}^* \end{matrix} \right\} + \left(1 - \frac{2}{\sqrt{n_g}} \right) \left\{ \begin{matrix} \tilde{a}^* \\ \tilde{b}^* \end{matrix} \right\} \tag{21}$$

for the thermal expansion and thermal bending coefficients of the satin composites.

Figure 5 shows numerical results of the analysis. The properties of constituent unidirectional laminae are given in Table 1 for a graphite/epoxy composite with a fiber volume fraction of 60% [1]. The general characteristics of the variations of thermal deformation coefficients with $1/n_g$ are very similar to those of the compliance constants $\bar{a}_{11}^*$ and $\bar{b}_{11}^*$ as discussed in Ref. [3]. For the thermal bending coefficients, there is considerable discrepancy between the results obtained from the one-dimensional models and the bridging model. Although the experimental results of Ref. [1] seem to be closer to the

Table 1. Material properties of a graphite/epoxy unidirectional lamina [1].

E_L	113. GPA
E_T	8.82 GPA
G_{LT}	4.46 GPa
ν_{TL}	0.3
ν_{TT}	0.5
α_L	0.0
α_T	$3.0 \times 10^{-5} °C^{-1}$

one-dimensional predictions. The data are insufficient to provide a valid comparison with the theoretical models.

The geometrical shape of the fiber undulation also affects $\bar{\bar{a}}_1^*$; this is demonstrated in Figure 6 where material properties of Table 1 have been used. The results indicate that the in-plane thermal expansion coefficient of satin weave composites is less sensitive to a_u/h than that of plain weave composites. Furthermore, the fiber undulaton model predicts a larger effect on $\bar{\bar{a}}_1^*$ due to a_u/h than the bridging model. In general, the bridging model predictions are also less sensitive to the n_g values than the undulation model. In both models, the maximum in $\bar{\bar{a}}_1^*$ occurs at $a_u/h \cong 1$.

5. COMPARISONS WITH EXPERIMENTS

Experimental data on thermal expansion coefficients of fabric composites are extremely limited. Rogers *et al.* [13–15] have performed measurements of thermal expansion of graphite fiber reinforced plastics. Their experiments [13], however, are based upon thick specimens with 15–25 plies. Due to the constraint of the neighboring layers, an individual ply in the laminate is not free to bend. As a result, modifications to the present theory are necessary so a more meaningful comparison with experiments can be made.

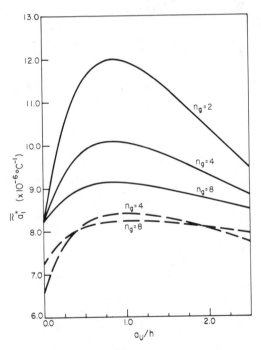

Figure 6. The effect of fiber undulation on $\bar{\bar{a}}_1^*$, fiber undulation model: ———, bridging model: ———.

It is assumed that the bending-free thermal expansion of a lamina can be realized if there exists a bending moment M_i under a temperature change ΔT, and no in-plane force is allowed. Thus

$$N_i = 0, \quad \kappa_i = 0 \tag{22}$$

From Eq. (2.1) and $\kappa_i = 0$

$$[d^*]\,\{M\} + \Delta T\,\{\widetilde{b}^*\} = 0 \tag{23}$$

Hence

$$\{M\} = -\Delta T\,[d^*]^{-1}\{\widetilde{b}^*\} \tag{24}$$

Substituting Eq. (24) into Eq. (2.1), and from the expression of ε^0, a modified in-plane thermal expansion coefficient under the bending-free condition can be defined as

$$\{\widetilde{a}^{**}\} = \{\widetilde{a}^*\} - [b^*]\,[d^*]^{-1}\{\widetilde{b}^*\} \tag{25}$$

Equation (25) can be evaluated for the mosaic, undulation and bridging models provided that the appropriate constants are given for a particular model. Also note the presence of elastic compliance constants in Eq. (25). Thus it is necessary to evaluate, for instance, $\bar{b}^{*S}_{ij}$ and $\bar{d}^{*S}_{ij}$ for calculating $\widetilde{a}^{**S}_i$ and $\bar{b}^{*U}_{ij}$ and $\bar{d}^{*U}_{ij}$ for $\widetilde{a}^{**U}_i$. The above modifications are of practical significance because it is desirable to overcome the anti-symmetrical behavior such as $\widetilde{b}^*_i$ by suitable stacking in laminate constructions.

Figure 7 gives the variation of $\widetilde{a}^{**}_1$ with $1/n_g$. The theoretical predictions are based upon both undulation and bridging models using the thermoelastic properties of unidirectional graphite/epoxy composite of Table 2. The experimental results for 5-harness satin composite, which has the same properties as given in Table 2 are also shown in Figure 7. Since the thickness of lamina is not given in Ref. [13], two estimated values for a/h were used for the analysis and a/a_u is assumed to be unity. The bridging model prediction coincides fairly well with experiments. It is also obvious that the in-plane thermal expansion coefficients are more sensitive to n_g in the bending-free case (Fig. 7) than the bending-unconstrained case (Fig. 5).

6. CONCLUSIONS

1. The mosaic model provides a simple means for estimating thermal expansion and thermal bending coefficients.
2. The one-dimensional fiber undulation model predicts slightly higher in-plane thermal expansion coefficients and the same thermal bending coefficients as compared to those obtained from the mosaic model. The limited experimental data on thermal bending coefficient coincides

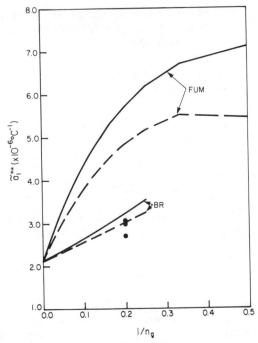

Figure 7. *Comparison of theoretical predictions with the experimental results of Ref. [13] for 5 harness-satin graphite/epoxy composites. $a/h = 3.75$; ——, $a/h = 7.5$: — — — —, and $a_u/a = 1.0$. FUM and BR indicate fiber undulation and bridging models, respectively.* ●: *experimental results at 300°K.*

rather well with the predictions of the mosaic and undulational models.
3. The bridging model is particularly suited for the prediction of thermal expansion constants for satin composites. The experimental results on in-plane thermal expansion coefficients for a 5-harness satin composite agree well with the theory.

Table 2. Material properties of a graphite/epoxy unidirectional lamina.

E_L	148 GPa†
E_T	7.35 GPa†
G_{LT}	3.92 GPa†
ν_{TL}	0.31†
ν_{TT}	0.52†
α_L	$3.1 \times 10^{-7}°C^{-1*}$
α_T	$3.1 \times 10^{-5}°C^{-1*}$

*Ref. [13].
†Values estimated based upon Refs. [13, 16, 17, 18] and longitudinal modulus of 237 GPa and transverse modulus of 12. GPa for Grafil XAS fibers.

7. ACKNOWLEDGMENT

This work is supported by the U.S. Army Research Office.

REFERENCES

1. Ishikawa, T., "Anti-Symmetric Elastic Properties of Composite Plates of Satin Weave Cloth," *Fib. Sci. Tech.,* Vol. 15 (1981), p. 127.
2. Ishikawa, T and Chou, T.-W., "Elastic Behavior of Woven Hybrid Composites," *J. Composite Materials,* Vol. 16 (1982), p. 2.
3. Ishikawa, T. and Chou, T.-W., "One-dimensional Analysis of Woven Fabric Composites," submitted for publication.
4. Ishikawa, T. and Chou, T.-W., "Stiffness and Strength Behavior of Woven Fabric Composites," *J. Materials Science,* Vol. 17 (1982), p. 3211.
5. Ishikawa, T. and Chou, T.-W., "Stiffness and Strength Properties of Woven Fabric Composites," Proceedings of ICCM 4 (Tokyo), (1982), p. 489.
6. Miller, E., "Textiles: Properties and Behavior," (rev. edn.) B. T. Batsford, London, (1976).
7. Zweben, C. and Norman, J. C., "Kevlar 49/Thornel 300 Hybrid Fabric Composites for Aerospace Applications," *SAMPE Quarterly,* July (1976), p. 1.
8. Tsai, S. W., "Structural Behavior of Composite Materials," NASA CR-71, July (1964).
9. Whitney, J. M. and Leissa, A. W., "Analysis of Heterogeneous Anisotropic Plate," *J. Applied Mec.,* Vol. 36 (1969), p. 261.
10. Jones, R. M., "Mechanics of Composite Materials," *Scripta,* Washington, D.C.. 1975.
11. Kimpara, I. and Takehana, M., "Analysis of Acoustic Emission from Internal Failure of Glass Fiber Reinforced Composites," 2nd Acoustic Emission Symp., Tokyo, 1974 Ses. 9/2-20.
12. Kavelka, J., "Thermal Expansion of Composites with Canvas-Type Reinforcement and Polymer Matrix," Proceedings of ICCM 3, Paris (1980), p. 770.
13. Rogers, K. F., Kingston-Lee, D. M., Phillips, L. N., Yates, B., Chandra, M., and Parker, S. F. H., "The Thermal Expansion of Carbon-fibre Reinforced Plastics, Part 6 The influence of fiber weave in fabric reinforcement," *J. Materials Science,* Vol. 16 (1981),p. 2803.
14. Rogers, K. F., Phillips, L. N., Kingston-Lee, D. M., Yates, B., Overy, M. J., Sargent, J. P., and McCalla, B. A., "The Thermal Expansion of Carbon Fibre-reinforced Plastics, Part 1 The influence of fibre type and orientation," *J. Materials Science,* Vol.12 (1977), p. 718.
15. Yates, B., Overy, M. J., Sargent, J. P., McCalla, B. A., Kingston-Lee, D. M., Phillips, L. N., and Rogers, K. F., "The Thermal Expansion of Carbon Fibre-reinforced Plastics, Part 2 The influence of fiber volume fraction," *J. Materials Science,* Vol. 13 (1978), p. 433.
16. Ishikawa, T., Koyama, K. and Kobayashi, S., "Elastic Moduli of Carbon-Epoxy Composites and Carbon Fibers," *J. Composite Materials,* Vol. 11, (1977), p. 332.
17. Ishikawa, T., Koyama K. and Kobayashi, S., "Thermal Expansion Coefficients of Unidirectional Composites," *J. Composite Materials,* Vol. 12, (1978), p. 153.
18. Dean, G. and Turner, G., "The Elastic Properties of Carbon Fibers and Their Composites," *Composites,* Vol. 4, (1973), p. 174.

27

Coefficient of Thermal Expansion for Composites with Randomly Oriented Fibers

W. J. CRAFT AND R. M. CHRISTENSEN

INTRODUCTION

THE PRESENT WORK IS CONCERNED WITH THE PREDICTION OF THE COEF-ficient of thermal expansion for composite materials. The primary objective is to predict from the individual thermal and mechanical properties of the separate phases the effective (average) coefficients of thermal expansion of the complete composite material. Information of this type is basic for design purposes.

Predictions of the effective coefficients of thermal expansion are available for aligned fiber systems that possess transverse isotropy; for example, see Reference [1]. Our interest in the present case is on the development of the corresponding predictive formulae for fiber systems that are isotropic in a plane or are three-dimensionally isotropic. The two-dimensionally isotropic or quasi-isotropic fiber system can be obtained by fusion of orthotropic laminae aligned at specific angles to obtain planar isotropy or by the impregnation of random fiber mat by a continuous matrix phase. This type of composite construction is of great technological importance. The three-dimensional isotropic composite can be obtained by randomizing fiber orientation in three dimensions and suspending the fibers in a matrix phase. While this latter case is less prevalent than the two-dimensional quasi-isotropic form, it nevertheless is of growing importance. These latter systems can be obtained by extrusion of a chopped-fiber dispersion in a polymer melt. Often partial orientation of the fiber phase is obtained. However, concern here is only with the random case.

Results corresponding to those sought here for the coefficient of thermal expansion are already available for the prediction of effective moduli [1]. The thermal expansion problem is expected to be more complicated than that of

330

effective moduli because of the coupling between thermal and mechanical effects. Nevertheless, a completely practical approach has been found for the problem.

The level of idealization is that of linear thermoelasticity theory wherein both phases in the fiber-matrix composite system are taken to be isotropic. Also, in possible application to chopped-fiber systems, end effects are neglected; thus the fiber aspect ratio must be very large.

The present analysis is divided into two sections: first, the two-dimensional quasi-isotropic case and, second, the fully isotropic three-dimensional case. The starting point in the two-dimensional case is the derivation of the in-plane value of the quasi-isotropic coefficient of thermal expansion, α_{2D}, in terms of the properties of an aligned fiber lamina. This derivation leads to the same result as that of Halpin and Pagano [2] and is included here for completeness.

Next the composite cylinders model of Hashin and Rosen [3] is used to predict α_{2D} in terms of the properties of the two phases. Finally, the theoretical solution for α_{2D} is put into the form of an asymptotic expansion, and the first two terms are explicitly obtained. The expansion parameter is that appropriate to a fiber phase that is very stiff compared with the matrix phase. The practical application of these results is discussed in the last section after the corresponding three-dimensional results are obtained.

In the three-dimensional case, first a new result is found whereby the coefficient of thermal expansion, α_{3D}, is given in terms of the thermal-mechanical properties of the corresponding aligned fiber system. This result is derived through a three-dimensional randomizing process, comparable to that employed in the mechanical case with no thermal effects. Thence the composite-cylinders model is used to relate α_{3D} to the individual properties of the fiber and matrix phases. Finally, these results are put into asymptotic form for ease of interpretation. The complete results are discussed in the last section.

TWO-DIMENSIONAL ANALYSIS

A lamina is taken with orthotropic properties. Interest in this section centers upon two-dimensional, plane stress conditions. Relative to rectangular Cartesian coordinate system, the linear stress-strain, temperature equation is given by:

$$\sigma_{ij} = C_{ijkl} \ (\varepsilon_{kl} - \alpha_{kl} \Delta T), \tag{1}$$

where C_{ijkl} and α_{kl} are the moduli tensor and the coefficients of thermal expansion tensor, respectively, and ΔT is the small deviation from the uniform reference temperature. Under plane stress conditions,

$$\sigma_{33} = \sigma_{31} = \sigma_{23} = 0 \quad .$$

Then (1) takes the form:

$$
\begin{bmatrix} \sigma_1 \\ \\ \sigma_2 \\ \\ \sigma_{12} \end{bmatrix} = \begin{bmatrix} \dfrac{E_{11}}{1 - \nu_{21}\nu_{12}} & \dfrac{\nu_{12}E_{22}}{1 - \nu_{12}\nu_{21}} & 0 \\ \dfrac{\nu_{12}E_{22}}{1 - \nu_{12}\nu_{21}} & \dfrac{E_{22}}{1 - \nu_{12}\nu_{21}} & 0 \\ 0 & 0 & \mu_{12} \end{bmatrix} \begin{bmatrix} \varepsilon_1 - \alpha_1 \Delta T \\ \\ \varepsilon_2 - \alpha_2 \Delta T \\ \\ 2\varepsilon_{12} \end{bmatrix}
$$

$$(2a)$$

or

$$
\begin{bmatrix} \sigma_1 \\ \\ \sigma_2 \\ \\ \sigma_{12} \end{bmatrix} = \begin{bmatrix} Q_{11} & Q_{12} & 0 \\ Q_{12} & Q_{22} & 0 \\ 0 & 0 & Q_{66} \end{bmatrix} \begin{bmatrix} \varepsilon_1 - \alpha_1 \Delta T \\ \\ \varepsilon_2 - \alpha_2 \Delta T \\ \\ 2\varepsilon_{12} \end{bmatrix} ,
$$

$$(2b)$$

where the 1,2 coordinate directions are taken to lie in planes of symmetry for an orthotropic material.

If the expansion of the orthotropic lamina is due strictly to thermal growth, then:

$$\varepsilon_1 = \alpha_1 \Delta T, \ \varepsilon_2 = \alpha_2 \Delta T, \text{ and } \varepsilon_{12} = 0 \ .$$

The effective expansions along some orientation θ to the material axes (see Figure 1) are given by:

$$\alpha_x = \alpha_1 \cos^2\theta + \alpha_2 \sin^2\theta \ ,$$

$$\alpha_y = \alpha_1 \sin^2\theta + \alpha_2 \cos^2\theta \ , \tag{3}$$

$$\alpha_{xy} = 2(\alpha_1 - \alpha_2)\sin\theta \cos\theta \ .$$

In matrix form, (3) becomes:

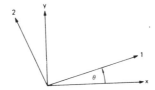

Figure 1. *Two-dimensional rotated coordinates.*

$$\begin{bmatrix} \alpha_x \\ \alpha_y \\ \alpha_{xy} \end{bmatrix} = \begin{bmatrix} L(\theta) \end{bmatrix}_{3 \times 2} \begin{bmatrix} \alpha_1 \\ \alpha_2 \end{bmatrix} . \qquad (4)$$

When the stress-strain relations are modified to allow the thermal strain, then:

$$\begin{bmatrix} \sigma_x \\ \sigma_y \\ \sigma_{xy} \end{bmatrix} = \begin{bmatrix} \bar{Q} \end{bmatrix} \begin{bmatrix} \varepsilon_x - \alpha_x \Delta T \\ \varepsilon_y - \alpha_y \Delta T \\ \varepsilon_{xy} - \alpha_{xy} \Delta T \end{bmatrix} ,$$

which becomes:

$$\begin{bmatrix} \sigma_x \\ \sigma_y \\ \sigma_{xy} \end{bmatrix} = \begin{bmatrix} \bar{Q} \end{bmatrix} \begin{bmatrix} \varepsilon_x \\ \varepsilon_y \\ \varepsilon_{xy} \end{bmatrix} - \Delta T \begin{bmatrix} \bar{Q} \end{bmatrix} \begin{bmatrix} L(\theta) \end{bmatrix} \begin{bmatrix} \alpha_1 \\ \alpha_2 \end{bmatrix} , \qquad (5)$$

where $[\bar{Q}]$ is the rotated coordinate form of $[Q]$ in (2b) and $[L]$ is as defined by (3).

The same expression (5) applies to the in-plane stresses of the Kth layer of composite laminate, i.e.:

$$\begin{bmatrix} \sigma_x \\ \sigma_y \\ \sigma_{xy} \end{bmatrix}_k = \begin{bmatrix} \bar{Q} \end{bmatrix}_k \left\{ \begin{bmatrix} \varepsilon_x \\ \varepsilon_y \\ \varepsilon_{xy} \end{bmatrix} - (\Delta T) \begin{bmatrix} L(\theta) \end{bmatrix} \begin{bmatrix} \alpha_1 \\ \alpha_2 \end{bmatrix} \right\}_k . \qquad (6)$$

Use will now be made of classical plate-theory kinematics as,

$$
\begin{aligned}
\varepsilon_x &= u_{0,x} - z\kappa_x \ , \\
\varepsilon_y &= v_{0,y} - z\kappa_y \ , \\
2\varepsilon_{ey} &= u_{0,y} + v_{0,x} - z\kappa_{xy},
\end{aligned}
\tag{7}
$$

where u_0 and v_0 are the in-plane displacements and the κ's are the curvatures. Force resultants are given by:

$$
\begin{bmatrix} N_x \\ N_y \\ N_{xy} \end{bmatrix}_k = \int_{-\frac{h}{2}}^{+\frac{h}{2}} \begin{bmatrix} \sigma_x \\ \sigma_y \\ \sigma_{xy} \end{bmatrix}_k dz = \sum_{k=1}^{N} \int_{z_{k-1}}^{z_k} \begin{bmatrix} \sigma_x \\ \sigma_y \\ \sigma_{xy} \end{bmatrix} dz
\tag{8}
$$

over each of N equally stiff layers of total thickness h.

With the thermal terms, force resultants become:

$$
\begin{bmatrix} N_x \\ N_y \\ N_{xy} \end{bmatrix} = [A] \begin{bmatrix} \varepsilon_x^0 \\ \varepsilon_y^0 \\ \varepsilon_{xy}^0 \end{bmatrix} - \Delta T \begin{bmatrix} E_1 \\ E_2 \\ E_3 \end{bmatrix} + [B] \begin{bmatrix} \kappa_x \\ \kappa_y \\ \kappa_{xy} \end{bmatrix} .
\tag{9}
$$

Similarly, the moment resultants become:

$$
\begin{bmatrix} M_x \\ M_y \\ M_{xy} \end{bmatrix} = [B] \begin{bmatrix} \varepsilon_x^0 \\ \varepsilon_y^0 \\ \varepsilon_{xy}^0 \end{bmatrix} - \Delta T \begin{bmatrix} F_1 \\ F_2 \\ F_3 \end{bmatrix} + [D] \begin{bmatrix} \kappa_x \\ \kappa_y \\ \kappa_{xy} \end{bmatrix} ,
\tag{10}
$$

where A_{ij}, B_{ij}, and D_{ij} are the appropriate forms of the extensional, coupling, and bending stiffnesses, respectively, and E_i and F_i are given by the following:

$$
E_i = \sum_{k=1}^{N} \left(\Delta_i \right)_k \left(z_k - z_{k-1} \right) ,
\tag{11}
$$

$$F_i \;=\; 1/2 \sum_{k=1}^{N} \left(\Delta_i \right)_k \left(z_k^2 - z_{k-1}^2 \right) \quad ,$$

where

$$\left(\Delta_i \right)_k \;=\; \begin{bmatrix} \Delta_1 \\ \Delta_2 \\ \Delta_3 \end{bmatrix}_k \;=\; \left[\bar{Q} \right]_k \; \left[L(\theta) \right]_k \begin{bmatrix} \alpha_1 \\ \alpha_2 \end{bmatrix}_k . \qquad (12)$$

It is convenient to define the product,

$$\left[J(\theta) \right]_k \;=\; \left[\bar{Q} \right]_k \; \left[L(\theta) \right]_k \quad , \qquad (13)$$

which involves products of $m = cos\theta$ and $n = sin\theta$ to power six (powers of four from Q_{ij} and powers of two from $L(\theta)$). These products may be expanded using the multiple angle summations listed below:

$$32m^6 \;=\; 10 + 15 \, cos \, 2\theta + 6 \, cos \, 4\theta + cos \, 6\theta$$

$$32m^5 n \;=\; 5 \, sin \, 2\theta + 4 \, sin \, 4\theta + sin \, 6\theta$$

$$32m^4 n^2 \;=\; 2 + cos \, 2\theta - 2 \, cos \, 4\theta - cos \, 6\theta$$

$$32m^3 n^3 \;=\; 3 \, sin \, 2\theta - sin \, 6\theta$$

$$32m^2 n^4 \;=\; 2 - cos \, 2\theta - 2 \, cos \, 4\theta + cos \, 6\theta$$

$$32mn^5 \;=\; 5 \, sin \, 2\theta - 4 \, sin \, 4\theta + sin \, 6\theta$$

$$32n^6 \;=\; 10 - 15 \, cos \, 2\theta + 6 \, cos \, 4\theta - cos \, 6\theta \quad .$$

$$(14)$$

The complete form of $[J(\theta)]_k$, after expansion, is given by:

$$[J(\theta)]_k \;=\; \begin{bmatrix} 1/2 \, (Q_{11} + Q_{12}) + 1/2 \, (Q_{11} - Q_{12}) \, cos \, 2\theta & 1/2 \, (Q_{12} + Q_{22}) + 1/2 \, (Q_{12} - Q_{22}) \, cos \, 2\theta \\ 1/2 \, (Q_{11} + Q_{12}) + 1/2 \, (Q_{12} - Q_{11}) \, cos \, 2\theta & 1/2 \, (Q_{12} + Q_{22}) + 1/2 \, (Q_{22} - Q_{12}) \, cos \, 2\theta \\ 1/2 \, (Q_{11} - Q_{12}) \, sin \, 2\theta & 1/2 \, (Q_{12} - Q_{22}) \, sin \, 2\theta \end{bmatrix}_k \qquad (15)$$

Thus all angular dependence above 2θ has vanished in the $[J(\theta)]_k$ expression that links the force and moment resultants to the coefficients of expansion in the material directions 1 and 2 (see Figure 1).

To investigate quasi-isotropy, we must examine conditions under which the resultants of equations (9) and (10) are not functions of the ply-layer orientations. It is sufficient to consider the case of a laminate of N layers with orientation angles:

$$\theta_i = \frac{i-1}{N} \, \pi \; ; i = 1, 2 ..., N.$$

In a well-known solution, quasi-isotropy exists if $N \geqslant 3$ for the material stiffness matrix, A_{ij} [Tsai and Pagano, 4]. Following that work, the effect of the E_i matrix must be considered along with A_{ij} in the force resultant (9).

Expansion of $[J(\theta)]_k$ involves both terms independent of θ and functions of $sin\,(2\theta)$ or $cos\,(2\theta)$. By making the substitution $\theta_k - \phi$ for θ_k, we can express the effect of the discrete lamina orientations in relation to the macroscopic properties of a laminate along an arbitrary direction ϕ. Thus if:

$$W_1 \;\; = \;\; 1/2\,(Q_{11} + Q_{12})\,\alpha_1 + 1/2\,(Q_{12} + Q_{22})\,\alpha_2 \quad,$$

$$W_2 \;\; = \;\; 1/2\,(Q_{11} - Q_{12})\,\alpha_1 + 1/2\,(Q_{12} - Q_{22})\,\alpha_2 \quad,$$

$$W_3 \;\; = \;\; 1/2\,(Q_{12} - Q_{11})\,\alpha_1 + 1/2\,(Q_{22} - Q_{12})\,\alpha_2 \quad,$$

and

$$W_4 \;\; = \;\; 1/2\,(Q_{11} - Q_{12})\,\alpha_1 + 1/2\,(Q_{12} - Q_{22})\,\alpha_2 \quad, \qquad\qquad (16)$$

then

$$E_1 \;\; = \;\; \left(\frac{h}{N}\right) \sum_{k=1}^{N} \{\, W_1 + W_2\, cos\, 2\,(\theta_k - \phi)\,\} \quad,$$

$$E_2 \;\; = \;\; \left(\frac{h}{N}\right) \sum_{k=1}^{N} \{\, W_1 + W_3\, cos\, 2\,(\theta_k - \phi)\} \quad,$$

$$E_3 \;\; = \;\; \left(\frac{h}{N}\right) \sum_{k=1}^{N} W_4\, sin\, 2\,(\theta_k - \phi) \quad.$$

$$\qquad\qquad (17)$$

The summations in equations (17) are constant for $N \geqslant 2$, because all 4θ dependency vanishes. While no angular dependency occurs in the E_i vector, the stiffness matrix A_{ij}, because of a 4θ component, is anisotropic in the $N = 2$ case.

If, on the other hand, $N \geqslant 3$, both A_{ij} and E_i are independent of θ and the corresponding form of equation (9) without curvature becomes:

$$\frac{1}{h} \begin{bmatrix} N_x \\ N_y \\ N_{xy} \end{bmatrix} = \begin{bmatrix} U_1 & U_4 & 0 \\ U_4 & U_1 & 0 \\ 0 & 0 & U_5 \end{bmatrix} \begin{bmatrix} \varepsilon_x^0 \\ \varepsilon_y^0 \\ 2\varepsilon_{xy}^0 \end{bmatrix} - \Delta T W_1 \begin{bmatrix} 1 \\ 1 \\ 0 \end{bmatrix} , \tag{18}$$

where

$$U_1 = (3Q_{11} + 3Q_{22} + 2Q_{12} + 4Q_{66})/8 ,$$

$$U_4 = (Q_{11} + Q_{22} + 6Q_{12} - 4Q_{66})/8 ,$$

$$U_5 = (Q_{11} + Q_{22} - 2Q_{12} + 4Q_{66})/8 . \tag{19}$$

If the material is allowed to expand free of applied in-plane stresses (through a change of temperature), then the expansion strain along any axis is given by ε as follows:

$$\varepsilon = \varepsilon_x^0 = \varepsilon_y^0 = \alpha_{2D} \, \Delta T, \tag{20}$$

where α_{2D} defines the quasi-isotropic planar coefficient of expansion. Applying equations (18) and (20), we get:

$$\alpha_{2D} = \frac{W_1}{U_1 + U_4} . \tag{21}$$

Applying equations (16) and (19), we get:

$$Q_{11} + Q_{12} = \frac{1 + \nu_{21}}{1 - \nu_{12}\,\nu_{21}} E_{11}$$

and

$$Q_{12} + Q_{22} = \frac{1 + \nu_{12}}{1 - \nu_{12}\,\nu_{21}} E_{22} .$$

Then (21) becomes:

$$\alpha_{2D} = \frac{(1 + \nu_{21})E_{11}\alpha_1 + (1 + \nu_{12})E_{22}\alpha_2}{(1 + \nu_{21})E_{11} + (1 + \nu_{12})E_{22}} \quad . \tag{22}$$

Now, using:

$$\nu_{21} = \left(\frac{E_{22}}{E_{11}}\right)\nu_{12}$$

in equation (22), we obtain:

$$\alpha_{2D} = \frac{E_{11}\alpha_1 + E_{22}(\alpha_1 + \alpha_2)\nu_{12} + E_{22}\alpha_2}{E_{11} + E_{22}(1 + 2\nu_{12})} \quad . \tag{23}$$

The result, (23), for α_{2D} is restricted at this point to the case $N \geqslant 3$ because of the corresponding restriction on (18). It is necessary here to consider the case $N = 2$ as a separate matter. For a two-layer (0°–90°) laminate, a simple computation shows that the same formula, (23), results for α_{2D} in the directions of material orthotropy. However, the transformation law (3) for the second-order tensor, α_{ij}, shows that if α_{2D} has equal values in two orthogonal directions, then it is quasi-isotropic. Thus the result (23) is now seen to apply for $N \geqslant 2$. The corresponding problem of effective moduli does not admit quasi-isotropy at the $N = 2$ level because the moduli tensor is not of second order. Again, it should be noted that the formula (23) was first derived by Halpin and Pagano [2]. The derivation given here is included for completeness in presenting the corresponding three-dimensional result, and for use with the derivation of some asymptotic results.

The question arises now as to what value of α_{2D} would be obtained as a function of constituent material properties for a composite material behaving quasi-isotropically. Let us take the notation, E_f, ν_f, μ_f, k_f, and α_f as referring to an isotropic fiber, and E_m, ν_m, μ_m, k_m, and α_m as referring to an isotropic matrix. Here, E = Young's modulus, ν = Poisson's ratio, μ = shear modulus, k = bulk modulus, and α = coefficient of linear expansion. Also, if c refers to the fiber-phase volume fraction, then $c = c_f$ and $c_m = 1-c$. To investigate the behavior of the factors that contribute materially to the effective in-plane coefficient of expansion, we will find the asymptotic behavior of (22) in terms of relative properties of fiber and matrix phases. Motivation for taking this approach comes from the corresponding isothermal results obtained by Christensen [5].

First, the appropriate property forms will be stated for the composites cylinder model, i.e.:

$$E_{11} = cE_f + \hat{E}_{11} \quad , \tag{24}$$

where

$$\hat{E}_{11} = (1-c)E_m + 4c(1-c)\mu_m \left[\cfrac{(\nu_f - \nu_m)^2}{\cfrac{(1-c)\mu_m}{k_f + \mu_m/3} + \cfrac{c\mu_m}{k_m + \mu_m/3} + 1} \right] , \tag{25}$$

with

$$\nu_{12} = c\nu_f + (1-c)\nu_m + \cfrac{c(1-c)(\nu_f - \nu_m)\left(\cfrac{\mu_m}{k_m + \mu_m/3} - \cfrac{\mu_m}{k_f + \mu_f/3} \right)}{\cfrac{(1-c)\mu_m}{k_f + \mu_f/3} + \cfrac{c\mu_m}{k_m + \mu_m/3} + 1} \quad , \tag{26}$$

$$K_{23} = k_m + \frac{\mu_m}{3} + \cfrac{c}{\cfrac{1}{k_f - k_m + (\mu_f - \mu_m)/3} + \cfrac{(1-c)}{k_m + 4\mu_m/3}} \quad , \tag{27}$$

$$\frac{\mu_{12}}{\mu_m} = \frac{\mu_f(1+c) + \mu_m(1-c)}{\mu_f(1-c) + \mu_m(1+c)} \quad , \tag{28}$$

and

$$\frac{\mu_{23}}{\mu_m} = 1 + \cfrac{c}{\cfrac{\mu_m}{\mu_f - \mu_m} + \cfrac{[k_m + 7\mu_m/3](1-c)}{2(k_m + 4\mu_m/3)}} \quad . \tag{29}$$

The above equations, (24) to (29), represent the five independent properties of a transversely isotropic material in terms of c and of the fiber and matrix properties. Actually, relation (29) is a bound; see Reference [1] for the relation of this bound to exact results. Index 1 refers to the fiber direction and index 2 refers to any direction perpendicular to the fiber direction.

The expansions above must be augmented by similar expressions for α_1 and α_2 below [1]:

$$\alpha_1 = \overline{\overline{\alpha}} + \frac{(\alpha_f - \alpha_m)}{\left(\dfrac{1}{k_f} - \dfrac{1}{k_m}\right)} \left[\frac{3(1 - 2\,\nu_{12})}{E_{11}} - \left(\overline{\overline{\dfrac{1}{k}}}\right) \right] \quad , \tag{30}$$

$$\alpha_2 = \overline{\overline{\alpha}} + \frac{(\alpha_f - \alpha_m)}{\left(\dfrac{1}{k_f} - \dfrac{1}{k_m}\right)} \left[\frac{3}{2\kappa_{23}} - \frac{3\nu_{12}(1 - 2\,\nu_{12})}{E_{11}} - \left(\overline{\overline{\dfrac{1}{k}}}\right) \right] \quad , \tag{31}$$

where

$$\overline{\overline{\alpha}} = c\alpha_f + (1 - c)\alpha_m, \left(\overline{\overline{\frac{1}{k}}}\right) = \frac{c}{k_f} + \frac{(1 - c)}{k_m} \quad . \tag{32}$$

Properties k_m and μ_m are taken to be of the same order of magnitude; further, it is assumed that:

$$\left. \begin{array}{l} \dfrac{E_m}{cE_f} \ll 1 \\[2ex] \dfrac{k_m}{k_f} \ll 1 \end{array} \right\} \quad \cdots \cdots \quad , \tag{33}$$

with cE_f and k_f being of the same magnitude. Expanding relations (25) to (31), retaining only terms of consistent order in E_m/cE_f and k_m/k_f and substituting them into the effective coefficient of thermal expansion formula (23) and expanding it to consistent order, we finally obtain the asymptotic result:

$$\alpha_{2D} = \alpha_f + (\alpha_m - \alpha_f)k_m \left\{ \left[\frac{3(1 + \tilde{\nu}_{12})\tilde{E}_{22}}{2\tilde{K}_{23}} + 3(1 - 2\tilde{\nu}_{12}) \right] \frac{1}{cE_f} - \frac{1}{k_f} \right\}$$

$$+ 0\left(\left(\frac{E_m}{cE_f}\right)^2 \right) \quad , \tag{34}$$

where

$$\tilde{E}_{22} = \frac{4\tilde{\mu}_{23}\tilde{K}_{23}}{\tilde{K}_{23} + \tilde{\mu}_{23}} \quad , \tag{35}$$

$$\tilde{\mu}_{23} = \mu_m + \frac{2c(k_m + 4\mu_m/3)}{(1-c)(k_m + 7\mu_m/3)} \mu_m \quad , \tag{36}$$

$$\tilde{K}_{23} = \frac{k_m}{1-c} + \frac{(1+3c)}{3(1-c)} \mu_m \quad , \tag{37}$$

$$\tilde{v}_{12} = cv_f + (1-c)v_m + \frac{c(1-c)(v_f - v_m)\left(\dfrac{\mu_m}{k_m + \mu_m/3}\right)}{1 + \dfrac{c\mu_m}{k_m + \mu_m/3}} \quad . \tag{38}$$

It is of interest to apply the asymptotic result for particular values of the Poisson's ratios. In particular, if we take $v_f = v_m = 1/4$, then (34) becomes:

$$\alpha_{2D} \cong \alpha_f + (\alpha_m - \alpha_f)\left(\frac{8 - 3c}{3}\right)\frac{E_m}{cE_f} \quad . \tag{39}$$

Relation (39) shows very simply the scale of interaction between the coefficients of thermal expansion of the two phases. In general, α_{2D} depends on the properties of both phases in a manner that could not be anticipated by any simple rules such as the rule of mixtures.

It should be noted that the result (34) and the reduced form (39) are not valid at the volume fraction $c = 1$. This restriction follows from the approximations introduced through the restrictions, (33), in the effective stiffness formulas (25) to (29). The application of these results to typical systems is discussed after the comparable three-dimensional problem is considered.

THREE DIMENSIONAL ANALYSIS

In the preceding section, the thermal-expansion results derived were appropriate to quasi-isotropic two-dimensional conditions. In this section, comparable results are derived for three-dimensional conditions. Specifically, the effective coefficient of thermal expansion is sought for a fiber-reinforced medium that has a random orientation of fibers. First, results will be derived for the value of α_{3D} expressed in terms of the corresponding properties of a transversely isotropic medium. The composite cylinders model is then invoked for asymptotic results.

Relative to a rectangular Cartesian coordinate system, the governing linear constitutive equation of a transversely isotropic medium is given by:

$$\sigma_{11} = C_{11}(\varepsilon_{11} - \alpha_1 \Delta T) + C_{12}(\varepsilon_{22} - \alpha_2 \Delta T) + C_{12}(\varepsilon_{33} - \alpha_2 \Delta T) \quad,$$

$$\sigma_{22} = C_{12}(\varepsilon_{11} - \alpha_1 \Delta T) + C_{22}(\varepsilon_{22} - \alpha_2 \Delta T) + C_{23}(\varepsilon_{33} - \alpha_2 \Delta T) \quad,$$

$$\sigma_{33} = C_{12}(\varepsilon_{11} - \alpha_1 \Delta T) + C_{23}(\varepsilon_{22} - \alpha_2 \Delta T) + C_{22}(\varepsilon_{33} - \alpha_2 \Delta T) \quad,$$

$$\sigma_{23} = (C_{22} - C_{23})\varepsilon_{23} \quad,$$

$$\sigma_{13} = 2C_{66}\varepsilon_{13} \quad,$$

$$\sigma_{12} = 2C_{66}\varepsilon_{12} \quad, \tag{40}$$

where axis 1 is the axis of symmetry.

The three-dimensional expansion coefficient is developed by means of a randomizing procedure. Preliminary to the randomizing process, let us take a rotated coordinate system (see Figure 2) such that the strains in the primed directions are given by:

$$\varepsilon'_{11} = \varepsilon'_{22} = \varepsilon'_{33} = \varepsilon \quad,$$

$$\varepsilon'_{23} = \varepsilon'_{13} = \varepsilon'_{12} = 0 \quad. \tag{41}$$

Substitution of (41) into (40) gives:

$$\sigma_{11} = (C_{11} + 2C_{12})\varepsilon - (C_{11}\alpha_1 + 2C_{12}\alpha_2)\Delta T \quad,$$

$$\sigma_{22} = (C_{12} + C_{22} + C_{23})\varepsilon - (C_{12}\alpha_1 + C_{22}\alpha_2 + C_{23}\alpha_2)\Delta T \quad,$$

$$\sigma_{33} = (C_{12} + C_{22} + C_{23})\varepsilon - (C_{12}\alpha_1 + C_{23}\alpha_2 + C_{23}\alpha_2)\Delta T \quad,$$

$$\sigma_{ij} (i \neq j) = 0 \quad. \tag{42}$$

Now without less of generality, the angles θ and ϕ define orientation of direction 1 with axes $1'$, $2'$, and $3'$, while direction 3 lies in the $1'$, $2'$ plane. Application of the stress-tensor-transformation equation to determine the stress in the $3'$ direction that is due to the unprimed stresses gives:

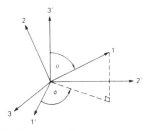

Figure 2. *Three-dimensional rotated coordinates.*

$$\sigma'_{33} = \sigma_{11} \cos^2 \theta + \sigma_{22} \sin^2 \theta \ . \tag{43}$$

A randomizing procedure, Reference [5], that averages the value of σ'_{33} over all possible fiber orientations in space requires the integration:

$$\sigma'_{33} \bigg|_{Random} = \frac{1}{2\pi} \int_0^\pi \int_0^\pi \sigma'_{33} \sin \theta \, d \, \theta \, d \phi \ . \tag{44}$$

Performing the integration of (44) using (43) and setting:

$$\sigma'_{33} \bigg|_{Random} = 0$$

gives

$$\sigma_{11} + 2\sigma_{22} = 0 \ . \tag{45}$$

The strain per unit temperature is defined as α_{3D}. When the expressions (42) are substituted into (45), the result is:

$$\alpha_{3D} = \frac{(C_{11} + 2C_{12})\alpha_1 + 2(C_{12} + C_{22} + C_{23})\alpha_2}{C_{11} + 4C_{12} + 2C_{22} + 2C_{23}} \ , \tag{46}$$

which can be expressed equivalently as:

$$\alpha_{3D} = \frac{[E_{11} + 4\nu_{12}(1 + \nu_{12})K_{23}]\alpha_1 + 4(1 + \nu_{12})K_{23}\alpha_2}{E_{11} + 4(1 + \nu_{12})^2 K_{23}} \ . \tag{47}$$

This result, (47), is the three-dimensional random-fiber-orientation form that correspondes to the two-dimensional form (23). Note that when $\alpha_1 = \alpha_2$, equation (47) correctly predicts that $\alpha_{3D} = \alpha_1$, even though the stiffness properties are not necessarily isotropic.

Knowledge of the properties of an aligned fiber system allow the prediction of the effective coefficient of thermal expansion for a three-dimensional randomized-fiber system. The properties α_1, α_2, E_{11}, ν_{12}, and K_{23} which enter (47) may either be known by direct measurement or may be predicted from relations (24) to (31). Motivated by the results of the preceding section, we now seek the asymptotic form of (47) that is appropriate to the stiff fiber case. If we use in relations (24) to (31) the restrictions (33) and the interrelations of the properties stated with (33), it follows that (47) takes the form:

$$\alpha_{3D} = \alpha_f + (\alpha_m - \alpha_f)\left(\frac{9}{cE_f} - \frac{1}{k_f}\right)k_m + 0\left(\left(\frac{E_m}{cE_f}\right)^2\right) \tag{48}$$

This simple expression offers an immediate determination of the scale of effects coming from the two different phases. It is of particular importance to evaluate the asymptotic result equation (48) against the original form (47).

DISCUSSION

The main results of the present work are the predictions (23) and (47) and the corresponding asymptotic forms (34) and (48) for the coefficients of thermal expansion of fiber composites in the quasi-isotropic two-dimensional case and the isotropic three-dimensional case. Equation (23) was previously derived by Halpin and Pagano [2]. The other three results are new.

As a first application of these results, a comparison with experimental data is made. A polyester-glass composite, which is quasi-isotropic in two dimensions, was measured to have a coefficient of thermal expansion α_{2D} of $16.7 \times 10^{-6}/°C$. The properties shown in Table 1 were used to obtain the theoretical prediction of α_{2D} from equation (23), with equations (24) to (31);

$$\alpha_{2D} = 17.7 \times 10^{-6}/°C.$$

The corresponding asymptotic prediction from equation (34) is:

Table 1. Properties of glass fibers and matrix phase.

Glass fibers	Matrix phase
E_G = 86,200 MPa	E_p = 2,800 MPa
ν_G = 0.2	ν_p = 0.35
α_G = 3 × 10^{-6}/°C	α_p = 8 × 10^{-5}/°C
c = 0.34	

$$\alpha_{2D} \cong 23.6 \times 10^{-6}/ \,^\circ C.$$

The theoretical prediction for α_{2D} of 17.7 and the measured value of 16.7 are in close agreement. The asymptotic value of 23.6 is in considerable error. This raises the question, which is considered next, of the range of practical use to which the asymptotic formulas can be put.

Figures 3 to 5 display predictions of the coefficient of thermal expansion in the two-dimensional case for a typical system involving stiff fibers in a polymeric matrix phase. The comparison is made between the analytical prediction (23) and the asymptotic form (34). The parameter that varies among Figures 3 to 5 is the volume fraction of fibers. First, it is seen that α_{2D} is a strong function of stiffness ratio E_f/E_m. Also, it is apparent that error in the asymptotic prediction increases with decreasing ratio of E_f/E_m. At a given value of E_f/E_m, the error decreases somewhat with increasing volume fraction. At a volume fraction c of 0.25, the error in the asymptotic prediction is larger than 25 percent if $E_f/E_m < 40$. If E_f/E_m is 100, the error is about 10 percent. Thus for the asymptotic result to give a reliable prediction of composite thermal expansion, the fibers must be very stiff compared with the matrix phase. The range of validity of the asymptotic formulas for α_{2D} is more restrictive than the corresponding range for effective moduli [6].

Figure 6 is related to Figure 4 except that in Figure 6, $\alpha_m < \alpha_f$ whereas in Figure 4, $\alpha_m > \alpha_f$. The restrictions upon E_f/E_m for the asymptotic formula to be valid are about the same as in the cases of Figures 3 to 5.

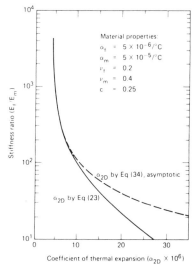

Figure 3. *Two-dimensional thermal expansion, $\alpha_f < \alpha_m$, $c = 0.25$.*

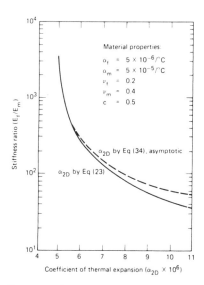

Figure 4. *Two-dimensional thermal expansion, $\alpha_f < \alpha_m$, $c = 0.5$.*

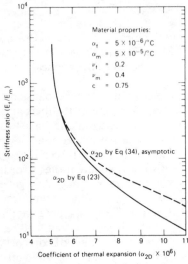

Figure 5. *Two-dimensional thermal expansion, $\alpha_f < \alpha_m$, $c = 0.75$.*

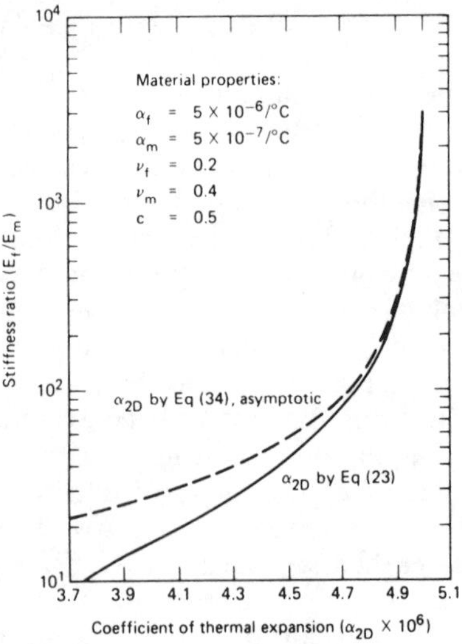

Figure 6. *Two-dimensional thermal expansion, $\alpha_f > \alpha_m$, $c = 0.50$.*

Figure 7 displays the coefficients of thermal expansion for the three-dimensional case, which may be compared with the two-dimensional case in Figure 3, because all specified properties and the volume fraction are the same. It is seen that the asymptotic three-dimensional formula has a more restrictive range of validity than does the two-dimensional result. This effect

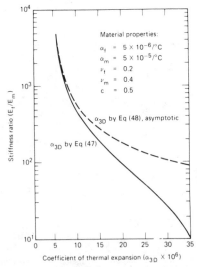

Figure 7. *Three-dimensional thermal expansion.*

is thought to be explained by the fact that at a given volume fraction, orienting fibers in three dimensions provide much less stiffening influence than orienting fibers in two dimensions.

The preceding examples show that the asymptotic formulas have a rather limited range of validity. However, they offer an immediate estimate of effects within that range. For more precise results, recourse must be made to the derived analytical forms (23) and (47). These forms can be used with either direct, experimentally determined properties of aligned fiber systems or predicted properties from a model, such as the composite cylinders model used here. Furthermore, two-dimensional laminates that are not quasi-isotropic are amenable to a predictive value of the anisotropic coefficient of thermal expansion. This result may be obtained from the forms derived herein if each lamina has a specified direction.

ACKNOWLEDGMENTS

This work was performed under the auspices of the U.S. Department of Energy by Lawrence Livermore National Laboratory under contract No. W-7405-Eng-48.

NOTICE

REFERENCES

1. Christensen, R. M. *Mechanics of Composite Materials.* New York: John Wiley (1979).
2. Halpin, J. C. and Pagano, N. J., "The Laminate Approximation for Randomly Oriented Fibrous Composites," *J. Comp. Materials, 3,* 720 (1969).
3. Hashin, Z. and Rosen, B. W., "The Elastic Moduli of Fiber-Reinforced Materials," *J. Appl. Mech., 31,* 223 (1964).
4. Tsai, S. W. and Pagano, N. J., "Invariant Properties of Composite Materials," in *Composite Materials Workshop,* S.W. Tsai, J.C. Halpin, and N.J. Pagano, ed. Lancaster, PA: Technomic Publishing Co., Inc. (1968).
5. Christensen, R. M. and Waals, F. M., "Effective Stiffness of Randomly Oriented Fiber Composites," *J. Comp. Materials, 6,* 518 (1972).
6. Christensen, R. M., "Asymptotic Modulus Results for Composites Containing Randomly Oriented Fibers," *Int. J. Solids Structures, 12,* 537 (1976).

28

Analysis of the Thermal Expansion Coefficients of Particle-Filled Polymers

K. TAKAHASHI, K. HARAKAWA, AND T. SAKAI

INTRODUCTION

IN RECENT YEARS, VARIOUS KINDS OF MOLDED PRODUCTS OF POLYMERS filled with short fibers or flakes have been developed for many practical applications; e.g., car parts, electrical appliances, or commodity plastics. In these composites, filler-particles have various types of orientation distributions. And the mechanical properties of composites, elastic moduli, strength, thermal expansion coefficients, etc., are affected considerably by the orientation distribution of particles. Therefore, the effect of the orientation distribution should be incorporated in the analysis of mechanical properties of particle-filled polymers. The present paper is concerned with the prediction of the effective thermal expansion coefficients of particle-filled polymers. From the engineering point of view, thermal expansion coefficients of composites are very important in relation to the dimensional stability and the mechanical compatibility when used with other materials.

The thermal expansion coefficient of a material filled with spherical particles have been formulated by Kerner [1]. Schapery [2] has derived the upper and lower bounds of isotropic composites by employing the energy principles. These bounds are independent of the shape of filler-particles. Wakashima et al. [3] and Ishikawa et al. [4] have analyzed the thermal expansions of the composites in which fibers or disks are oriented uniaxially.

Very few data can be found in the literature to estimate the effect of the orientation distribution of particles. Christensen and Waals [5] have derived simple formulas for the elastic moduli of a material with two- or three-dimensionally randomly oriented fibers, but the effect of the interaction among the randomly oriented fibers has not been considered. Fukuda and Kawata [6] have calculated the effect of fiber orientation on Young's modulus of a short fiber composite under the condition of plane stress assuming the interaction among the fibers to be negligible.

There is at present little hope of rigorously solving the problem of a composite in which there are many interacting particles. An approximate method which makes the estimation of the interaction among the particles possible is termed the self-consistent method or the smearing-out method. Summaries of the smearing-out method may be found in Hashin [7]. The analysis by Kerner [1] or Wakashima et al. [3] is one of them. For the composite in which many ellipsoidal particles are oriented uniaxially, Mori and Tanaka [8] have discussed theoretically the image strain which is produced by the interaction among the particles.

In this paper, we will develop a kind of self-consistent method to estimate the interaction among the ellipsoidal particles having various orientation distributions. Following Eshelby [9], the real particles are replaced by the equivalent particles whose elastic moduli are equal to those of the matrix. The transformation strain allocated to each equivalent particle is a function of the orientation of each particle, and it is determined in such a manner that the existence of the image strain is satisfied self-consistently. Here the composite is an assumed homogeneous body because of the introduction of equivalent particles, so it is not unreasonable to assume that the image strain produced by the interaction among equivalent particles is uniform throughout the composite. This assumption will give a good approximation, if attention is directed not to the detailed strain distribution but to a macroscopic average strain in the composite.

From the average of the transformation strain over the orientation distribution of particles, the effective thermal expansion coefficient of the composite can be derived. Numerical calculations are carried out for the glass particle-epoxy resin composites in which ellipsoidal particles are oriented uniaxially, plane-randomly, or space-randomly. And the results of calculations are compared with existing experimental data.

THEORETICAL

A self-consistent method will be developed for the calculation of the thermal expansion coefficient of a composite containing ellipsoidal particles with an arbitrary orientation distribution. Our theoretical analysis will be based on the following assumptions. 1) The matrix and the particles are ideally elastic, isotropic, and homogeneous. 2) The particles can be treated, on average, as similar ellipsoids (Figure 1).

$$\frac{x_1^2}{a^2} + \frac{x_2^2}{a^2} + \frac{x_3^2}{c^2} = 1 \tag{1}$$

3) The volume of each particle is sufficiently small compared with the total volume of the composite and the distribution of their position in space is

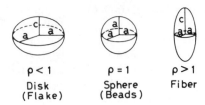

Figure 1. *Particles are assumed to be ellipsoids whose shape effect is characterized by the aspect ratio $\varrho = c/a$.*

statistically uniform, so that the composite may be regarded as a quasi-homogeneous body on a macroscopic scale. 4) Matrix and particles are perfectly bonded together at the interface. 5) When the real particles are replaced by the equivalent ones, which have the elastic moduli equal to those of the matrix and the same shape and size as the real ones, the image strain produced by the interaction among the particles is uniform throughout the composite.

The shape effect of the particles is characterized by the aspect ratio $\varrho = c/a$. If ϱ is larger than unity the particle is fiber-shaped, and if it is smaller than unity, the particle is a disk-shaped one. A spherical particle is given when ϱ is equal to unity.

In equation (1) the axes of the coordinate system x_i ($i = 1, 2, 3$) are taken along the principal axes of the ellipsoid, and x_i varies with the orientation of each ellipsoidal particle. Accordingly, it is necessary to introduce the coordinate system $\mathring{x}_i$ ($i = 1, 2, 3$) fixed in the composite specimen. The strain components referred to x_i and $\mathring{x}_i$ coordinate systems are expressed by e_{ij} and $\mathring{e}_{ij}$, respectively. The transformation law for these strains is given by:

$$e_{ij} = a_{im} a_{jn} \mathring{e}_{mn}$$

where a_{ij} is the direction cosine between the x_i and $\mathring{x}_j$ axes, and a repeated suffix is summed over the values 1,2,3. The orientation of the particle is evaluated by a_{ij} and the orientation distribution function.

The analysis consists of two parts: The first deals with the interaction among the equivalent particles, and the second with the thermal expansion coefficient of the composite.

Interaction Among Equivalent Particles

We consider first a region (the "equivalent particle") in an infinite homogeneous body is subjected to a transformation which, in the absence of the constraint imposed by its surroundings (the "matrix"), would be prescribed uniform strain e_{ij}^T. This is the transformation problem of Eshelby [9]. If the region is an ellipsoid, the strain inside it is uniform and can be expressed by:

$$e_{ij}^C = T_{ijkl}e_{kl}^T \tag{2}$$

where T_{ijkl} is Eshelby's tensor which is determined by the aspect ratio of the ellipsoidal particle and Poisson's ratio of the matrix. On the other hand, the average strain produced in the matrix can be regarded approximately to be zero when a single particle is an infinite matrix [10].

When certain fractions of particles are filled in a finite matrix, the additional elastic strain $\mathring{e}_{ij}^m$ is produced by the interaction among the particles and the presence of free boundary of the specimen. $\mathring{e}_{ij}^m$ is termed image strain [8]. $\mathring{e}_{ij}^m$ may be a function of the position in the composite specimen, but it is almost impossible to obtain the precise solution in general. Therefore, we assume $\mathring{e}_{ij}^m$ to be uniform throughout the homogeneous specimen containing the equivalent particles. This assumption will give a good approximation to the prediction of the effective thermal expansion coefficients of composites, since these are related to the macroscopic average strains.

As a result, total strain $(e_{ij}^C + e_{ij}^m)$ is produced in an equivalent particle, in which elastic comonent is $(e_{ij}^C + e_{ij}^m - e_{ij}^T)$. And the corresponding stress is expressed by $C_{pqij}(e_{ij}^C + e_{ij}^m - e_{ij}^T)$, where C_{pqij} is the elastic modulus of the matrix.

When a homogeneous specimen with the thermal expansion coefficient α undergoes a uniform temperature change ΔT, the thermal expansion is uniform ($\alpha\Delta T$) throughout the specimen and an internal elastic field is not generated. However, when the specimen includes particles with the thermal expansion coefficient α^* different from that of the matrix, the extra strain $(\alpha^* - \alpha)\Delta T\delta_{ij}$ is remained in the particles by the temperature change ΔT and an internal elastic field is generated. Here δ_{ij} is the Kronecker delta. This real particle can be replaced by an equivalent particle with transformation strain e_{ij}^T, applying the inhomogeneity problem of Eshelby [9], if the following relation is satisfied:

$$C_{pqij}(e_{ij}^C + e_{ij}^m - e_{ij}^T) = C_{pqij}^*[e_{ij}^C + e_{ij}^m - (\alpha^* - \alpha)\Delta T\delta_{ij}] \tag{3}$$

where C_{pqij}^* is the elastic modulus of the real particle. The left side member of equation (3) exhibits the stress in the equivalent particle, while the right side member exhibits the stress in the real particle, when these are subjected to the total strain $(e_{ij}^C + e_{ij}^m)$.

After eliminating e_{ij}^C with the help of equation (2), the transformation strain e_{ij}^T allocated to the equivalent particle can be obtained by solving the six equations (3). Substituting the solution e_{ij}^T again into equation (2), we obtain the constraint strain e_{ij}^C in the equivalent particle. It should be noted that up to this step of calculation the components of image strain $\mathring{e}_{ij}^m$ remain unknowns, and that equations (2) and (3) are only valid in a coordinate system x_i whose axes are parallel to the principal axes of an ellipsoidal particle in question.

e_{ij}^T and e_{ij}^C derived for each orientation of the particles are transformed to $\mathring{e}_{ij}^T$ and $\mathring{e}_{ij}^C$ which are the strain components with respect to the coordinate system x_i fixed in the specimen. The strain components $\mathring{e}_{ij}^T$ and $\mathring{e}_{ij}^C$ are then weighted according to the frequency of occurrence of each orientation, and averaged over the whole orientation distribution of particles. The resulting average strains are expressed by $<\mathring{e}_{ij}^T>$ and $<e_{ij}^C>$, respectively.

Here the composite is assumed as a homogeneous body by the introduction of the equivalent particles. For a homogeneous elastic body, the volume average of any elastic strain component associated with an internal elastic field should vanish provided that the averaging procedure is carried throughout the body. The elastic strain in an equivalent particle is expressed as $(\mathring{e}_{ij}^C + \mathring{e}_{ij}^m - \mathring{e}_{ij}^T)$, while the average elastic strain in the matrix is expressed by $\mathring{e}_{ij}^m$ alone because the average strain in the matrix is regarded approximately to be zero if the interaction among particles can be neglected [10]. Averaging the elastic strain throughout the composite, we have

$$v(<\mathring{e}_{ij}^C> + \mathring{e}_{ij}^m - <\mathring{e}_{ij}^T>) + (1-v)\mathring{e}_{ij}^m = 0$$

where v is the volume fraction of the particles.

Consequently, the image strain $\mathring{e}_{ij}^m$ produced by the interaction among equivalent particles can be found by solving the following six equations:

$$\mathring{e}_{ij}^m = -v(<\mathring{e}_{ij}^C> - <\mathring{e}_{ij}^T>) \tag{4}$$

Once the six components of $\mathring{e}_{ij}^m$ are found, $<\mathring{e}_{ij}^T>$ can be calculated in terms of C_{pqij}, C_{pjij}^*, α, α^*, ΔT, ϱ, and the orientation distribution function of the particles.

Thermal Expansion Coefficients of Composites

In the preceding analysis, it has been shown that an elastic state in the composite containing real particles can be duplicated by that in the homogeneous body containing the equivalent particles with e_{ij}^T determined properly by equation (3).

The overall strain generated in a homogeneous body containing the equivalent particles is easily calculated as the sum of two components: One is generated by the temperature change ΔT; $\alpha \Delta T \delta_{ij}$. The other is the average strain due to the transformation strain in the equivalent particles [3,8]; $v<e_{ij}^T>$. On the other hand, when the composite is regarded as a quasi-homogeneous body with the effective thermal expansion coefficient α_{ij}^{**}, the overall strain in the composite by temperature change ΔT is also expressed by $\alpha_{ij}^{**}\Delta T$.

Consequently, the effective thermal expansion coefficient of the composite α_{ij}^{**} is found to be:

$$\alpha_{ij}^{**} = \alpha\delta_{ij} + v<e_{ij}^T>/\Delta T \tag{5}$$

Explicit solutions for the general cases would be very complicated and in this section the results of only a few special cases are shown. When particles are distributed space-randomly, the thermal expansion coefficient of the composite is apparently isotropic.

$$\alpha_{ij}^{**} = \alpha^{**}\delta_{ij}$$

For a composite containing the spherical particles ($a = c$), we have

$$\alpha^{**} = \alpha + \frac{\kappa^*(3\kappa + 4\mu)(\alpha^* - \alpha)\nu}{\kappa^*(3\kappa + 4\mu\nu) + 4\mu\kappa(1 - \nu)} \tag{6}$$

where μ is the shear modulus of the matrix, and κ^* and κ are the bulk moduli of the particles and the matrix, respectively. Equation (6) agrees with that derived by Kerner [1] and with the upper bound of isotropic composites derived by Schapery [2].

For a composite in which fiber-shaped particles ($\varrho = \infty$) are oriented space-randomly, we have a new expression:

$$\alpha^{**} = \alpha + \frac{\kappa^*(\mu^* + 3\kappa + 3\mu)(\alpha^* - \alpha)\nu}{\kappa^*(\mu^*\nu + 3\kappa + 3\mu\nu) + \kappa(\mu^* + 3\mu)(1 - \nu)} \tag{7}$$

where μ^* is the shear modulus of the particles.

For a composite in which disk-shaped particles ($\varrho = 0$) are oriented space-randomly, we have

$$\alpha^{**} = \alpha + \frac{\kappa^*(4\mu^* + 3\kappa)(\alpha^* - \alpha)\nu}{\kappa^*(4\mu^*\nu + 3\kappa) + 4\mu^*\kappa(1 - \nu)} \tag{8}$$

Equation (8) agrees with the lower bound of isotropic composites derived by Schapery [2].

It is interesting to note that equation (6) for a sphere-composite is independent of μ^*, and equation (8) for a disk-composite is independent of μ.

NUMERICAL RESULTS

In this section, some results obtained from numerical calculations will be presented. Calculations are carried out for the glass particle-epoxy resin composites and the effects of the orientation distribution and the aspect ratio of the particles are investigated theoretically. And the results of calculations are also compared with the existing experimental data.

The properties of the constituents used in the calculations are listed in Table 1. For brevity, the thermal expansion coefficient of a composite is denoted simply by α^{**} hereafter.

Table 1. Properties of constituents.

	Thermal Expansion Coefficient $(10^{-6}/°C)$	Young's Modulus (kg/mm^2)	Poisson's Ratio
Epoxy resin	60	370	0.39
Glass particles	5	7400	0.22

The Effect of Orientation Distribution of Particles

Three typical cases of the orientation distribution of particles are considered:

1. Uniaxial orientation; c-axes of ellipsoidal particles are oriented parallel to each other.
2. Plane random orientation; c-axes of particles are oriented at random in the planes which are parallel to each other.
3. Space random orientation; c-axes of particles are oriented at random in space.

Figure 2 shows the effect of the orientation distribution of glass fibers with infinite aspect ratio ($\varrho = \infty$). Superscripts, U, P, and S indicate the uniaxial, plane random, and space random orientations, respectively.

For the uniaxial orientation of fibers, subscript L indicates the longitudinal direction (parallel to c-axes) of the fibers, and subscript T indicates the transverse direction (perpendicular to c-axes). For the plane random orientation of fibers, subscript L indicates the direction parallel to the planes in which c-axes of fibers are oriented at random, and subscript T indicates the direction perpendicular to the planes.

In the longitudinal direction, α^{**} decreases with an increase of volume fraction of fibers, and the uniaxial orientation is more effective on the decrease of α^{**} than plane random or space random orientation. While in the transverse direction, α^{**} increases initially by loading of glass fibers, and the increase is more remarkable in the case of plane random orientation. This increase of α^{**} can be interpreted in the following manner. A fiber constrains the thermal expansion of the matrix in the longitudinal direction. This constraint (compression) of the matrix in the longitudinal direction is accompanied by the transverse strain components (extensions) equal to the product of the compression strain and Poisson's ratio. In the range of small volume fraction of fibers, this extension of matrix is more effective than the small thermal expansion of fibers on the overall expansion of the composite in the transverse direction. Therefore in the transverse direction, α^{**} becomes effectively larger even than that of the matrix.

Figure 3 shows the calculated results for the glass flake (disk)-epoxy resin

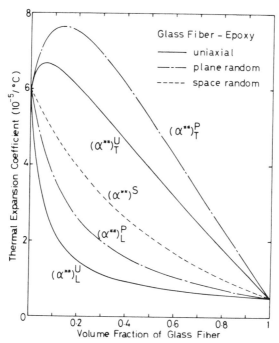

Figure 2. α^{**} *of a glass fiber-epoxy resin composite as a function of volume fraction of glass fibers. Aspect ratio of fibers is assumed infinite. Superscripts U,P, and S represent the orientation distribution of fibers and subscripts L and T represent the directions along which α^{**} are measured. U: uniaxial orientation, L: longitudinal direction (parallel to c-axis of the fiber). T: transverse direction (perpendicular to c-axis). P: plane random orientation; c-axes of fibers are oriented at random in the planes which are parallel to each other, L: parallel to the planes. T: perpendicular to the planes. S: space random orientation.*

composites. Aspect ratio of disks is assumed infinitesimal ($\varrho = 0$). Since it is practically impossible that c-axes of disks are oriented plane-randomly, the results for the plane random orientation are not shown in Figure 3.

For the uniaxial orientation of disks, subscript L indicates the direction parallel to the disk-plane (perpendicular to c-axis of a disk), and subscript T indicates the transverse direction of the disk-plane (parallel to c-axis of a disk).

The initial increase of $(\alpha^{**})_T^U$ also appears but it is larger than that of a fiber-composite. It is because the matrix is constrained two-dimensionally by disks while one-dimensionally by fibers. $(\alpha^{**})^S$ of a disk-composite (the broken line), which is lower than that of a fiber-composite, coincides with the lower bound of isotropic composites derived by Schapery.

Critical Aspect Ratio of Particles

The critical aspect ratio is interpreted as the aspect ratio which is required

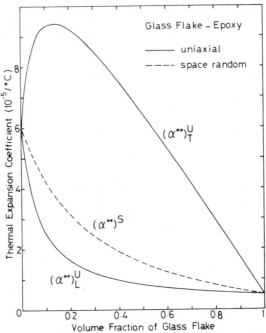

Figure 3. α^{**} *of a glass flake (disk)-epoxy resin composite as a function of volume fraction of glass flakes (disks). Aspect ratio of disks is assumed infinitesimal. Superscript U represents the uniaxial orientation; c-axes of disks are oriented uniaxially. Subscript L represents the direction parallel to the disk-plane (perpendicular to c-axis of the disk), and subscript T represents the direction perpendicular to the disk-plane (parallel to c-axis of the disk).*

for the particles to exhibit the shape-effect of a disk or a fiber almost sufficiently. Then setting the sphere ($\varrho = 1$) on the base, we define the critical aspect ratio ϱ_c of disks or fibers by:

$$\frac{\alpha^{**}(1) - \alpha^{**}(\rho_c)}{\alpha^{**}(1) - \alpha^{**}(0, \infty)} = 0.9 \tag{9}$$

where $\alpha^{**}(1)$ represents the α^{**} of a composite containing spherhical particles ($\varrho = 1$). And $\alpha^{**}(0, \infty)$ represents the α^{**} of a composite containing the particles with the ultimate aspect ratio; disks with $\varrho = 0$ or fibers with $\varrho = \infty$.

Figure 4 shows the α^{**}-ϱ relations of the uniaxially oriented glass particle-epoxy resin composites for various volume fractions of particles. Solid lines represent the α^{**} along the c-axes, and broken lines represent the α^{**} along the a-axes of the ellipsoidal particles. Maximum values of α^{**} is obtained by about 5 vol percent of fibers ($\varrho>1$), or about 10 vol percent of disks ($\varrho<1$). Open circles indicate the critical aspect ratios defined by equation (9). The

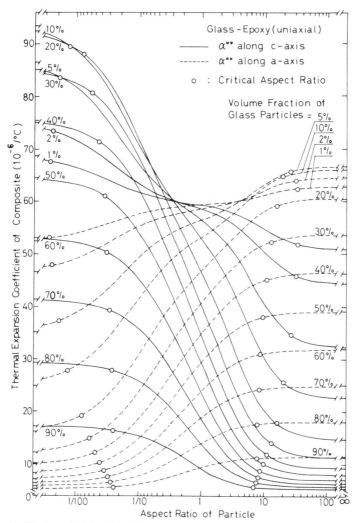

Figure 4. α^{**} *of a uniaxially oriented glass particle-epoxy resin composite as a function of the aspect ratio of particles. Solid lines represent α^{**} along c-axis, and broken lines along a-axis of the ellipsoidal particle. Open circles indicate the critical aspect ratio defined by Equation (9).*

critical aspect ratio decreases with an increase of the volume fraction of particles.

Figure 5 shows the α^{**}-ϱ relations of the space-randomly oriented glass particle-epoxy resin composites for various volume fraction of particles. α^{**} has a maximum at $\varrho = 1$ (spheres), and decreases when spheres ($\varrho = 1$) are changed into fibers ($\varrho > 1$) or disks ($\varrho < 1$). α^{**} of a disk-composite is lower

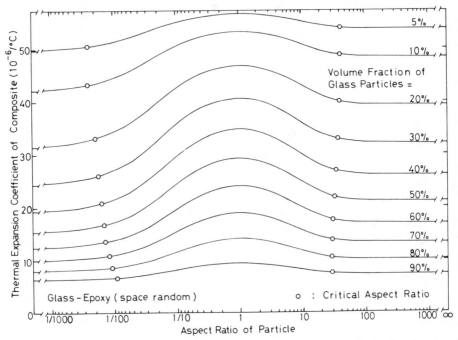

Figure 5. α^{**} of a space-randomly oriented glass particle-epoxy resin composite as a function of aspect ratio of particles. Open circles indicate the critical aspect ratios.

than that of a fiber-composite on the whole. The critical aspect ratio of space random orientation has a relatively large value for fibers ($\varrho > 1$), and a small value for disks ($\varrho < 1$), compared with that of the uniaxial orientation.

The effect of orientation distribution and volume fraction of particles on the critical aspect ratio is summarized in Table 2.

When a space-randomly oriented glass particle-epoxy resin composite is designed in relation to the thermal expansion behavior, a glass fiber with

Table 2. Critical aspect ratios of glass particles in an epoxy resin.

	Orientation Distribution	Volume Fraction of Particles		
		5%	10%	50%
Glass Fiber	Uniaxial	27	20	10
	Plane Random	34	30	20
	Space Random	40	40	32
Glass Flake	Uniaxial	1/180	1/130	1/40
	Space Random	1/270	1/250	1/150

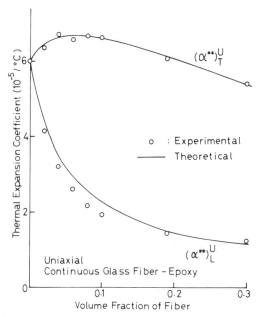

Figure 6. *Comparison of calculated results with experimental data by Sato et al. [11] for a uniaxially oriented continuous glass fiber-epoxy resin composite. Cross-sectional diameter of glass fibers is 13μm.*

$\varrho>40$ or a glass flake (disk) with $\varrho<1/270$ is recommended as an effective filler particle.

Comparison with Experimental Data

Now the numerical results obtained here will be compared with the experimental data conducted by Sato et al. [11]. Composite specimens prepared by them were epoxy resins filled with: 1) uniaxially oriented continuous glass fibers (13 μm cross-sectional diameter), 2) glass beads (30 μm average diameter), 3) glass flakes 325 mesh (3 μm thickness) and 4) glass flakes 150 mesh (3 μm thickness).

Figure 6 shows α^{**} as a function of volume fraction on the uniaxially oriented continuous glass fibers. Agreement between theoretical (solid lines) and experimental results (open circles) is satisfactory.

Figure 7 shows α^{**} as a function of volume fraction of glass particles. In the theoretical calculations, it is assumed that the aspect ratios of glass beads, glass flakes 325 mesh, and glass flakes 150 mesh are 1, 1/7, 1/30, respectively, and that these particles are oriented space-randomly. Because these glass particles have sufficiently small sizes compared with the thickness of the specimen (about 5 mm), it can be considered that these particles are oriented space-randomly in the specimen.

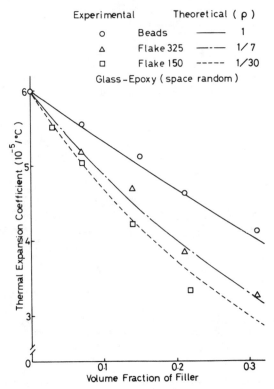

Figure 7. *Comparison of calculated results with experimental data by Sato et al. [11] for a glass particle-epoxy resin composite. Particles used in the experiment were: (O) glass beads of 30μm average diameter, (Δ) glass flake 325 mesh of 3 μm thickness, and (□) glass flake 150 mesh of 3 μm thickness.*

As shown in Figures 6 and 7, calculated results are almost in good agreement with the experimental results. It appears to indicate that the thermal expansion coefficients of particle-filled polymers can be estimated by the approximate method proposed in this paper.

CONCLUSIONS

The effective thermal expansion coefficients of particle-filled composites have been estimated by an approximate method that takes account of the interaction among the arbitrary oriented particles. The expression for a composite filled with spherical particles ($a = c$), equation (6), agrees with that derived by Kerner [1] and the upper bound of isotropic composites derived by Schapery [2]. The expression for a composite in which disk-shaped particles ($a \gg c$) are oriented space-randomly, equation (8), agrees with the lower bound derived by Schapery [2].

Numerical calculations were performed for the glass particle-epoxy resin composites, and the effects of the orientation distribution and the aspect ratio of particles on α^{**} were investigated theoretically. In the case of space random orientation, disk-shaped particles ($a>c$) are more effective on α^{**} than fiber-shaped particles ($a<c$). The critical aspect ratio ϱ_c, defined by equation (9), depends on the orientation distribution and the volume fraction of particles (Table 2).

The calculated results were compared with existing experimental data for the glass-epoxy composites, and it was found that the method of calculation proposed here provides a good estimation for the thermal expansion coefficients of particle-filled polymers.

NOMENCLATURE

a_{ij}	direction cosine between x_i and $\mathring{x}_j$ axes
C_{pqij}	elastic modulus of matrix
C^*_{pqij}	elastic modulus of particle
e_{ij}	strain component referred to x_i coordinate system
$\mathring{e}_{ij}$	strain component referred to $\mathring{x}_i$ coordinate system
e^T_{ij}	transformation strain allocated to the equivalent particle when the real particle is replaced by the equivalent particle whose elastic moduli are equal to those of the matrix
e^C_{ij}	constrained strain produced by e^T_{ij} in the equivalent inclusion
e^m_{ij}	image strain produced by the interaction among equivalent particles
$\langle e_{ij} \rangle$	volume average of e_{ij} over the orientation distribution of particles
T_{ijkl}	Eshelby's tensor
v	volume fraction of particles
$x_i(i=1,2,3)$	the coordinate system whose axes are parallel to the principal axes of an ellipsoidal particle
$\mathring{x}_i(i=1,2,3)$	the coordinate system fixed in the specimen
α	thermal expansion coefficient of matrix
α^*	thermal expansion coefficient of particles
α^{**} or α^{**}_{ij}	thermal expansion coefficient of composite
δ_{ij}	Kronecker delta
ΔT	temperature change to which specimen is subjected
$\varkappa$	bulk modulus of matrix
$\varkappa^*$	bulk modulus of particles
μ	shear modulus of matrix
μ^*	shear modulus of particles
$\varrho = c/a$	aspect ratio of particle
ϱ_c	critical aspect ratio

Superscript

U	uniaxial orientation

P	plane random orientation
S	space random orientation

Subscript

L	longitudinal direction
T	transverse direction

REFERENCES

1. Kerner, E. H., "The Elastic and Thermoelastic Properties of Composite Media," *Proc. Phys. Soc., 69B,* 808 (1956).
2. Schapery, R. A., "Thermal Expansion Coefficients of Composite Materials Based on Energy Principles," *J. Composite Materials, 2,* 380 (1968).
3. Wakashima, K., Otsuka, M., and Umekawa, S., "Thermal Expansion of Heteroegeneous Solids Containing Aligned Ellipsoidal Inclusions," *J. Composite Materials, 8,* 391 (1974).
4. Ishikawa, T., Koyama, K., and Kobayashi, S., "Thermal Expansion Coefficients of Unidirectional Composites," *J. Composite Materials, 12,* 153 (1978).
5. Christensen, R. M. and Waals, F. M., "Effective Stiffness of Randomly Oriented Fiber Composites," *J. Composite Materials, 6,* 518 (1972).
6. Fukuda, H. and Kawata, K., "On Young's Modulus of Short Fiber Composites," *Fibre Sci. & Tech., 7,* 207 (1974).
7. Hashin, Z., "Theory of Mechanical Behavior of Heterogeneous Media," *Appl. Mech. Rev., 17,* 1 (1964).
8. Mori, T. and Tanaka, K., "Average Stress in Matrix and Average Elastic Energy of Materials with Misfitting Inclusions," *Acta Metall., 21,* 571 (1973).
9. Eshelby, J. D., "The Determination of the Elastic Field of an Ellipsoidal Inclusion, and Related Problems," *Proc. Roy. Soc., 241A,* 376 (1957).
10. Tanaka, K. and Mori, T., "Note on Volume Integrals of the Elastic Field Around an Ellipsoidal Inclusion," *J. Elasticity, 2,* 199 (1972).
11. Sato, N., Sakai, T., and Kuze, E., "Thermal Expansion Characteristics of Glass-Epoxy Resin Composite in the Viscoelastic State (II)," *Trans. Japan Soc. Composite Materials, 4,* 5 (1978).

29

Analysis of the Viscoelastic Response of Composite Laminates During Hygrothermal Exposure

D. L. FLAGGS AND F. W. CROSSMAN

INTRODUCTION

THE EXPOSURE OF POLYMERS AND FIBER-REINFORCED POLYMERIC MATRIX composites to elevated hygrothermal conditions results in a plasticization* of the polymer which alters the mechanical response of the material. To account for this matrix plasticization phenomena, three distinctly different classes of constitutive models have been employed: elastic, elastic-plastic, and viscoelastic. Elastic constitutive laws, either linear or nonlinear, retain the assumption of a conservative or path-independent system. Elastic-plastic models are frequently based on metal plasticity relations, which assume time-independent plastic flow and a yield condition independent of the first stress invariant. In contrast, experimental studies of polymers have shown that polymer deformation is both history and time-dependent and that yielding is a function of both the hydrostatic and deviatoric stress components [1]. To properly account for the time-temperature/moisture and mechanical loading dependent history of the polymeric composite laminate, a viscoelastic constitutive model is required.

In past research, the time-dependent of viscoelastic nature of the polymeric matrix has often been neglected for applications near ambient temperature and moisture conditions. Under these conditions, the composite exhibits "elastic" or more accurately described, glassy behavior characteristic of viscoelastic materials. However, with exposure to an elevated hygrothermal environment, the material experiences a transition from glassy to leathery behavior with its strong time-temperature/moisture dependent response.

---

*A simplistic explanation for plasticization is that both increased thermal agitation and absorbed moisture induce matrix swelling which decreases the intermolecular attractions of adjacent polymer chains permitting greater chain mobility.

In this paper, we have chosen to explore the utility of a linear viscoelastic material description based upon a single hereditary integral formulation to describe the response of polymeric composite laminates to simultaneously hygrothermal exposure and mechanical loading. In the range of linear viscoelastic behavior, this has been shown to be a particularly useful approach requiring a minimum of experimental data to accurately predict material response under complex loading and environmental conditions. This approach has been used successfully by DeRuntz [2] to predict residual stress relaxation and creep response of graphite-epoxy laminates from unidirectional relaxation moduli, by Weitsman [3,4] to analyze the effects of time-varying moisture and temperature in a polymer slab and composite, by Flaggs [5] to analyze the response of a viscoelastic adhesive to hygrothermal exposure, and by Maksimov et al. [6–9] to predict the response of PN-3 polyester to uniaxial and biaxial stress states under stationary and cyclic hygrothermal environments.

VISCOELASTIC ANALYSIS FORMULATION

Consider a fiber-reinforced polymeric material system where due to time, temperature, and moisture effects, this class of material is assumed to exhibit a thermorheologically complex viscoelastic material behavior [10]. The constitutive law, based upon a single hereditary integral formulation is:

$$\sigma_i (T, M, t) = \int_{0^+}^{t} Q_{ij} (T, M, t-\tau) \frac{d}{d\tau} \overline{\epsilon}_j(\tau) \, d\tau \tag{1}$$

with

$$\overline{\epsilon}_j(\tau) = \epsilon_j(\tau) - \alpha_j \Delta T(\tau) - \beta_j \Delta M(\tau)$$

where σ_i and ϵ_j are the infinitesimal stress and engineering strain components, Q_{ij} are the relaxation moduli, α_j and β_j are the coefficients of thermal and hygroscopic expansion, and ΔT and ΔM are the changes in temperature and moisture from the stress-free state.

Introducing the concept of time-temperature/moisture superposition [11–13] permits the use of a master relaxation modulus curve representation of the experimental data, obtained from short-time stress relaxation tests, given by:

$$Q_{ij}(T, M, t) = a_v(T, M) \cdot Q_{ij}(T_o, M_o, \xi) \tag{2}$$

where T_o and M_o are the reference temperature and moisture conditions for the master curve, a_v is the vertical shift factor, and ξ is the reduced time defined as [14]:

$$\xi = \int_{0+}^{t} \frac{d\zeta}{a_{TM}\ (T,\ M)} \tag{3}$$

with a_{TM} being the horizontal temperature/moisture shift factor. Substituting equation (2) into equation (1) and integrating by parts yields [2,15]:

$$\sigma_i(T,\ M,\ t) = Q_{ij}\ (T,\ M,0\ )\overline{\epsilon}_j(t) - \int_{0+}^{t} \frac{d[a_v(T,\ M)\cdot Q_{ij}\ (T_o,M_o,\xi-\xi')]}{d\tau}\ \overline{\epsilon}_j\ (\tau)d\tau \tag{4}$$

where

$$\xi' = \int_{0+}^{\tau} \frac{d\zeta}{a_{TM}\ (T,\ M)} \tag{5}$$

To use equation (4) in a numerical scheme, it was necessary to simplfy the hereditary integral and, at the same time, express the relaxation moduli, Q_{ij}, in some functional form so that the necessary time derivative may be taken. To this end, the relaxation moduli are expanded in a finite exponential series of the form:

$$Q_{ij}(T_o,M_o,\xi) = Q_{ij}(T_o,M_o,0) - \sum_{k=1}^{n} F_{ij}^k(T_o,M_o\ )\left[1 - e^{-\lambda_{ij}^k\xi}\right] \tag{6}$$

In order to integrate the hereditary integral exactly, an incremental solution procedure is used in which temperature, moisture, and external boundary conditions are assumed to remain constant during viscoelastic time steps (Δt), while changes in hygrothermal conditions and/or boundary conditions are made in "elastic" steps during which time is held constant ($\Delta t = 0$).

With these simplifying assumptions, equation (4) becomes:

$$\sigma_i(T,\ M,\ t) = Q_{ij}(T,\ M,\ 0)\overline{\epsilon}_j(t) - \sum_{k=1}^{n} a_v(T,M)\ F_{ij}^k\ (T_o,M_o)\ e^{-\lambda_{ij}^k\xi}\ .$$

$$\left\{\frac{\lambda_{ij}^k}{a_{TM}\ (T,M)} \int_{0+}^{t} e^{\lambda_{ij}^k\xi'}\ \overline{\epsilon}_j(\tau)\ d\tau\right\} \tag{7}$$

If we consider an additional time increment, equation (7) would be updated as follows,

$$\sigma_i(T,M,t + \Delta t^+) = Q_{ij}(T,M,0)\,\overline{\epsilon}_j(t + \Delta t^+)$$

$$- \sum_{k=1}^{n} F_{ij}^k (T_o, M_o) e^{-\lambda_{ij}^k(\xi + \Delta\xi^+)} \left\{ I_{ij}^k + \Delta\, I_{ij}^k \right\} \qquad (8)$$

where ΔI_{ij}^k is the contribution to the hereditary integral over the current time step Δt^+ as given by:

$$\Delta I_{ij}^k = \frac{a_v^+(T,M)\lambda_{ij}^k}{a_{TM}^+(T,M)} \int_t^{t+\Delta t^+} e^{\lambda_{ij}^k \tau'}\,\overline{\epsilon}_j(\tau)\,d\tau \qquad (9)$$

The integrals in equation (8) are then easily updated at each time step and do not involve integration over the previous load/hygrothermal history.

In performing the integration indicated in equation (9), a quadratic strain variation in time is assumed as:

$$\overline{\epsilon}_j(\tau) = A_j + B_j\tau + C_j\tau^2 \qquad (10)$$

where A_j, B_j, and C_j are determined from the current and the two previous time steps. The resulting integration of equation (9) yields:

$$\Delta\, I_{ij}^k = e^{\lambda_{ij}^k\,(\xi + \Delta\xi^+)} D_{ij}^k\,(t + \Delta t^+) - e^{\lambda_{ij}^k\,\xi} D_{ij}^k\,(t) \qquad (11)$$

where

$$D_{ij}^k\,(\tau) = A_j + B_j \left(\tau - \frac{a_{TM}^+}{\lambda_{ij}^k}\right) + C_j \left\{\tau^2 - 2\left[\frac{a_{TM}^+}{\lambda_{ij}^k}\tau - \left(\frac{a_{TM}^+}{\lambda_{ij}^k}\right)^2\right]\right\} \qquad (12)$$

When a viscoelastic time step follows an elastic one, a linear strain variation is used in place of equation (10) so that the terms in equation (12) containing C_j are zero.

Since the viscoelastic constitutive law has been posed in terms of relaxation moduli, problems which specify stress or mixed boundary conditions are solved by a modified Newton technique with over/under relaxation for a set of compatible strains at each time step.

The moisture distribution within the laminate at a specified time is calculated assuming exposure on both laminate surfaces $[z = \pm (h/2)]$ by [16]:

$$M(z,t) = \left(\frac{M_u - M_1}{h} \right) z + \frac{1}{2} (M_u + M_1) + \sum_{r=0}^{\infty} m_r \cos a_r z \qquad (13)$$

with

$$a_r = \frac{(2r + 1) \pi}{h}$$

$$m_r = \left\{ \frac{2}{h} \int_{-h/2}^{h/2} M(z) \cos a_r z \, dz - \frac{2(M_u + M_1)(-1)^r}{(2r + 1)\pi} \right\} exp \left[-a_r^2 Dt \right]$$

where M_u and M_1 are the moisture boundary concentrations on the upper and lower laminate surfaces, $M(z)$ is the initial moisture distribution, and D is the hygroscopic diffusivity, which is assumed in this analysis to be independent of moisture concentration.

VISCOELASTIC MATERIAL DATA

In this paper, the viscoelastic material response of both GY70/339 and T300/934 composite laminates are investigated. The experimentally determined elastic material properties used in the viscoelastic analysis are [13,17]:

	GY70/339	*T300/934*
E_{11} (MSI)	42.0	21.0
E_{22} (MSI)	.88	1.5
G_{12} (MSI)	.6	.7
ν_{12}	.31	.29
α_1 ($\mu\epsilon/°$F)	$-.4$	.05
α_2 ($\mu\epsilon/°$F)	14.0	16.5
β_1 ($\mu\epsilon/\%$M)	0	0
β_2 ($\mu\epsilon/\%$M)	5400*	5400*
D(in^2/sec)	4.65×10^{-11} (75°F/95% RH)	5.27×10^{-10} (160°F/95% RH)
	1.55×10^{-10} (125°/95%)	1.25×10^{-9} (200°/95%)
	1.5×10^{-10} (160°/21%)	
	3.8×10^{-10} (160°/51%)	
	5.19×10^{-10} (160°/71%)	
	9.3×20^{-10} (160°/95%)	

*No expansion below 0.4% absorbed moisture is assumed to conform with experimental observations.

The master relaxation modulus curves for both materials, obtained from
$(\pm 45)_s$ tensile stress relaxation tests [13], with the necessary temperature/
moisture shift factors are presented in Figures 1-3. As described in Reference
[13], horizontal shifting alone was considered sufficient to bring the experimental data into coincidence for construction of the master relaxation
curves; thus, for the remainder of this paper, $a_v(T,M) = 1$. This is equivalent
to assuming a thermorheologically simple material behavior.

With the computer solution time being directly proportional to the number
of terms needed in equation (6) to represent the experimental data, a
numerical optimization procedure based upon the method of Hooke and
Jeeves [21] was implemented to obtain the "best" analytic fit with the fewest
terms. The nonlinear unconstrained minimization problem which is solved to
determine the unknown coefficients F_{ij}^k and λ_{ij}^k may be stated as:

$$min \; \epsilon^2 = \left\{ Q_{ij}(T_o,M_o,\xi) - Q_{ij}(T_o,M_o,0) + \sum_{k=1}^{n} F_{ij}^k(T_o,M_o) \left[1 - e^{-\lambda_{ij}^k \xi} \right] \right\}^2$$

(14)

For the two material systems modeled in this paper, ten term approximations
were used in the subsequent analyses.

PREDICTION OF NONSYMMETRIC LAMINATE WARPING—
CORRELATION TO EXPERIMENT

To determine the usefulness of the linear viscoelastic material model to
describe composite laminate behavior, the warping of nonsymmetric $(0_4/90_4)_T$
GY70/339 laminates during prolonged hygrothermal exposure was analyzed
and compared to experimental data obtained from laminates exposed to a
series of constant temperature and humidity environments [13]. This technique provides a convenient experimental means of predicting the magnitude
of laminate residual stresses.

The stress-free temperatures used in this and subsequent sections were
determined by separate viscoelastic analyses. The procedure, described in
more detail in Reference [5], involves the simulation of a two-hour exponential cool-down from the cure temperature to room temperature. The "effective" stress-free temperature at that time is then calculated by determining the
change in temperature during an elastic step ($\Delta t = 0$) which reduces the
residual thermal stress state to zero. The stress-free temperatures calculated in
this manner were 193 °F for GY70/339 and 317 °F for T300/934.

Figure 4 contrasts our experimental results with those for both elastic and
viscoelastic analyses. Ratios of final to initial curvature measured at room
temperature are plotted versus equilibrium moisture content for specimens
exposed to 75 °F (95 percent RH), 125 °F (95 percent RH), and 160 °F (21, 51,

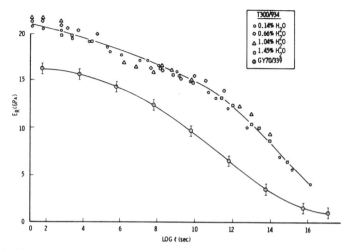

Figure 1. *Master relaxation modulus curves for [±45]$_s$ T300/934 and GY70/339 laminates. Horizontally shifted T300/934 data shown (Ref. 13).*

71, and 95 percent RH). Comparison of the experimentally measured curvatures (open symbols) to the viscoelastic predictions (filled symbols) indicates that slightly more stress relaxation occurred than was predicted by the linear viscoelastic analysis. Nonlinear or stress-dependent viscoelastic effects may account for the differences between the predicted and experimentally deter-

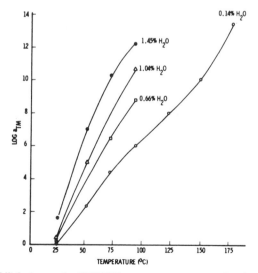

Figure 2. *Shift factor a$_{TM}$ for T300/934 versus temperature and moisture (Ref. 13).*

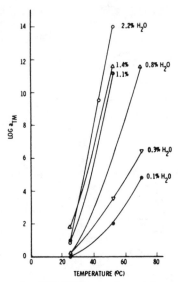

Figure 3. *Shift factor a_{TM} for GY70/339 versus temperature and moisture (Ref. 13).*

mined curves. It is, however, more likely that these differences resulted from the experimental test procedures. To ensure an equilibrium moisture content, experimental specimens were often maintained at the elevated environmental conditions for a time period longer than that predicted analytically to reach an equilibrium moisture content, thus permitting additional stress relaxation to occur.

The large deviation of the elastic analysis (dashed line) from both the experimental and viscoelastic results (see Figure 4) underscores the need to ac-

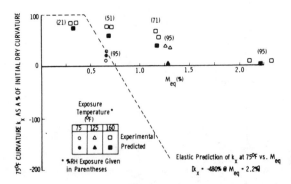

Figure 4. *Ratios of final to initial curvature at 75° F of $(0_4/90_4)_T$ GY 70/339 laminates versus equilibrium moisture content for different hygrothermal exposure conditions.*

count for the history dependence of polymeric composite laminates exposed to elevated hygrothermal environments. Once thermorheologically simple material (TSM) behavior is assumed, the correlation between theory and experiment is good. In the sections which follow, the TSM model is used to analytically investigate the influence of hygrothermal exposure on inplane dimensional changes and stresses, the strain rate dependence of inplane shear properties, and, finally, the influence of laminate residual stress relaxation on the creep response of laminated composite plates under a constant applied mechanical load.

PREDICTION OF INPLANE DIMENSIONAL CHANGES DURING ABSORPTION/DESORPTION

In this section, the results of both viscoelastic and elastic analyses are examined which predict the dimensional changes in a symmetric $(0/\pm45/90)_s$ GY70/339 laminate during absorption and desorption at elevated hygrothermal conditions. The initial exposure at 125°F/95 percent RH to an equilibrium moisture content of 1.275 percent is followed by desorption at 125°F/0 percent RH until dry. The average moisture content in the laminate as a function of exposure time is shown in Figure 5. The time increments used in the analysis are also evident in Figure 5.

Significant differences between the inplane strain responses ε_x and ε_y are seen for both the elastic and viscoelastic analyses in Figures 6 and 7. In the elastic case, the difference between ε_x and ε_y is a result of the transient moisture distribution within the laminate. In the adsorption phase, the outermost 0° lamina are the first to be affected by adsorbed moisture; hence, the large transverse ε_y strain. Not until moisture diffuses into the 45° layers is an

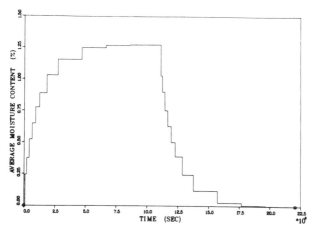

Figure 5. *Average moisture content of $(0/\pm45/90)_s$ GY70/339 laminate as a function of exposure time.*

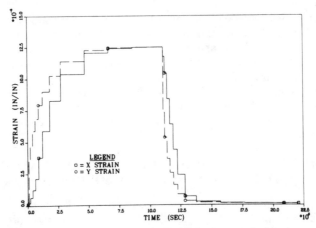

Figure 6. *Elastic predictions of (0/±45/90)ₛ GY70/339 laminate strain εₓ and εᵧ as a function of exposure time.*

appreciable ε_x strain induced by hygroscopic expansion. As equilibrium is approached, the differences between ε_x and ε_y diminish as moisture is absorbed in the innermost 90° lamina. As expected for this quasi-isotropic laminate, $\varepsilon_x = \varepsilon_y$ at saturation. The effect of desorption is essentially opposite to that encountered in absorption, with ε_x and ε_y returning to their initial dry value.

The predicted inplane strain history, based on a viscoelastic analysis, is shown in Figure 7. As in the elastic case, ε_y is initially larger than ε_x due to the transient moisture distribution. However, during each time increment, stress relaxation (see Figure 9) is permitted to occur, thereby reducing by some por-

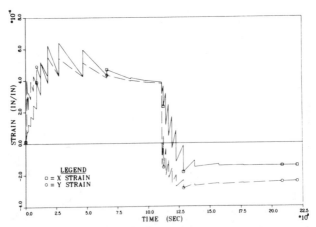

Figure 7. *Viscoelastic predictions of (0/±45/90)ₛ GY70/339 laminate strain εₓ and εᵧ as a function of exposure time.*

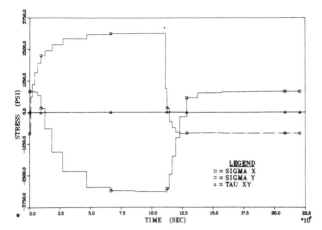

Figure 8. *Elastic predictions of the 90° lamina stress distribution in the (0/±45/90)ₛ GY70/339 laminate as a function of exposure time.*

tion the total elastic change which took place during the preceding elastic step due to the updating of the moisture distribution. Greater stress relaxation occurs in the outer lamina because an initially higher moisture content acts over a period of time longer than that experienced by laminae farther from the surface. Less relaxation is thus predicted in the innermost 90° lamina which dominates the ε_x response, resulting in an ε_x strain which exceeds ε_y before the equilibrium moisture content is reached in the laminate. The large changes seen in both strain components are a result of the large temperature-moisture shift factor which shifts the real time behavior of the material far to the right on the master relaxation curve. This results in the reduction of any differences between strains to a point where their valves are almost identical at equilibrium.

As in the elastic case, ε_y leads ε_x during desorption. This time, however, less

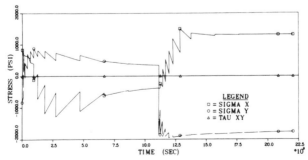

Figure 9. *Viscoelastic predictions of the 90° lamina stress distribution in the (0/±45/90)ₛ GY70/339 laminate as a function of exposure time.*

relaxation occurs in the outer lamina because these lamina are at a high moisture content for only a short time during the desorption phase. This results in a final strain state where ε_y is approximately 40 percent greater than ε_x, a value significantly different from that predicted by the elastic analysis.

The stress distributions in the 90° lamina, as determined by the elastic and viscoelastic analyses, are shown in Figures 8 and 9, respectively. In the visco-elastic case, large alteration of the lamina stress state due to stress relaxation is a result of the simulated hygrothermal exposure. During the absorption phase, stress relaxation reduces the magnitude of residual stresses in the laminate. Once desorption is completed, however, the magnitude of the stresses as calculated by the viscoelastic analysis is significantly higher than that predicted by the elastic analysis. In this instance, the altered residual stresses due to hygroscopic swelling coupled with the simultaneous visco-elastic stress relaxation act together to increase the final stress levels about those predicted by the elastic analysis.

The effect of stress relaxation on the other plies of this quasi-isotropic laminate is similar to that seen in Figure 9. It is interesting to note that upon desorption, an elastic analysis predicts $\sigma_x = \sigma_y$ for the 0° and 90° laminate (see Figure 8) with $\sigma_x = \sigma_y = 0$ for the ±45° laminae. For the viscoelastic case, residual σ_x and σ_y stresses are observed to occur in the ±45° lamina with the result that $\sigma_x \neq \sigma_y$ in the 0 and 90° laminae.

In summary, viscoelastic stress relaxation during a complete moisture absorption-desorption cycle at constant temperature can lead to non-recoverable dimensional and in-plane stress changes in a quasi-isotropic com-posite laminate which would not be predicted by an elastic analysis. Similar nonrecoverable dimensional changes have been observed experimentally in both GY70/339 and T300/5209 laminates which absorbed moisture at one temperature and desorbed at a different temperature [17].

VISCOELASTIC EFFECTS IN CONSTANT STRAIN RATE TENSILE TESTING

This section investigates the effect of linear viscoelastic material behavior on the tensile strain rate response of $[\pm 45]_{2s}$ and $[0/\pm 45/90]_s$ T300/934 laminates. Constant strain rate tensile loading at 0.001, 0.01, and 0.1 in/in/min is simulated for specimens which are assumed to be at equilibrated hygrothermal conditions of 77°F/0.14 percent, 200°F/0.14 percent, and 200°F/1.4 percent prior to testing. The analytically predicted σ_x versus ε_x responses for the different hygrothermal test conditions of the two laminates up to 1 percent strain are shown in Figures 10 and 11.

As a result of the fiber-dominated elastic response of the 0° lamina, the quasi-isotropic $[0/\pm 45/90]_s$ laminate (see Figure 11) exhibits, as expected, little strain rate-dependent behavior, even at the elevated hygrothermal test condi-tions. This is not the case for the $[\pm 45]_{2s}$ laminate (see Figure 10) where the matrix-dominated tensile response shows pronounced time-temperature/ moisture dependent behavior. Here, strain rate insensitivity occurs only at

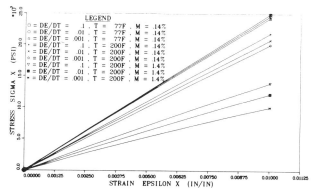

Figure 10. *Strain rate dependent response of (±45)₂ₛ T300/934 laminate for different hygrothermal test conditions.*

ambient conditions where the material behaves in a glassy or nearly "elastic" manner. To facilitate comparison with the elastic value of $E_x = 2.502$ *Msi*, the initial and final ($\varepsilon_x = 1$ percent) tangent moduli of the $[\pm 45]_{2s}$ laminate at the different simulated test conditions are tabulated in Table 1.

It is interesting to note that, as a result of the assumed linear viscoelastic behavior, the degree of nonlinearity in the stress-strain curves is very slight, even for elevated hygrothermal conditions. Recent experimental $[\pm 45]_s$ tensile tests conducted on AS/3501-6 at elevated temperature and moisture conditions by Augl [18] confirm the quasi-linear stress-strain behavior predicted by the linear viscoelastic analysis for tensile strains up to $\varepsilon_x = 0.01$.

At higher load levels, a highly nonlinear stress-strain curve develops in testing of $[\pm 45]_s$ composite laminates. These nonlinearities require a stress-

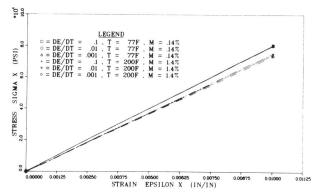

Figure 11. *Strain rate dependent response of (0/±45/90)ₛ T300/934 laminate for different hygrothermal test conditions.*

Table 1. Tangent modulus values for $(\pm 45)_{2s}$ T300/934 laminate
($E_{x_{elastic}}$ = 2.502 Msi).

Test Environment (T°F/M%)	Strain Level (%)	Strain Rate (in./in./min.)		
		0.1	0.01	0.001
77°/.14%	0	2.5	2.483	2.441
	1	2.485	2.451	2.418
200°/.14%	0	2.184	2.130	1.999
	1	2.165	2.039	1.982
200°/1.4%	0	1.434	1.273	1.090
	1	1.342	1.160	.912

dependent viscoelastic constitutive description which is not accounted for in this linear viscoelastic material model. However, most structural applications of composite laminates limit the inplane strain allowables to less than .01 in/in making the assumption of linear viscoelasticity a viable one.

Comparison of the linear viscoelastic predictions with experimental data indicates that this material model is sufficient for describing the macroscopic ply-by-ply time dependent mechanical response of a composite laminate. A non-linear or stress-dependent constitutive model is needed for applications where the shear stress in the off-axis laminae approaches the inplane shear strength.

EFFECT OF RESIDUAL THERMAL STRESS RELAXATION ON CREEP BEHAVIOR

Due to the significant residual thermal and hygroscopic stresses which exist in composite laminates, the viscoelastic response of the laminate to an applied mechanical load is necessarily coupled to the simultaneous relaxation of these residual streses. As a result of this as yet undetermined contribution to the creep response of a laminate, Chamis [19] has suggested that the 10° off-axis tensile specimen, rather than the more commonly used $[\pm 45]_s$ specimen, be employed to determine the inplane viscoelastic response.

In this section, we have analyzed the creep behavior of a $[\pm 45]_s$ GY70/339 laminate when residual thermal stresses are allowed to relax prior to a series of simulated constant load creep tests for 50 hours at 122°F/0 percent RH. Table 2 shows the relative magnitudes of the residual thermal stresses, before and after stress relaxation at 122°F, and the three levels of mechnically applied load employed in this creep simulation. The maximum shear stress of τ_{xy} = 947 psi resulting from an applied uniaxial stress of σ_x = 2000 psi is on the order of 15 percent of the ultimate laminate strength at the 122°F test temperature.

The laminate axial strain, ε_x, normalized to the initial mechanically applied strain, is plotted versus time for applied loads of σ_x = 20, 200, and 2000 psi in Figure 12. Before the simulated creep test was started, the specimen was pre-

Table 2. **Predicted shear stress,** τ_{xy}, **$(\pm 45)_{2s}$ GY70/339 laminate during conditioning and creep testing at 122°F.**

Condition	τ_{xy}(psi)
1. Elastic prediction for $\Delta T = -71°F$	870
2. After 1800 sec. conditioning	762
3. After 180,000 sec. conditioning	670
4. After 360,000 sec. conditioning	650
5. Shear stress due to 20 psi tensile load	9.47
6. Shear stress due to 200 psi tensile load	94.7
7. Shear stress due to 2000 psi tensile load	947

conditioned at each load level by holding at the test temperature of 122 °F for either 1800 s. (30 min) or 180,000 s. (50 H), thus permitting stress relaxation to occur. The results indicate that the limiting case of pure mechanical loading (zero residual stresses) is approached with increasing load and longer preconditioning exposure times. In all cases, the effect of simultaneous residual stress relaxation is to increase the normalized creep strain above that due to mechanical loading alone.

Table 2 indicates that the alteration in the measured creep response is small when the absolute change in residual stresses in the time period associated with creep testing is small with respect to the mechanically applied creep stress. The results show that the resultant creep curve is characteristic of mechanical load alone when the residual stress change during the time of creep measurement is less than ~1/10 of the applied mechanical load.

It is important to note that, in this example, the magnitude of the elastic residual shear stress (τ_{xy} = 870 psi) is of the same order of magnitude as that of the maximum applied mechanical shear stress of 947 psi. This does not,

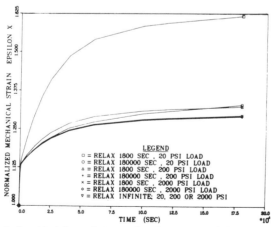

Figure 12. *Effect of residual thermal stress relaxation on creep behavior of $(\pm 45)_s$ GY70/339 laminate at different mechanical load levels.*

however, imply that significant alteration of the predicted creep response will occur. Rather, it is the magnitude of the residual thermal stress relaxation during the time of creep testing which influences the laminate creep response to a mechanical load.

The amount of residual stress relaxation during a particular time period is typically not known prior to the generation of the appropriate creep or relaxation master curve. Yet the experimentalist must isolate the response due to mechanical loading from that due to residual stress relaxation in order to determine the true relaxation behavior. Two procedures are suggested here. The first involves holding the test specimen at the test temperature/moisture conditions for a period several times longer than the period of the subsequent mechanical load application. This method, however, is in practice limited to the generation of short-term relaxation data. The second procedure makes use of previously measured creep or relaxation data on a similar composite material system to estimate the applied stress level needed to overpower the influence of residual stress relaxation. This latter method may require such a high stress level that nonlinear viscoelastic deformation is introduced in the testing. A combination of these two procedures may be necessary to generate material response curves which have not been significantly influenced by residual stress relaxation.

By testing uniaxially reinforced composites, it is possible to eliminate from consideration lamina residual stress. However, the microresidual stress relaxation in the matrix and fiber components will influence the response of the unidirectional composite to mechanically applied loads in the same manner as residual stresses in a multiaxially-reinforced laminate. The relaxation of residual thermal stresses may also be thought of as taking place in an unreinforced, but microscopically heterogeneous polymer.

It is interesting to note that the analytically predicted effects of residual stress relaxation on creep behavior have also been attributed to some type of undefined chemical or physical aging process. Current experimental studies at Texas A&M [20] characterizing the physical behavior of neat 3502 epoxy resin show enhanced creep similar to the composite behavior described above. Creep/recovery tests conducted at 300 °F ($T_g = 375$ °F) immediately after quenching from 425 °F show a dramatic increase in compliance as compared to results of creep/recovery tests conducted on specimens "aged" at ambient temperature for several months prior to testing. In the absence of any large free volume changes associated with the quench prior to loading, the investigators concluded that relaxation of microscopic thermal residual stresses may account for the observed "aging" phenomenon.

In summary, it was found that when the absolute change in residual stresses in the time period associated with creep testing is small compared to the applied mechanical creep stress, then the alteration in the measured creep response is also small (see Table 2). For changes in the residual stress state during creep testing of the order of 10 percent, we find that a creep curve is obtained which is characteristic of the material due to mechanical loading alone.

CONCLUSION

A linear viscoelastic constitutive model and computational procedure developed within the framework of lamination theory has been shown to accurately predict the curvature changes of nonsymmetric GY70/339 laminates after exposures to elevated hygrothermal environments for up to one year. Application of the analysis to problems involving dimensional stability, strain rate effects, and the effect of residual stress relaxation on creep properties has been described to emphasize the need to account for the interaction of time, temperature, and moisture, and additionally, to explore the bounds of the linear viscoelastic model to describe composite laminate behavior.

ACKNOWLEDGMENTS

The research described in this report was funded in part by the Air Force Office of Scientific Research under Contract F49620-77-C-0122 and in part by the Lockheed Independent Research Program "Structural Composite Materials." The authors are grateful to Mr. J. R. Zumsteg for his assistance in the preparation of this paper.

REFERENCES

1. Sternstein, S. S., "Mechanical Properties of Glassy Polymers, in *Treaties on Materials Science and Technology,* Vol. 10., J.M. Schultz, ed. Academic Press (1977).
2. DeRuntz, J. A. and Crossman, F. W., "Time and Temperature Effects in Laminated Composites," *Nuclear Metallurgy, 20,* 1085 (1976).
3. Weitsman, Y., "Hygrothermal Viscoelastic Analysis of a Resin Slab Under Time-Varying Moisture and Temperature," AIAA Structures, Dynamics and Materials Conference, San Diego (March 1977).
4. Weitsman, Y., "Residual Thermal Stresses Due to Cool-Down of Epoxy-Resin Composites," *J. of Applied Mechanics, 46,* 563 (Sept. 1979).
5. Flaggs, D. L. and Crossman, F. W., "Viscoelastic Response of a Bonded Joint due to Transient Hygrothermal Exposure," in *Modern Developments in Composite Materials and Structures,"* J.R. Vinson, ed. ASME (1979).
6. Maksimov, R. D., Sokolov, E. A., and Mochalov, V. P., *Mechanical Polymerov, 3,* 393 (1975).
7. Maksimov, R. D., Mochalov, V. P., and Sokolov, E. A., *Mechanica Polymerov, 4,* 627 (1975).
8. Maksimov, R. D., Sokolov, E. A., and Mochalov, V. P., *Mechanica Polymerov, 6,* 976 (1975).
9. Maksimov, R. D., Mochalov, V. P., and Sokolov, E. A., *Mechanica Polymerov, 6,* 982 (1975).
10. Schapery, R. A., "Viscoelastic Behavior and Analysis of Composite Materials," in *Composite Materials,* Broutman and Krock, eds. Academic Press, p. 85 (1974).
11. Yeow, Y. T., Morris, D. H., and Brinson, H. F., "The Time-Temperature Behavior of an Unidirectional Graphite/Epoxy Composite," ASTM Symposium on Composite Materials: Testing and Design, Fifth Conference, New Orleans, Louisiana (March 20–22, 1978).
12. Griffith, W. I., Morris, D. H., and Brinson, H. F., "The Accelerated Characterization of Viscoelastic Composite Materials," VPI-E-80-15, Department of Engineering Science and Mechanics, Virginia Polytechnic Institute (April 1980).
13. Crossman, F. W., Mauri, R. E., and Warren, W. J., "Moisture Altered Viscoelastic

Response of Graphite-Epoxy Composites," ASTM Symposium on Environmental Effects on Advanced Composite Materials, ASTM-STP 658, Dayton, Ohio (September 29–30, 1977).
14. Morland, L. W. and Lee, E. H., "Stress Analysis for Linear Viscoelastic Materials with Temperature Variation," *Trans. Soc. Rheology, 4,* 223 (1960).
15. Taylor, R. L., Pister, K. S., and Goudreau, G. L., "Thermomechanical Analysis of Viscoelastic Solids," Structural Engineering Laboratory Report 68-7, Department of Civil Engineering, University of California at Berkeley (June 1968).
16. Flaggs, D. L. and Vinson, J. R., "Hygrothermal Effects on the Buckling of Laminated Composite Plates," *Fibre Science and Technology, 11,* 353 (1978).
17. Mauri, R. E., Crossman, F. W., and Warren, W. J., "Assessment of Moisture Altered Dimensional Stability of Structural Composites," Proc. 23rd National SAMPE Symposium, Anaheim, CA, 1202 (May 1978).
18. Augl, J. M., "Prediction and Verification of Moisture Effects on Carbon Fiber-Epoxy Composites," NSWC TR79-43 (March 30, 1979).
19. Chamis, C. C. and Sinclair, J. H., "10-Degree Off-Axis Test for Shear Properties of Fiber Composites," SESA Spring Meeting, Dallas, TX (May 1977).
20. Schapery, R. A., et al., "Composite Materials for Structural Design," AFOSR TR-79-0347 (March 1979).
21. Kuester, J. L. and Mize, J. H. *Optimization Techniques with Fortran.* McGraw-Hill, p. 309 (1973).

30

Time-Temperature Effect in
Adhesively Bonded Joints

F. Delale and F. Erdogan

INTRODUCTION

IN MOST PRACTICAL APPLICATIONS OF ADHESIVELY BONDED JOINTS AND epoxy-based composites the operating temperature is such that in the stress analysis of the structure any viscoelastic behavior which may be exhibited by the adhesive or the epoxy matrix may be neglected. On the other hand, depending on the time-temperature behavior of the particular epoxy, time-history of loading, and the level of accuracy required of the analysis, even at moderately low temperatures the viscoelastic effects may have to be taken into account in analyzing the structure. In particular, if the structure is subjected to a loading with a relatively high frequency cyclic component, temperature rise may occur due to internal heat generation and it may become necessary to investigate the time-temperature effects in the stress analysis. The objective of this paper is to study these effects by considering a relatively simple geometry. The corresponding elastic problem for the adhesively bonded joints have been studied quite extensively. Some typical models used in these studies may be found in [1–4].

FORMULATION OF THE PROBLEM

For constant temperature the basic formulation of the adhesively bonded joints was considered in a previous paper [5] where it was assumed that the adhesive is a linear viscoelastic material which can be modeled by using differential operators. In this paper the hereditary integrals will be used to model the adhesive and the solution will be given for various temperatures in order to give some idea about the relative importance of the temperature changes or of the operating temperatures. It is assumed that the stress relaxation process in the viscoelastic adhesive takes place in a much slower rate than the heat conduction process in the bonded joint. Therefore, the spatial variation of temperature and its effect on the stress distribution may be neglected and it may be assumed that the adherends and the adhesive have the same temperature which is a function of the time only. Thus, the general formulation given in this section includes thermal stresses coming from differential thermal expansion only.

The problem under consideration is described in Figure 1. In this study, the adherends are treated as "plates" in which the transverse shear effects are taken into account. The adhesive is assumed to be a viscoelastic solid under inplane deformations in which the thickness variation of stresses is neglected. The assumptions regarding the mechanical modeling of the adherends and the adhesive may be justified on the basis of the fact that generally the thickness of the adherends is one order of magnitude and that of the adhesive is approximately two orders of magnitude smaller than the characteristic "length" dimension of the joint. In the analysis it is also assumed that the dimension of the joint in z-direction is relatively large and the external loads are independent of z, meaning that the problem may be approximated by one of plane strain (Figure 1).

Referring to Figure 1, the equilibrium conditions for the plate elements representing the adherends 1 and 2 may be expressed as

$$\frac{\partial N_{1x}}{\partial x} = \tau, \qquad \frac{\partial Q_{1x}}{\partial x} = \sigma, \qquad \frac{\partial M_{1x}}{\partial x} = Q_{1x} - \frac{h_1 + h_o}{2}\tau , \qquad (1\text{a-c})$$

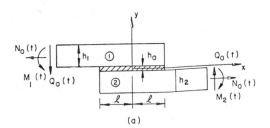

(a)

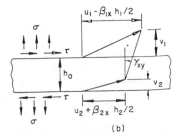

(b)

Figure 1. *The geometry of the bonded joint.*

$$\frac{\partial N_{2x}}{\partial x} = -\tau, \quad \frac{\partial Q_{2x}}{\partial x} = -\sigma, \quad \frac{\partial M_{2x}}{\partial x} = Q_{2x} - \frac{h_2 + h_o}{2} \tau, \quad \text{(2a-c)}$$

where N_{ix}, Q_{ix}, M_{ix}, ($i = 1,2$) are respectively the membrane, the transverse shear, and the moment resultants for the adherends 1 and 2, and σ and τ are the normal and the shear stress in the adhesive.

Assume that at (the homogeneous) temperature T_o the joint is free from temperature-induced stresses. If the temperature is raised to T at time t_1, the stress-displacement relations can be written as

$$\frac{\partial u_1}{\partial x} = C_1 N_{1x} + (1 + v_1)\alpha_1 (T - T_o)H(t - t_1),$$

$$\frac{\partial \beta_{1x}}{\partial x} = D_1 M_{1x}, \quad \frac{\partial v}{\partial x} + \beta_{1x} = Q_{1x}/B_1, \quad \text{(3a-c)}$$

$$\frac{\partial u_2}{\partial x} = C_2 N_{2x} + (1 + v_2)\alpha_2 (T - T_o)H(t - t_1),$$

$$\frac{\partial \beta_{2x}}{\partial x} = D_2 M_{2x}, \quad \frac{\partial v_2}{\partial x} + \beta_{2x} = Q_{2x}/B_2. \quad \text{(4a-c)}$$

where u_i, v_i, β_{ix}, ($i = 1,2$) and the x and y-components of the displacement vector, and the rotation of the normal at the midplane of the plates, α_i, ($i = 1,2$) is the coefficient of thermal expansion, $H(t)$ is the Heaviside function, and

$$C_i = \frac{1 - v_i^2}{E_i h_i}, \quad D_i = \frac{12(1 - v_i^2)}{E_i h_i^3}, \quad B_i = \frac{5}{6}\mu_i h_i, \quad (i = 1,2) \quad \text{(5a-c)}$$

E_i, μ_i, and v_i ($i = 1,2$) being the elastic constants of the adherends.

To complete the formulation of the problem, the continuity conditions of the displacements in the region of adhesion have to be considered. Again, referring to Figure 1 the strains in the adhesive averaged over the thickness may be expressed as

$$\gamma_{xy} = 2\varepsilon_{xy} = \frac{1}{h_o}\left(u_1 - \frac{h_1}{2}\beta_{1x} - u_2 - \frac{h_2}{2}\beta_{2x}\right),$$

$$\varepsilon_y = \frac{1}{h_o}(v_1 - v_2), \quad \text{(6a-c)}$$

$$\varepsilon_x = \frac{1}{2}\left(\frac{\partial u_1}{\partial x} - \frac{h_1}{2}\frac{\partial \beta_{1x}}{\partial x} + \frac{\partial u_2}{\partial x} + \frac{h_2}{2}\frac{\partial \beta_{2x}}{\partial x}\right)$$

where the remaining components of the adhesive strains are zero. The nonzero stress components in the adhesive, again averaged over the thickness, are σ_x, $\tau_{xy} = \tau$, $\sigma_y = \sigma$, and σ_z. Noting that $\varepsilon_z = 0$, in the standard fashion the hydrostatic and deviatoric components of the strain and the stress tensor for the adhesive may be defined as

$$e = (\varepsilon_x + \varepsilon_y)/3, \; e_{ij} = \varepsilon_{ij} - e\delta_{ij}, \; (i,j = x,y,z) \qquad \text{(7a,b)}$$

$$s = (\sigma_x + \sigma + \sigma_z)/3, \; s_{ij} = \sigma_{ij} - s\delta_{ij}, \; (i,j = x,y,z) \qquad \text{(8a,b)}$$

Using now the hereditary integrals and observing that for the practical range of temperature and stress levels under hydrostatic stress, most viscoelastic materials behave elastically, the constitutive equations of the adhesive may be written as follows [6-8]:

$$s_{ij} = 2\int_{-\infty}^{t} G(T, t-\zeta)\; \frac{\partial e_{ij}}{\partial \xi}\, d\xi, \; (i,j = x,y,z) \qquad \text{(9)}$$

$$e = \frac{s}{3K(T)} + \alpha_3(T)(T-T_o)H(t-t_1), \qquad \text{(10)}$$

where the known functions $G(T,t)$, $K(T)$, and $\alpha_3(T)$ are respectively the relaxation modulus, the bulk modulus, and the coefficient of thermal expansion of the adhesive. Equations (9) and (10) may also be expressed in the following explicit form:

$$2\sigma_x - \sigma - \rho_z = 2\int_{-\infty}^{t} G(T, t-\xi)\left(2\frac{\partial \varepsilon_x}{\partial \xi} - \frac{\partial \varepsilon_y}{\partial \xi}\right)d\xi, \qquad \text{(11)}$$

$$2\sigma - \sigma_x - \sigma_z = 2\int_{-\infty}^{t} G(T, t-\xi)\left(2\frac{\partial \varepsilon_y}{\partial \xi} - \frac{\partial \varepsilon_x}{\partial \xi}\right)d\xi, \qquad \text{(12)}$$

$$2\sigma_z - \sigma_x - \sigma = -2\int_{-\infty}^{t} G(T, t-\xi)\left(\frac{\partial \varepsilon_x}{\partial \xi} + \frac{\partial \varepsilon_y}{\partial \xi}\right)d\xi, \qquad \text{(13)}$$

$$\tau = \int_{-\infty}^{t} G(T, t-\xi) \ \frac{\partial \gamma_{xy}}{\partial \xi} d\xi, \tag{14}$$

$$\sigma_x + \sigma + \sigma_z = 3K(T) \ [\varepsilon_x + \varepsilon_y - 3\alpha_3(T)(T-T_o)H(t-t_1)]. \tag{15}$$

It may be noted that since $\Sigma s_{ii} = 0$ and $\Delta e_{ii} = 0$, equations (11–13) are not linearly independent; (12), for example, can be obtained by adding (11) and (13) and can, therefore, be ignored. Eliminating σ_x and σ_z, from (11), (13), and (15) it follows that

$$\sigma = K(T)(\varepsilon_x + \varepsilon_y) - 3K(T)\alpha_3(T)(T-T_o)H(t-t_1)$$

$$- \frac{2}{3} \int_{-\infty}^{t} G(T, t-\xi) \ \left(\frac{\partial \varepsilon_x}{\partial \xi} - 2 \ \frac{\partial \varepsilon_y}{\partial \xi} \right) d\xi. \tag{16}$$

Equations (14) and (16) with (6) provide the needed constitutive equations for the adhesive. Thus, (1–4), (14), and (16) (with (6)) give a system of fourteen equations to determine the fourteen unknown functions σ, τ, u_i, v_i, β_{ix}, N_{ix}, M_{ix}, and Q_{ix} ($i = 1,2$). Solution of the system of differential equations (1–4) contains twelve integration "constants" which are functions of time and are determined by using the boundary conditions for plates (1) and (2) at $x = \pm \ell$.

SOLUTION FOR THE SINGLE LAP JOINT

In the general formulation given in the previous section by eliminating all unknown functions other than $\sigma(x,t)$ and $\tau(x,t)$, it is possible to reduce the problem to a system of equations for σ and τ only. Even though relatively straightforward, this process is quite lengthy and the resulting equations are coupled. On the other hand, if the lap joint consists of two identical adherends, then the elimination process is somewhat simpler and the resulting equations for σ and τ are uncoupled. For this case, the problem is considerably simplified and at the same time still yields the main features of the solution. For the identical adherends, the material constants become

$$C_1 = C_2 = C = \frac{1-v^2}{Eh}, D_1 = D_2 = D = \frac{12(1-v^2)}{Eh^3} ,$$

$$B_1 = B_2 = B = \frac{5}{6}\mu h, \alpha_1 = \alpha_2 = \alpha , \tag{17}$$

where $h = h_1 = h_2, E = E_1 = E_2, v = v_1 = v_2$, and $\mu = \mu_1 = \mu_2$.

For $t<t_1$, let the joint be at a homogeneous temperature $T = T_o$ and be stress-free. If the temperature is raised to T at $t = t_1<0$ and is held constant for $t>t_1$, then the particular adhesive model used in the analysis described in the previous section would give no temperature induced stresses. Therefore, if the external loads are applied at $t = 0$, in the constitutive equations (14) and (16) the lower limit of the integrals may be taken as zero. Needless to say, if the adherends are not identical, for $t_1<t<0$, the stress state in the joint is not zero. In this case, one can still use, for example, one-sided Laplace transform technique to solve the problem by shifting the time such that $t_1 = 0$.

In the problem under consideration, it is assumed that the external loads are applied at the ends of the composite plate. In particular, no external transverse shear load is applied to the plate in $-\ell<x<\ell$ (Figure 1). Thus, from the equilibrium of transverse shear resultants it follows that:

$$Q_1(x,t) + Q_2(x,t) \cong Q_o(t) \tag{18}$$

where $Q_o(t)$ is the transverse shear resultant applied to the ends of the plate (Figure 1). Using the relations (6), (17), and (18), the equations (1–4), (14), and (16) may now be reduced to

$$\frac{\partial^2 \tau}{\partial x^2} - \frac{4C + h(h+h_o)D}{2h_o} \int_o^t G(T, t-\xi) \frac{\partial \tau}{\partial \xi} d\xi \tag{19}$$

$$= -\frac{hD}{2h_o} \int_{o^-}^t G(T, t-\xi) \frac{dQ_o(\xi)}{d\xi} d\xi ,$$

$$\frac{\partial^4 \sigma}{\partial x^4} - K(T) \left[\left(\frac{2}{Bh_o} - \frac{hD}{2} \right) \frac{\partial^2 \sigma}{\partial x^2} - \frac{2D}{h_o} \sigma \right] \tag{20}$$

$$-\frac{2}{3} \int_{o^-}^t G(T, t-\xi) \left[\left(\frac{hD}{2} + \frac{4}{Bh_o} \right) \frac{\partial^3 \sigma}{\partial x^2 \partial \xi} - \frac{4D}{h_o} \frac{\partial \sigma}{\partial \xi} \right] d\xi = 0.$$

First, we observe that if T is constant when the bulk modulus $K(T)$ is constant and the relaxation modulus G is a function of time. Hence, the equations (19) and (20) have "constant coefficients" and a convolution type kernel. Therefore, the equations may be solved by using the standard Laplace transform technique. Let $F_1(x,s)$ and $F_2(x,s)$ be the Laplace transform of $\tau(x,t)$ and $\sigma(x,t)$, respectively. Noting that for $t<0$ the composite plate is stress-free, from (19) and (20) we obtain

$$\frac{d^2F_1}{dx^2} - \gamma_1^2 F_1 = \beta \ , \tag{21}$$

$$\frac{d^4F_2}{dx^4} - 2\gamma_2^2 \frac{d^2F_2}{dx^2} + \gamma_3^4 F_2 = 0 \tag{22}$$

where

$$\gamma_1^2 = \frac{1}{2h_o} [4C + hD(h + h_o)]s\, G_1(s) \tag{23}$$

$$\beta = - \frac{hD}{2h_o} s\, G_1(s)\, q_o(s) \tag{24}$$

$$\gamma_2^2 = \left(\frac{1}{h_o B} - \frac{hD}{4}\right) K(T) + \left(\frac{hD}{6} + \frac{4}{3h_o B}\right)s\, G_1(s) \ , \tag{25}$$

$$\gamma_3^4 = \frac{8D}{3h_o} s\, G_1(s) + \frac{2D}{h_o} K(T) \ , \tag{26}$$

and $G_1(s)$ and $q_o(s)$ are the Laplace transforms of G and Q_o, respectively. The solution of (21) and (22) is

$$F_1(x,s) = A_1 \sinh(\gamma_1 x) + A_2 \cosh(\gamma_1 x) - \frac{\beta}{\gamma_1^2} \tag{27}$$

$$F_2(x,s) = A_3 \sinh(\phi_1 x) + A_4 \cosh(\phi_1 x) + A_5 \sinh(\phi_2 x) + A_6 \cosh(\phi_2 x) \tag{28}$$

where

$$\phi_1 = [\gamma_2^2 + (\gamma_2^4 - \gamma_3^4)^{1/2}]^{1/2} \ , \ \phi_2 = [\gamma_2^2 - (\gamma_2^4 - \gamma_3^4)^{1/2}]^{1/2} \ . \tag{29}$$

The functions $A_1(s),\dots,A_6(s)$ are unknown and are determined from the boundary conditions.

EXAMPLES

The solution of the lap joint problem shown in Figure 1 is obtained for three separate loading conditions. The solution given by (27) and (28) is derived under the assumption that the adherends are identical, the operating temperature T is constant, and the external loads are applied at $t = 0$ ($T \neq T_o$ where T_o is a base temperature corresponding to zero stress and deformation state).

Membrane Loading

Let the composite medium be subjected to a constant membrane loading N_o for $t \geqslant 0$. Reduced to the end points $x = +\ell$ the boundary conditions for the plates 1 and 2 may be expressed as

$$N_{1x}(\ell,t) = 0, \; M_{1x}(\ell,t) = 0, \; Q_{1x}(\ell,t) = 0, \; N_{1x}(-\ell,t) = N_oH(t) \, ,$$

$$M_{1x}(-\ell,t) = -N_o\frac{h_o+h}{2}H(t), \; Q_{1x}(-,t) = 0, \tag{30}$$

$$N_{2x}(\ell,t) = N_oH(t), \; M_{2x}(\ell,t) = N_o\frac{h+h_o}{2}H(t), \; Q_{2x}(\ell,t) = 0,$$

$$N_{2x}(-\ell,t) = 0, \; M_{2x}(-\ell,t) = 0, \; Q_{2x}(-\ell,t) = 0 \, , \tag{31}$$

where $H(t)$ is the Heaviside function. From the symmetry of the problem it can be shown that the conditions (30) and (31) are equivalent to

$$\tau(x,t) = \tau(-x,t) \, , \quad \int_{-\ell}^{\ell} \tau(x,t)dx = -N_oH(t) \, , \tag{32a,b}$$

$$\sigma(x,t) = \sigma(-x,t) \, , \int_{-\ell}^{\ell} \sigma(x,t)dx = 0 \tag{33a,b}$$

$$\frac{\partial^2}{\partial x^2}\sigma(\ell,t) = K(T)\left[\left(\frac{2}{h_oB} - \frac{hD}{2}\right)\sigma(\ell,t) + \frac{h+h_o}{2h_o}DN_oH(t)\right]$$

$$\tag{34}$$

$$+ \frac{2}{3}\int_{o^-}^{t} G(T, t-\xi)\left[\left(\frac{hD}{2} + \frac{4}{Bh_o}\right)\frac{\partial}{\partial \xi}\sigma(\ell,\xi) + \frac{h+h_o}{h_o}DN_o\delta(\xi)\right]d\xi \, .$$

Note that for the loading under consideration $\beta = 0$. Thus, taking the Laplace transforms and substituting from (27) and (28), from (32–34) we obtain

$$A_1(s) = 0, \; A_3(s) = 0, \; A_5(s) = 0, \; A_2(s) = -\frac{\gamma_1N_o}{2s \, sinh \, (\gamma_1\ell)} \, , \tag{35}$$

$$A_4(s) = -\frac{(h+h_o)N_o\gamma_3{}^4 \sinh(\phi_2\ell)}{4s\phi_2\Delta_a(s)} \,,\quad A_6(s) = \frac{(h+h_o)N_o\gamma_3{}^4 \sinh(\phi_1\ell)}{4s\phi_1\Delta_a(s)} \,,$$

where

$$\Delta_a(s) = \phi_2\cosh(\phi_1\ell)\sinh(\phi_2\ell) - \phi_1\sinh(\phi_1\ell)\cosh(\phi_2\ell). \tag{36}$$

The adhesive stresses $\tau(x,t)$ and $\sigma(x,t)$ are obtained by substituting from (27), (28), and (35) into the corresponding inversion integrals.

Bending

In this case of the twelve stress and moment resultants which are prescribed on the boundaries (see (30) and (31)), the following are the only nonzero components:

$$M_{1x}(-\ell,t) = M_oH(t) \,,\; M_{2x}(\ell,t) = M_oH(t) \tag{37}$$

Again, it can be shown that the boundary conditions are equivalent to

$$\tau(x,t) = -\tau(-x,t) \,, \tag{38}$$

$$\frac{\partial}{\partial x}\tau(\ell,t) = -\frac{hD}{2h_o}M_o\int_{o^-}^t G(T,\, t-\xi)\delta\,(\xi)d\xi \tag{39}$$

$$\sigma(x,t) = -\sigma(-x,t),\quad \int_{-\ell}^t \sigma(x,t)xdx = M_oH(t) \,, \tag{40a,b}$$

$$\frac{\partial^2}{\partial x^2}\sigma(\ell,t) = K(T)\left[\left(\frac{2}{Bh_o} - \frac{hD}{2}\right)\sigma(\ell,t) + \frac{D}{h_o}M_oH(t)\right]$$

$$+ \frac{2}{3}\int_{o^-}^t G(T,\, t-\xi)\left[\left(\frac{hD}{2} + \frac{4}{h_oB}\right)\frac{\partial}{\partial\xi}\sigma(\ell,y) + \frac{2D}{h_o}M_o\delta(\xi)\right]d\xi. \tag{41}$$

Substituting now from (27) and (28) into the Laplace transforms of (38–41) we find

$$A_2(s) = 0,\; A_4(s) = 0,\; A_6(s) = 0,\; A_1(s) = -\frac{hDM_oG_1(s)}{2h_o\gamma_1\cosh(\gamma_1\ell)} \,, \tag{42}$$

$$A_3(s) = -\frac{\gamma_3{}^4 M_o \cosh(\phi_2 \ell)}{2s\phi_2 \Delta_b(s)}, \quad A_5(s) = \frac{\gamma_3{}^4 M_o \cosh(\phi_1 \ell)}{2s\phi_1 \Delta_b(s)},$$

where

$$\Delta_b(s) = \phi_2 \sinh(\phi_1 \ell)\cosh(\phi_2(s)) - \phi_1 \cosh(\phi_1 \ell)\sinh(\phi_2 \ell). \tag{43}$$

Transverse Shear Loading

For this loading condition the following are the only nonhomogeneous boundary conditions:

$$Q_{1x}(-\ell,t) = Q_o H(t), \; M_x(-\ell,t) = -Q_o \ell H(t),$$

$$Q_{2x}(\ell,t) = Q_o H(t), \; M_{2x}(\ell,t) = Q_o \ell H(t). \tag{44}$$

The equivalent conditions in terms of τ and σ may be shown to be

$$\tau(x,t) = \tau(-x,t), \quad \int_{-\ell}^{\ell} \tau(x,t)dx = 0, \tag{45a,b}$$

$$\sigma(x,t) = \sigma(-x,t), \quad \int_{-\ell}^{\ell} \sigma(x,t)dx = -Q_o H(t), \tag{46,a,b}$$

$$\frac{\partial^2}{\partial x^2}\sigma(\ell,t) = K(T)\left[\left(\frac{2}{Bh_o} - \frac{hD}{2}\right)\sigma(\ell,t) + \frac{D\ell}{h_o}Q_o H(t)\right]$$

$$+ \frac{2}{3}\int_{o^-}^{\ell} G(T, t-\xi)\left[\left(\frac{hD}{2} + \frac{4}{Bh_o}\right)\frac{\partial}{\partial \xi}\sigma(\ell,\xi) + \frac{2D\ell}{h_o}Q_o \delta(\xi)\right]d\xi. \tag{47}$$

In this case

$$\beta = -\frac{hD}{2h_o}Q_o G_1(s) \tag{48}$$

and the functions $A_1(s),..,A_6(s)$ are obtained as

$$A_1(s) = A_3(s) = A_5(s) = 0, \; A_2 s) = \frac{\beta\ell}{\gamma_1 \sinh(\gamma_1 \ell)},$$

$$A_4(s) = \frac{Q_o[\gamma_3{}^2\phi_1\cosh(\phi_2\ell) - \gamma_3{}^4\ell \, \sinh(\phi_2\ell)}{2s5f_2\Delta_c(s)}$$

$$A_6(s) = -\frac{Q_o[\gamma_3{}^2\phi_2\cosh(\phi_1\ell) - \gamma_3{}^4\ell \, \sinh(\phi_1\ell)}{2s\phi\Delta_c(s)} ,$$ (49)

where

$$\Delta_c(s) = \phi_2\cosh(\phi_1\ell)\sinh(\phi_2\ell) - \phi_1\sinh(\phi_1\ell)\cosh(\phi_2\ell) .$$ (50)

THE GENERAL PROBLEM

As pointed out earlier, the general problem for dissimilar adherends under arbitrary temperature and loading conditions for $t>0$ may be reduced to a system of coupled equations of the form (19) and (20). If the temperature T is constant for $t>0$, then the equations have constant coefficients and may easily be solved by using Laplace transforms. On the other hand, if T is not constant, the equations would have time-dependent coefficients, and the Laplace transforms would not be applicable. In this case, to solve the problem, one may have to use a purely numerical technique.

There is one special case for which the solution may be obtained by following the procedure outlined in the previous section. Let the external loads be zero for $t<0$, and be given arbitrary functions of t for $t>0$. Also, let the temperature be a piecewise constant function of time, i.e., let

$$T(t) = T_i, \quad t_{i-1}<t < t_i, \quad t_o = 0, \quad (i=1,...,n), \quad T(t) = T_{n+1} , \quad t > t_n \quad (51)$$

Then, for $T = T_1$, the solution given in the previous section is valid in $0<t<t_1$. After obtaining this solution the functions $\tau(x,t_1)$ and $\sigma(x,t_1)$ can be calculated. Using now this information as the "initial conditions" in time shifted to t_1, assuming $T = T_2$, and repeating the procedure of the previous section, the solution may be obtained which is valid for $t_1<t<t_2$. The complete solution is obtained by repeating this process for the intervals $t_2<t<t_3,...,t_n<t$.

NUMERICAL RESULTS

Once the relaxation modulus G and the bulk modulus K are specified, the solution may be expressed in terms of Laplace inversion integrals. These integrals are much too complicated for closed form evaluation. However, they can easily be expressed in terms of real integrals and can be evaluated numerically [5]. The relaxation modulus of the adhesive is obtained from a torsion relaxation test. In practice $G(T,t)$ is generally measured in an interval $t_1<t<t_2$ for different temperature levels. If the material is thermorheologically simple, the function G for $0<t<\infty$ may then be obtained by using the time-temperature shift factor $a(T)$. For a reference temperature T_o, G is expressed as

$$G(T,t) = G(T_o,\eta), \quad \eta = \frac{t}{a(T)}, \tag{52}$$

η being the reduced time.

For the numerical calculations the relaxation modulus of the epoxy is assumed to have the following form:

$$G(T,t) = \{[\mu_o(T) - \mu_\infty(T)]e^{-t/\varepsilon(T)} + \mu_\infty(T)\}H(t), \tag{53}$$

where

$$\varepsilon(T) = \frac{\mu_\infty(T)}{\mu_o(T)} t_o(T) . \tag{54}$$

$\mu_o(T)$ represents the shear modulus at $t=0$, $\mu_\infty(T)$ at $T=\infty$, and $t_o(T)$ corresponds to the retardation time. For such a material the bulk modulus may be expressed as

$$K(T) = \frac{E_o(T)\mu_o(T)}{3[3\mu_o(T) - E_o(T)]} \tag{55}$$

The constants E_o, μ_o, μ_∞, and t_o for the temperature levels used in the calculations are given in Table 1. The adherends are assumed to be aluminum plates for which $E = 10^7$ psi, $v = 0.3$. The table also shows the dimensions of the lap joint.

For the adhesive model used the Laplace transform of the relaxation modulus is

$$G_1(s) = \frac{\mu_o(T) - \mu_\infty(T)}{s + 1/\varepsilon(T)} + \frac{\mu_\infty(T)}{s} . \tag{56}$$

The calculated results are shown in Tables 2–7. To show the trends some limited results are also shown in Figures 2–4. For a lap joint under membrane loading Figure 2 shows the distribution of the shear stress in the adhesive at

Table 1. Viscoelastic constants of the adhesive.

$T(^\circ F)$	E_o(psi)	μ_o(psi)	μ_∞(psi)	t_o(hours)
70	4.65×10^5	1.80×10^5	0.8×10^5	0.5
100	4.40×10^5	1.70×10^5	0.7×10^5	0.5
140	4.10×10^5	1.58×10^5	0.58×10^5	0.5
180	3.85×10^5	1.50×10^5	0.50×10^5	0.5

various times and for various operating temperatures. Figures 3 and 4 show the "relaxation" of the adhesive stresses $\tau(x,t)$ and $\sigma(x,t)$ for various operating temperatures. These figures indicate that after approximately one hour the stress state in the joint reaches a steady-state. From the figures, it may be observed that the peak values of σ and τ (which are at the end points $x = \pm \ell$) at $t=0$ may be considerably greater than the corresponding peak stresses at steady-state. Except for the variation with time and temperature, as expected the distribution of stresses in the adhesive has the same trends as those obtained from the related elastic problem [4,5].

Table 2. Variation of $\tau(x,t)/N_o/\ell)$ for the case of membrane loading (t in hours).

x/ℓ	t = 0.01	t = 0.05	t = 0.1	t = 0.5	t = 1	t = 2
			T = 70°F			
0.	-6.98×10^{-4}	-9.10×10^{-4}	-1.27×10^{-3}	-4.27×10^{-3}	-7.17×10^{-3}	-9.54×10^{-3}
0.1	-1.04×10^{-3}	-1.33×10^{-3}	-1.82×10^{-3}	-5.62×10^{-3}	-9.06×10^{-3}	-0.012
0.2	-2.40×10^{-3}	-2.97×10^{-3}	-3.91×10^{-3}	-0.011	-0.016	-0.019
0.3	-6.10×10^{-3}	-7.32×10^{-3}	-9.30×10^{-3}	-0.022	-0.030	-0.035
0.4	-0.016	-0.018	-0.023	-0.046	-0.060	-0.066
0.5	-0.041	-0.046	-0.055	-0.098	-0.119	-0.127
0.6	-0.106	-0.116	-0.132	-0.205	-0.234	-0.243
0.7	-0.273	-0.292	-0.318	-0.425	-0.458	-0.465
0.8	-0.708	-0.732	-0.764	-0.871	-0.890	-0.888
0.9	-1.831	-1.837	-1.829	-1.761	-1.714	-1.695
1.0	-4.740	-4.603	-4.358	-3.501	-3.277	-3.233
			T = 100°F			
0.	-9.06×10^{-4}	-1.21×10^{-3}	-1.73×10^{-3}	-6.11×10^{-3}	-0.010	-0.014
0.1	-1.32×10^{-3}	-1.72×10^{-3}	-2.41×10^{-3}	-7.85×10^{-3}	-0.013	-0.016
0.2	-2.94×10^{-3}	-3.71×10^{-3}	-4.98×10^{-3}	-0.014	-0.021	-0.025
0.3	-7.26×10^{-3}	-8.85×10^{-3}	-0.011	-0.028	-0.039	-0.044
0.4	-0.018	-0.022	-0.027	-0.056	-0.073	-0.080
0.5	-0.046	-0.052	-0.063	-0.114	-0.138	-0.147
0.6	-0.115	-0.128	-0.146	-0.230	-0.261	-0.269
0.7	-0.289	-0.310	-0.340	-0.456	-0.488	-0.493
0.8	-0.727	-0.754	-0.788	-0.895	-0.908	-0.903
0.9	-1.828	-1.831	-1.818	-1.728	-1.673	-1.653
1.0	-4.594	-4.440	-4.173	-3.279	-3.063	-3.024
			T = 140°F			
0.	-1.26×10^{-3}	-1.74×10^{-3}	-2.58×10^{-3}	-9.81×10^{-3}	-0.017	-0.022
0.1	-1.79×10^{-3}	-2.41×10^{-3}	-3.49×10^{-3}	-0.012	-0.020	-0.025
0.2	-3.81×10^{-3}	-4.95×10^{-3}	-6.85×10^{-3}	-0.020	-0.031	-0.037
0.3	-9.04×10^{-3}	-0.011	-0.015	-0.038	-0.053	-0.060
0.4	-0.022	-0.026	-0.033	-0.073	-0.094	-0.102

(continued)

Table 2. (continued).

x/ℓ	t = 0.01	t = 0.05	t = 0.1	t = 0.5	t = 1	t = 2
			T = 140°F			
0.5	−0.053	−0.062	−0.075	−0.140	−0.168	−0.176
0.6	−0.128	−0.144	−0.167	−0.266	−0.298	−0.305
0.7	−0.310	−0.335	−0.371	−0.498	−0.527	−0.529
0.8	−0.752	−0.781	−0.819	−0.921	−0.924	−0.917
0.9	−1.821	−1.819	−1.800	−1.674	−1.609	−1.589
1.0	−4.408	−4.229	−3.928	−2.985	−2.784	−2.752
			T = 180°F			
0.	-1.59×10^{-3}	-2.26×10^{-3}	-3.45×10^{-3}	−0.014	−0.024	−0.030
0.1	-2.21×10^{-3}	-3.07×10^{-3}	-4.57×10^{-3}	−0.017	−0.027	−0.034
0.2	-4.58×10^{-3}	-6.09×10^{-3}	-8.65×10^{-3}	−0.027	−0.040	−0.048
0.3	−0.011	−0.013	−0.018	−0.048	−0.066	−0.074
0.4	−0.025	−0.030	−0.039	−0.088	−0.112	−0.120
0.5	−0.058	−0.069	−0.085	−0.162	−0.192	−0.199
0.6	−0.138	−0.156	−0.184	−0.295	−0.327	−0.332
0.7	−0.326	−0.354	−0.395	−0.529	−0.553	−0.553
0.8	−0.769	−0.801	−0.841	−0.936	−0.931	−0.921
0.9	−1.814	−1.808	−1.782	−1.626	−1.555	−1.534
1.0	−4.277	−4.075	−4.744	−2.770	−2.582	−2.555

Table 3. Variation of $o(x,t)/N_o/\ell$ for membrane loading (t in hours).

x/ℓ	t = 0.01	t = 0.05	t = 0.1	t = 0.5	t = 1	t = 2
			T = 70°F			
0.	1.93×10^{-4}	2.18×10^{-4}	2.55×10^{-4}	4.07×10^{-4}	4.57×10^{-4}	4.70×10^{-4}
0.1	2.30×10^{-4}	2.67×10^{-4}	3.24×10^{-4}	5.51×10^{-4}	6.24×10^{-4}	6.41×10^{-4}
0.2	9.78×10^{-5}	1.70×10^{-4}	2.81×10^{-4}	7.15×10^{-4}	8.45×10^{-4}	8.71×10^{-4}
0.3	-1.69×10^{-3}	-1.60×10^{-3}	-1.44×10^{-3}	-8.77×10^{-4}	-7.31×10^{-4}	-7.10×10^{-4}
0.4	−0.011	−0.011	−0.011	−0.012	−0.012	−0.012
0.5	−0.050	−0.051	−0.052	−0.056	−0.057	−0.057
0.6	−0.174	−0.177	−0.181	−0.194	−0.197	−0.198
0.7	−0.495	−0.502	−0.507	−0.526	−0.530	−0.531
0.8	−1.032	−1.034	−1.030	−1.013	−1.006	−1.005
0.9	−0.640	−0.621	−0.587	−0.473	−0.448	−0.445
1.0	7.870	7.806	7.649	7.151	7.044	7.033
			T = 100°F			
0.	2.41×10^{-4}	2.72×10^{-4}	3.18×10^{-4}	5.01×10^{-4}	5.57×10^{-4}	5.70×10^{-4}
0.1	3.01×10^{-4}	3.47×10^{-4}	4.17×10^{-4}	6.87×10^{-4}	7.67×10^{-4}	7.83×10^{-4}
0.2	2.33×10^{-4}	3.21×10^{-4}	4.53×10^{-4}	9.54×10^{-4}	1.09×10^{-3}	1.11×10^{-3}
0.3	-1.52×10^{-3}	-1.41×10^{-3}	-1.23×10^{-3}	-6.19×10^{-4}	-4.83×10^{-4}	-4.68×10^{-4}

(continued)

Table 3. (continued).

x/ℓ	t = 0.01	t = 0.05	t = 0.1	t = 0.5	t = 1	t = 2
			T = 100°F			
0.4	−0.011	−0.012	−0.012	−0.012	−0.012	−0.012
0.5	−0.051	−0.052	−0.054	−0.058	−0.060	−0.060
0.6	−0.179	−0.183	−0.187	−0.201	−0.205	−0.205
0.7	−0.503	−0.509	−0.515	−0.535	−0.539	−0.539
0.8	−1.026	−1.028	−1.022	−1.002	−0.995	−0.994
0.9	−0.596	−0.574	−0.537	−0.419	−0.396	−0.394
1.0	7.673	−7.599	7.429	6.920	6.823	6.816
			T = 140°F			
0.	3.10×10^{-4}	3.52×10^{-4}	4.13×10^{-4}	6.45×10^{-4}	7.09×10^{-4}	7.20×10^{-4}
0.1	4.03×10^{-4}	4.64×10^{-4}	5.55×10^{-4}	8.91×10^{-4}	9.79×10^{-4}	9.93×10^{-4}
0.2	4.21×10^{-4}	5.35×10^{-4}	7.04×10^{-4}	1.30×10^{-3}	1.44×10^{-3}	1.45×10^{-3}
0.3	-1.29×10^{-3}	-1.16×10^{-3}	-9.46×10^{-4}	-2.86×10^{-4}	-1.71×10^{-4}	-1.66×10^{-4}
0.4	−0.012	−0.012	−0.012	−0.013	−0.013	−0.013
0.5	−0.053	−0.055	−0.056	−0.062	−0.064	−0.064
0.6	−0.186	−0.190	−0.195	−0.211	−0.214	−0.214
0.7	−0.512	−0.519	−0.526	−0.546	−0.549	−0.549
0.8	−1.018	−1.018	−1.011	−0.985	−0.978	−0.977
0.9	−0.541	−0.515	−0.472	−0.350	−0.331	−0.330
1.0	7.428	7.338	7.150	6.631	6.549	6.546
			T = 180°F			
0.	4.42×10^{-4}	5.04×10^{-4}	5.95×10^{-4}	9.35×10^{-4}	1.02×10^{-3}	1.04×10^{-3}
0.1	5.91×10^{-4}	6.80×10^{-4}	8.12×10^{-4}	1.28×10^{-3}	1.39×10^{-3}	1.41×10^{-3}
0.2	7.48×10^{-4}	9.05×10^{-4}	1.13×10^{-3}	1.90×10^{-3}	2.05×10^{-3}	2.07×10^{-3}
0.3	-9.54×10^{-4}	-7.89×10^{-4}	-5.38×10^{-4}	1.56×10^{-4}	2.36×10^{-4}	2.31×10^{-4}
0.4	−0.012	−0.012	−0.013	−0.014	−0.014	−0.014
0.5	−0.057	−0.059	−0.061	−0.069	−0.071	−0.071
0.6	−0.196	−0.202	−0.208	−0.227	−0.230	−0.231
0.7	−0.526	−0.534	−0.542	−0.562	−0.564	−0.564
0.8	−1.002	−1.000	−0.989	−0.954	−0.946	−0.946
0.9	−0.454	−0.423	−0.374	−0.246	−0.229	−0.229
1.0	7.053	6.944	6.733	6.195	6.124	6.123

Table 4. Variation of $\tau(x,t)/M_o/\ell^2)$ for bending (t in hours).

x/ℓ	t = 0.01	t = 0.05	t = 0.1	t = 0.5	t = 1	t = 2
			T = 70°F			
0.	0.	0.	0.	0.	0.	0.
0.1	-6.21×10^{-3}	-7.80×10^{-3}	−0.010	−0.030	−0.045	−0.055
0.2	−0.018	−0.023	−0.030	−0.078	−0.113	−0.134

(continued)

Table 4. (continued).

x/ℓ	t = 0.01	t = 0.05	t = 0.1	t = 0.5	t = 1	t = 2
			T = 70°F			
0.3	−0.049	−0.059	−0.074	−0.173	−0.238	−0.272
0.4	−0.127	−0.148	−0.181	−0.372	−0.481	−0.531
0.5	−0.329	−0.372	−0.440	−0.788	−0.957	−1.022
0.6	−0.851	−0.936	−1.064	−1.653	−1.889	−1.960
0.7	−2.204	−2.353	−2.565	−3.429	−3.696	−3.749
0.8	−5.706	−5.906	−6.162	−7.028	−7.177	−7.161
0.9	−14.77	−14.81	−14.75	−14.20	−13.82	−13.67
1.0	−38.22	−37.12	−35.15	−28.23	−26.42	−26.07
			T = 100°F			
0.	0.	0.	0.	0.	0.	0.
0.1	-7.75×10^{-3}	-9.92×10^{-3}	−0.014	−0.040	−0.060	−0.073
0.2	−0.023	−0.028	−0.038	−0.102	−0.148	−0.173
0.3	−0.058	−0.071	−0.091	−0.219	−0.300	−0.340
0.4	−0.147	−0.173	−0.215	−0.453	−0.583	−0.638
0.5	−0.369	−0.422	−0.505	−0.921	−1.113	−1.179
0.6	−0.928	−1.028	−1.179	−1.853	−2.103	−2.168
0.7	−2.333	−2.502	−2.743	−3.681	−3.939	−3.976
0.8	−5.864	−6.080	−6.356	−7.217	−7.319	−7.283
0.9	−14.74	−14.76	−14.66	−13.94	−13.49	−13.33
1.0	−37.04	−35.81	−33.66	−26.44	−24.70	−24.39
			T = 140°F			
0.	0.	0.	0.	0.	0.	0.
0.1	−0.010	−0.013	−0.019	−0.059	−0.088	−0.103
0.2	−0.029	−0.037	−0.051	−0.145	−0.208	−0.239
0.3	−0.072	−0.090	−0.119	−0.298	−0.405	−0.450
0.4	−0.176	−0.212	−0.269	−0.586	−0.746	−0.803
0.5	−0.426	−0.496	−0.604	−0.128	−1.346	−1.408
0.6	−1.033	−1.159	−1.347	−2.143	−2.402	−2.452
0.7	−2.503	−2.704	−2.990	−4.019	−4.247	−4.260
0.8	−6.062	−6.302	−6.606	−7.428	−7.452	−7.390
0.9	−14.68	−14.67	−14.51	−13.50	−12.98	−12.81
1.0	−35.55	−34.11	−31.68	−24.08	−22.45	−22.19
			T = 180°F			
0.	0.	0.	0.	0.	0.	0.
0.1	−0.012	−0.017	−0.024	−0.078	−0.115	−0.132
0.2	−0.035	−0.046	−0.064	−0.188	−0.266	−0.299
0.3	−0.084	−0.107	−0.145	−0.373	−0.500	−0.547
0.4	−0.199	−0.245	−0.317	−0.706	−0.887	−0.942
0.5	−0.471	−0.556	−0.688	−1.305	−1.536	−1.591
0.6	−1.113	−1.262	−1.484	−2.376	−2.629	−2.665
0.7	−2.628	−2.857	−3.182	−4.268	−4.459	−4.451
0.8	−6.200	−6.459	−6.786	−7.547	−7.503	−7.423
0.9	−14.63	−14.58	−14.37	−13.11	−12.54	−12.37
1.0	−34.49	−32.86	−30.19	−22.34	−20.82	−20.60

(continued)

Table 5. Variation of $\sigma(x,t)/M_o/\ell^2)$ for bending (t in hours).

x/ℓ	t = 0.01	t = 0.05	t = 0.1	t = 0.5	t = 1	t = 2
			T = 70 °F			
0.	0.	0.	0.	0.	0.	0.
0.1	1.71×10^{-3}	2.03×10^{-3}	2.51×10^{-3}	4.47×10^{-3}	5.09×10^{-3}	5.23×10^{-3}
0.2	8.28×10^{-4}	1.58×10^{-3}	2.73×10^{-3}	7.25×10^{-3}	8.61×10^{-3}	8.89×10^{-3}
0.3	−0.018	−0.017	−0.015	-9.38×10^{-3}	-7.83×10^{-3}	-7.60×10^{-3}
0.4	−0.120	−0.121	−0.121	−0.124	−0.126	−0.126
0.5	−0.529	−0.539	−0.550	−0.594	−0.607	−0.610
0.6	−1.854	−1.888	−1.925	−2.064	−2.100	−2.106
0.7	−5.269	−5.335	−5.393	−5.599	−5.640	−5.646
0.8	−10.98	−11.00	−10.95	−10.77	−10.71	−10.70
0.9	−6.813	−6.606	−6.241	−5.032	−4.768	−4.737
1.0	83.72	83.04	81.37	76.07	74.93	74.82
			T = 100 °F			
0.	0.	0.	0.	0.	0.	0.
0.1	2.31×10^{-3}	2.71×10^{-3}	3.30×10^{-3}	5.62×10^{-3}	6.30×10^{-3}	6.43×10^{-3}
0.2	2.23×10^{-3}	3.14×10^{-3}	4.52×10^{-3}	9.74×10^{-3}	0.011	0.011
0.3	−0.016	−0.015	−0.013	-6.64×10^{-3}	-5.18×10^{-3}	-5.02×10^{-3}
0.4	−0.122	−0.123	−0.123	−0.128	−0.130	−0.130
0.5	−0.545	−0.557	−0.570	−0.621	−0.635	−0.638
0.6	−1.905	−1.944	−1.987	−2.141	−2.176	−2.181
0.7	−5.349	−5.420	−5.484	−5.695	−5.730	−5.735
0.8	−10.92	−10.94	−10.88	−10.65	−10.58	−10.57
0.9	−6.343	−6.108	−5.707	−4.455	−4.213	−4.191
1.0	81.63	80.84	79.03	73.61	72.58	72.51
			T = 140 °F			
0.	0.	0.	0.	0.	0.	0.
0.1	3.18×10^{-3}	3.70×10^{-3}	4.48×10^{-3}	7.34×10^{-3}	8.06×10^{-3}	8.18×10^{-3}
0.2	4.18×10^{-3}	5.36×10^{-3}	7.12×10^{-3}	0.013	0.015	0.015
0.3	−0.014	−0.012	−0.010	-3.09×10^{-3}	-1.86×10^{-3}	-1.80×10^{-3}
0.4	−0.124	−0.125	−0.126	−0.134	−0.136	−0.137
0.5	−0.567	−0.582	−0.599	−0.661	−0.676	−0.678
0.6	−1.973	−2.019	−2.071	−2.244	−2.277	−2.281
0.7	−5.448	−5.526	−5.599	−5.813	−5.841	−5.843
0.8	−10.83	−10.83	−10.75	−10.48	−10.41	−10.40
0.9	−5.751	−5.474	−5.021	−3.726	−3.522	−3.511
1.0	79.02	78.07	76.06	70.54	69.67	69.64
			T = 180 °F			
0.	0.	0.	0.	0.	0.	0.
0.1	4.76×10^{-3}	5.51×10^{-3}	6.62×10^{-3}	0.011	0.011	0.012
0.2	7.58×10^{-3}	9.20×10^{-3}	0.012	0.020	0.021	0.021
0.3	−0.010	-8.45×10^{-3}	-5.78×10^{-3}	1.64×10^{-3}	2.51×10^{-3}	2.46×10^{-3}

(continued)

397

Table 5. (continued).

x/ℓ	t = 0.01	t = 0.05	t = 0.1	t = 0.5	t = 1	t = 2
			T = 180°F			
0.4	−0.129	−0.132	−0.135	−0.148	−0.153	−0.153
0.5	−0.609	−0.628	−0.653	−0.736	−0.754	−0.756
0.6	−2.089	−2.146	−2.212	−2.417	−2.450	−2.453
0.7	−5.600	−5.685	−5.766	−5.981	−6.000	−6.001
0.8	−10.65	−10.63	−10.52	−10.15	−10.06	−10.06
0.9	−4.828	−4.497	−3.980	−2.617	−2.439	−2.434
1.0	75.03	73.87	71.63	65.91	65.14	65.13

Table 6. Variation of $\tau(x,t)/Q_o/\ell)$ for transverse shear loading (t in hours).

x/ℓ	t = 0.01	t = 0.05	t = 0.1	t = 0.5	t = 1	t = 2
			T = 70°F			
0.	4.027	4.025	4.022	3.998	3.974	3.955
0.1	4.024	4.022	4.018	3.987	3.959	3.938
0.2	4.013	4.008	4.001	3.948	3.906	3.878
0.3	3.983	3.973	3.957	3.856	3.788	3.750
0.4	3.905	3.884	3.851	3.659	3.548	3.496
0.5	3.703	3.660	3.592	3.244	3.073	3.007
0.6	3.181	3.096	2.968	2.379	2.143	2.071
0.7	1.828	1.680	1.467	0.603	0.336	0.283
0.8	−1.674	−1.873	−2.129	−2.996	−3.145	−3.129
0.9	−10.74	−10.78	−10.71	−10.17	−9.792	−9.637
1.0	−34.19	−33.09	−31.11	−24.20	−22.39	−22.04
			T = 100°F			
0.	4.025	4.023	4.018	3.983	3.949	3.922
0.1	4.022	4.018	4.013	3.969	3.929	3.901
0.2	4.009	4.002	3.992	3.919	3.863	3.828
0.3	3.974	3.961	3.940	3.808	3.721	3.675
0.4	3.885	3.859	3.816	3.577	3.444	3.385
0.5	3.663	3.610	3.527	3.110	2.917	2.848
0.6	3.104	3.004	2.852	2.179	1.928	1.862
0.7	1.699	1.531	1.289	0.351	0.093	0.055
0.8	−1.832	−2.048	−2.324	−3.185	−3.287	−3.252
0.9	−10.71	−10.73	−10.63	−9.905	−9.463	−9.299
1.0	−33.01	−31.78	−29.62	−22.41	−20.67	−20.35
			T = 140°F			
0.	4.022	4.018	4.011	3.953	3.898	3.858
0.1	4.018	4.013	4.004	3.934	3.872	3.830

(continued)

Table 6. (continued).

x/ℓ	t = 0.01	t = 0.05	t = 0.1	t = 0.5	t = 1	t = 2
			T = 140°F			
0.2	4.002	3.992	3.977	3.867	3.785	3.737
0.3	3.959	3.941	3.912	3.724	3.607	3.550
0.4	3.856	3.820	3.762	3.442	3.276	3.211
0.5	3.606	3.536	3.428	2.902	2.681	2.614
0.6	2.999	2.874	2.686	1.888	1.628	1.574
0.7	1.530	1.329	1.042	0.013	−0.217	−0.231
0.8	−2.030	−2.269	−2.574	−3.396	−3.421	−3.360
0.9	−10.65	−10.64	−10.48	−9.470	−8.947	−8.779
1.0	−31.52	−30.07	−27.64	−20.04	−18.42	−18.16
			T = 180°F			
0.	4.019	4.014	4.004	3.920	3.843	3.791
0.1	4.014	4.007	3.995	3.896	3.811	3.758
0.2	3.995	3.983	3.963	3.814	3.707	3.649
0.3	3.947	3.924	3.885	3.643	3.499	3.436
0.4	3.832	3.787	3.714	3.319	3.127	3.061
0.5	3.561	3.476	3.344	2.724	2.486	2.424
0.6	2.919	2.770	2.548	1.654	1.398	1.357
0.7	1.405	1.175	0.850	−0.236	−0.430	−0.425
0.8	−2.168	−2.427	−2.754	−3.515	−3.473	−3.394
0.9	−10.59	−10.55	−10.34	−9.082	−8.507	−8.340
1.0	−30.46	−28.83	−26.16	−18.31	−16.79	−16.57

Table 7. Variation of $\sigma(x,t)/Q_o/\ell)$ for transverse shear loading (t in hours).

x/ℓ	t = 0.01	t = 0.05	t = 0.1	t = 0.5	t = 1	t = 2
			T = 70°F			
0.	1.94×10^{-3}	2.19×10^{-3}	2.56×10^{-3}	4.05×10^{-3}	4.54×10^{-3}	4.65×10^{-3}
0.1	2.37×10^{-3}	2.74×10^{-3}	3.31×10^{-3}	5.58×10^{-3}	6.30×10^{-3}	6.47×10^{-3}
0.2	1.34×10^{-3}	2.09×10^{-3}	3.23×10^{-3}	7.69×10^{-3}	9.02×10^{-3}	9.30×10^{-3}
0.3	−0.015	−0.014	−0.013	-6.49×10^{-3}	-4.85×10^{-3}	-4.59×10^{-3}
0.4	−0.109	−0.109	−0.109	−0.109	−0.110	−0.111
0.5	−0.487	−0.495	−0.504	−0.540	−0.551	−0.554
0.6	−1.727	−1.757	−1.789	−1.912	−1.944	−1.950
0.7	−4.980	−5.041	−5.094	−5.283	−5.321	−5.327
0.8	−10.67	−10.70	−10.66	−10.51	−10.45	−10.44
0.9	−8.223	−8.048	−7.720	−6.638	−6.400	−6.373
1.0	71.59	70.97	69.47	64.74	63.73	63.63

(continued)

399

Table 7. (continued).

x/ℓ	t = 0.01	t = 0.05	t = 0.1	t = 0.5	t = 1	t = 2
			T = 100°F			
0.	2.42×10^{-3}	2.72×10^{-3}	3.18×10^{-3}	4.97×10^{-3}	5.51×10^{-3}	5.63×10^{-3}
0.1	3.08×10^{-3}	3.54×10^{-3}	4.23×10^{-3}	6.93×10^{-3}	7.72×10^{-3}	7.88×10^{-3}
0.2	2.74×10^{-3}	3.64×10^{-3}	5.01×10^{-3}	0.010	0.012	0.012
0.3	−0.014	−0.012	−0.010	-3.55×10^{-3}	-1.98×10^{-3}	-1.79×10^{-3}
0.4	−0.109	−0.110	−0.110	−0.112	−0.113	−0.113
0.5	−0.499	−0.510	−0.520	−0.563	−0.575	−0.578
0.6	−1.772	−1.806	−1.844	−1.980	−2.011	−2.016
0.7	−5.053	−5.119	−5.177	−5.372	−5.405	−5.409
0.8	−10.62	−10.64	−10.59	−10.41	−10.35	−10.34
0.9	−7.802	−7.601	−7.242	−6.120	−5.903	−5.884
1.0	69.71	68.99	67.38	62.54	61.63	61.56
			T = 140°F			
0.	3.10×10^{-3}	3.51×10^{-3}	4.11×10^{-3}	6.37×10^{-3}	6.98×10^{-3}	7.09×10^{-3}
0.1	4.10×10^{-3}	4.71×10^{-3}	5.62×10^{-3}	8.95×10^{-3}	9.81×10^{-3}	9.95×10^{-3}
0.2	4.68×10^{-3}	5.85×10^{-3}	7.59×10^{-3}	0.014	0.015	0.015
0.3	−0.011	-9.50×10^{-3}	-7.15×10^{-3}	3.28×10^{-4}	1.69×10^{-3}	1.79×10^{-3}
0.4	−0.110	−0.111	−0.112	−0.116	−0.117	−0.118
0.5	−0.519	−0.531	−0.545	−0.597	−0.610	−0.612
0.6	−1.832	−1.873	−1.919	−2.072	−2.101	−2.105
0.7	−5.144	−5.217	−5.283	−5.482	−5.508	−5.510
0.8	−10.55	−10.56	−10.49	−10.26	−10.20	−10.19
0.9	−7.272	−7.033	−6.627	−5.467	−5.283	−5.273
1.0	67.38	66.51	64.72	59.80	59.03	59.00
			T = 180°F			
0.	4.39×10^{-3}	4.99×10^{-3}	5.88×10^{-3}	9.18×10^{-3}	0.010	0.010
0.1	5.98×10^{-3}	6.86×10^{-3}	8.17×10^{-3}	0.013	0.014	0.014
0.2	8.06×10^{-3}	9.68×10^{-3}	0.012	0.020	0.022	0.022
0.3	-7.11×10^{-3}	-5.22×10^{-3}	-2.37×10^{-3}	5.81×10^{-3}	6.90×10^{-3}	6.89×10^{-3}
0.4	−0.114	−0.116	−0.117	−0.127	−0.130	−0.130
0.5	−0.554	−0.570	−0.591	−0.662	−0.676	−0.678
0.6	−1.935	−1.985	−2.044	−2.226	−2.256	−2.258
0.7	−5.284	−5.364	−5.439	−5.639	−5.658	−5.659
0.8	−10.40	−10.39	−10.29	−9.980	−9.908	−9.904
0.9	−6.443	−6.157	−5.693	−4.472	−4.311	−4.306
1.0	63.80	62.75	60.75	55.66	54.98	54.97

400

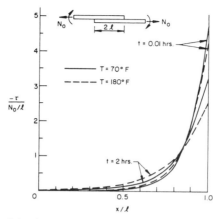

Figure 2. Distribution of the shear stress in the adhesive in a bonded joint under uniform membrane loading.

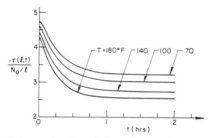

Figure 3. Relaxation of the peak value of the adhesive shear stress for various operating temperatures.

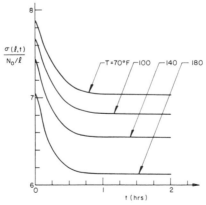

Figure 4. Relaxation of the peak value of the normal stress in the adhesive for various operating temperatures.

401

REFERENCES

1. Goland, M. and Reissner, E., "The Stresses in Cemented Joints," *Journal of Applied Mechanics, 1,* Trans. ASME, A17–A27 (1944).
2. Erdogan, F. and Civelek, M. B., "Contact Problem for an Elastic Reinforcement Bonded to an Elastic Plate," *Journal of Applied Mechanics, 41,* Trans. ASME, 1014–1018 (1974).
3. Williams, J. H., Jr., "Stresses in Adhesive Between Dissimilar Adherends," *Journal of Adhesions, 7,* 97–107 (1975).
4. Yuceoglu, U. and Updike, D. P., "Stress Analysis of Bonded Plates and Joints," *Journal of the Engineering Mechanics Division, ASCE, 106,* 37–56 (1980).
5. Delale, F. and Erdogan, F., "Viscoelastic Analysis of Adhesively Bonded Joint," *Journal of Applied Mechanics, 48,* ASME, 331–338 (1981).
6. Findley, W. N., Lai, J. S., and Onaran, K. *Creep and Relaxation of Nonlinear Viscoelastic Materials.* North-Holland Publ. Co. (1976).
7. Pipkin, A. C. *Lectures on Viscoelastic Theory.* Springer-Verlag (1972).
8. Boley, B. A. and Weiner, J. H. *Theory of Thermal Stresses.* John Wiley and Sons (1962).

31

Adhesive Joints Involving Composites and Laminates. Measurement of Stresses Caused by Resin Swelling

J. P. SARGENT AND K. H. G. ASHBEE

INTRODUCTION

IN ORDER TO DETERMINE THE STRESSES CREATED NORMAL TO AN ADHESIVE joint during swelling associated with water uptake at the joint edge, Sargent and Ashbee [1] have refined an optical interference method first used by Farrar and Ashbee [2] to measure the displacement field during swelling of an adhesive layer that is only lightly constrained not to deform. The method consists of bonding a thin, and therefore flexible, microscope cover slip to an adherend such as phosphoric anodised aluminum, and bringing an optical flat into close proximity to the free surface of the cover slip in order to create a pattern of interference fringes, the precise geometry of which changes as the glue line thickness changes. Full details of the experimental method are given in Reference [1]. Using thin plate elasticity theory, the measured displacements have been converted into stresses normal to the joint, the magnitudes of which vary from as much as 1 kb compressive near the edge to tensile stresses of up to 0.5 kb deeper inside.

The same optical interference method has now been used to study the nature and magnitude of the self-stresses created by resin-swelling in joints involving fiber composites and laminates.

EXPERIMENTAL

Joints between a thin microscope cover slip and each of two different materials have been studied. One material is a unidirectional composite manufactured from "S" glass and RAC 7250 epoxy resin. The other material is a 0 ± 63° 57-ply laminate manufactured from Modmor type I carbon fiber and MY720 epoxy resin.

403

The adhesive used for both sets of experiments is Redux 312/5. This material was selected because it retains good bonding properties when subjected to warm aqueous atmospheres. It also exhibits relatively small dimensional changes during water uptake. In addition, Redux 312/5 is characterized by a low cure temperature (120 °C).

Joints between cover slips and specimens of each of the composite and laminate were mounted so that the free surface of the cover slip was in close proximity to an optical flat. The interference pattern formed in the gap between optical flat and cover slip, and changes in that pattern during swelling of the adhesive, were recorded using a 35 mm SLR camera. Numerical data in respect of swelling normal to the joint was obtained from Moire fringes generated by superimposing photographed images on the image obtained prior to water uptake.

RESULTS

Unidirectional S2 Glass Fiber/SP280 Epoxy Composite

Figure 1 shows a sequence of Moire patterns for a joint manufactured using a square microscope cover slip with edges oriented parallel and perpendicular

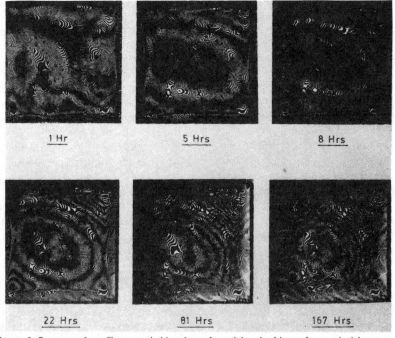

Figure 1. Progress of swelling revealed by circumferential moire fringes for a uniaxial composite joint during water uptake at 60°C. The fiber direction is vertical.

to the fiber direction, that has been immersed in distilled water at 60 °C. The thickness of the composite is 1.31 mm, that of the adhesive layer is 0.2 mm and that of the cover slip is 150 μm.

Figure 2 shows the rates of migration of the first Moire fringe in directions parallel and perpendicular to the fiber direction. It is evident that anisotropy of water uptake by the composite influences the overall water uptake and hence influence the nature of the swelling of the adhesive.

A characteristic feature of Figure 1, and of similar figures for adhesively bonded metal joints [1], is the occurrence of Moire fringes which form closed loops. This is a consequence of there being a region of compression ahead of the boundary between unswollen and swollen adhesive.

Figure 3 shows the variation across the joint of the normal displacement

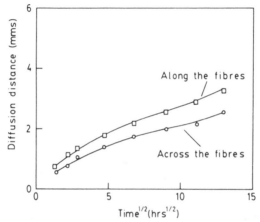

Figure 2. *Migration distance of the first Moire fringe for the specimen from Figure 1 as a function of the square root of time. $\square$ parallel to the fiber axis, $\bigcirc$ perpendicular to the fiber axis.*

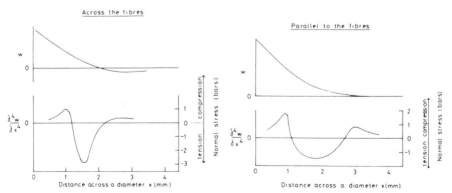

Figure 3. *Normal displacement (w) and its fourth differential ($\nabla^4 w$) across two directions in the plane of the joint shown in Figure 1.*

(w) for a trace oriented parallel and another for a trace oriented perpendicular to the fiber direction. Using thin plate elasticity theory, these displacements can be converted to stress normal to the joint. Love [3] shows that the pressure drop across a deformed membrane is given by:

$$p = D\nabla^4 w$$

The magnitudes of the normal stresses so calculated are not particularly large (varying from 2 bars compression near the joint edge to 2 bars tensile deeper inside) but do reflect the anisotropic contribution to swelling arising from water diffusion through the composite.

Carbon Fiber MY720 Epoxy Laminate

Figure 4 shows a sequence of Moire patterns from a joint manufactured with a cover slip mounted parallel to the laminae and immersed in water at 60 °C. The free surface of the laminate was bonded to a rigid block of plate glass.

It is evident from Figure 4 that water uptake is very rapid. This is demonstrated more rigorously in Figure 5, which shows the position of the

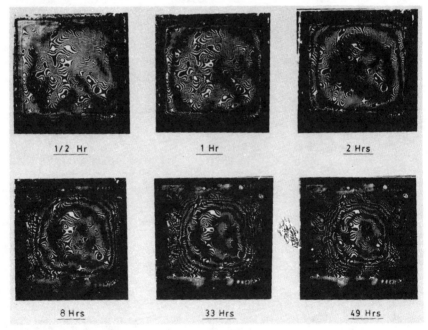

Figure 4. Sequence of Moire patterns for a 0 ± 63° 57 ply CFRP specimen immersed in distilled water at 60°C. The cover slip lies parallel to the plane of the laminate.

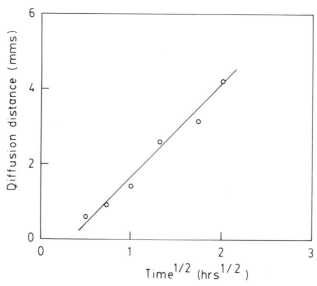

Figure 5. *Migration distance of the first Moire fringe for the specimen from Figure 4 as a function of the square root of time.*

first Moire fringe as a function of the square root of time of exposure of the specimen to water. Detectable swelling (.004 percent) at the center of the specimen has occurred after only nine hours' exposure. Figure 4 also reveals an anomalously different rate of water diffusion from one edge of the specimen. This is attributed to a delamination defect in the composite.

Laminate Edge Joints

Figure 6 shows a sequence of photographs on the changing pattern of interference fringes during hot water uptake of a joint between a cover slip and the edge of the 57-ply laminate used in the previous section. In this case, cracks in the cover slip attributed to anisotropy of in-plane laminate swelling, occur after one hour and continue to propagate by a tearing mode due to differential uplift by the swelling adhesive/composite combination. This in turn is a consequence of inclination of the crack to the plane of the cover slip, as indicated by the sketch in Figure 7 which was constructed from optical microscope examination of one such crack. Water access to the adhesive is aided by the slope on one side of the crack and restricted by the slope on the other. Figure 8 is a Moire image of crack region photographed after 23.5 hours. The relative uplift either side of the crack is clearly shown by the mis-match of Moire fringes. During propagation of such a crack, the excess of fringes on one side often changes to a deficit. In fact, the crack profile, obtained by counting fringes on neighboring points either side of a crack, can vary considerably, as shown in Figure 9.

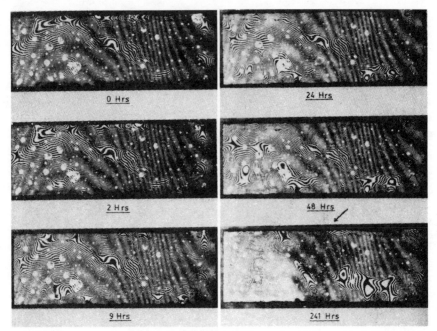

Figure 6. *Interference patterns for a 0 ± 63° 57 ply CFRP specimen immersed in distilled water at 60°C. The cover slip is bonded to the edge of the laminate.*

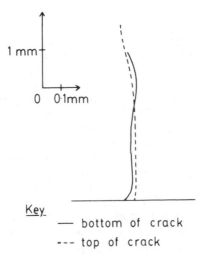

1 mm

0   0·1mm

Key

—— bottom of crack

--- top of crack

Figure 7. *Sketch of the crack indicated by an arrow in Figure 6.*

Figure 8. Moire pattern of the same crack, generated by superimposing photographs of the interference pattern recorded after 0 and 23.5 hours' immersion of the joint in 60°C water.

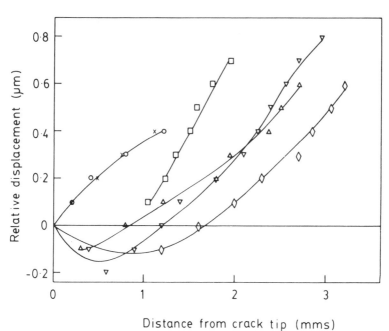

Distance from crack tip (mms)

Figure 9. Crack opening displacement data taken from Figure 8 and from patterns similar to Figure 8 that were obtained after different exposure times. 0 defines the crack tip. X = 8.5 hrs, O = 23.5 hrs, □ = 48 hrs, Δ = 73 hrs, ∇ = 98 hrs, ◇ = 168 hrs.

It is evident that the crack profiles in Figure 9 do not approximate to the parabolic profile assumed in linear elastic fracture mechanics (nor for that matter do they approximate to the elliptic profile assumed in Griffith's theory). Nevertheless, the relative displacement after 8.5 hours and after 23.5 hours does vary as the square root of distance from the crack tip and the gradient of that relationship has been used to evaluate K_{III}. According to linear elastic fracture mechanics

$$\Delta w = \frac{4\,K_{III}\,r^2\,\mathrm{Sin}\,1/2\theta}{\mu\,(2\pi)^{1/2}} \tag{1}$$

In Figure 10, the values of K_{III} so obtained have been plotted with Freiman's [4] data of crack velocity versus K_I for soda-lime-silicate glass of composition similar to that of the cover slips used for the present work.

Figure 8 affords a method for measuring the displacement field local to the crack tip. Thus, by subtracting displacement to one side of the tip from displacements at mirror image points on the other side of the tip, the data points shown in Figure 11 have been generated. (The measured displacements are relative to the tip as origin). For comparison, crack profiles computed using equation (1) have also been reproduced.

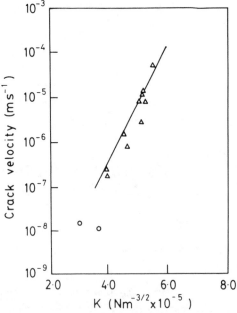

Figure 10. *The crack velocity as a function of stress intensity factor (K) for commercial soda-lime-silicate glass tested in water. Δ K_I from Freiman [4], ○ present K_{III} data.*

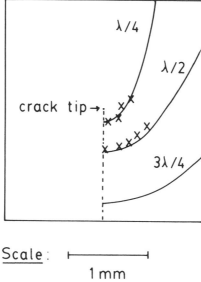

Scale: ├────────────────┤
1 mm

Figure 11. *λ/4, λ/2 and 3λ/4 relative displacement contours near the tip of a mode III crack (broken lines) computed using equation 1. Each cross denotes the position on one side of the crack for which the displacement relative to its mirror image points on the other side of the crack is λ/4, λ/2.*

CONCLUSIONS

A high resolution optical interference method has been developed to study the geometry of swelling associated with water uptake by adhesive joints formed with composite materials. By generating Moire fringes from images of such interference patterns, it is possible to obtain a precise picture of the microscope cover slip used as one adherend and by appling thin plate elasticity theory to the measured deformation of the cover slip, it is possible to determine the swelling stresses that would be encountered in a real joint. One noteworthy feature of this stress field is its compressive sign in a wide annular region adjacent to the rim of the joint. It has also been possible to detect and explore the consequence of anisotropic water uptake by composites and laminates themselves.

ACKNOWLEDGMENTS

The authors gratefully acknowledge financial support from the U.S. Army (Grant No. DA-ERO-78-G-117) and from the Science Research Council (Grant No. GR/A/8261-1). Some of the materials were supplied by Ciba-Geigy, some by the U.S. Army Materials and Mechanics Research Center and other by Rotorway Composites, Ltd.

NOMENCLATURE

w	The normal displacement
D	The flexural rigidity
p	The pressure drop across the membrane
K_{III}	Mode III stress intensity factor
Δw	The relative displacement
r	The distance from the crack tip
θ	The angle subtended by the point
μ	The shear modulus

REFERENCES

1. Sargent, J. P. and Ashbee, K. H. G., "High Resolution Optical Interference Investigation of Swelling Due to Water Uptake by Model Adhesive Joints," *J. Adhesion, 11,* 175 (1980).
2. Farrar, N. R. and Ashbee, K. H. G., "Self-stress-enhanced Water Migration in Composites," ACS Symposium Resins for Aerospace, #132, 435 (1980).
3. Love, A. E. H. *Treatise on the Mathematical Theory of Elasticity.* Cambridge University Press, Chapter XXII, p. 455 (1959).
4. Freiman, S. W., *J. Am. Ceram. Soc., 57,* 350 (1974).

32

Authors and References Cited in Volumes 1 and 2

Abbott, N. J., Donovan, J. G., Schoppee, M. M., and Skelton, J., "Some Mechanical Properties of Kevlar and Other Heat Resistant Nonflammable Fibers, Yarns and Fabrics," Fabric Research Laboratories Report AFML-TR-74-65, Part III, Dedham, MA (March 1975).

Adams, D. F.
  See Cairns, D. S. and Adams, D. F.
    Crane, D. A. and Adams, D. F.
    Miller, A. K. and Adams, D. F.
    Schafter, B. G. and Adams, D. F.
    Tsai, S. W. and Adams, D. F.
    Walrath, D. E. and Adams, D. F.

Adams, D. F. and Doner, D. R., "Longitudinal Shear Loading of a Unidirectional Composite," *J. Composite Materials, 1,* 4–17 (1967).

Adams, D. F. and Miller, A. K., "Hygrothermal Microstresses in a Unidirectional Composite Exhibiting Inelastic Material Behavior," *Journal of Composite Materials, 11,* 285–299 (1977).

Adams, D. F. and Miller, A. K., "The Influence of Material Variability on the Predicted Environmental Behavior of Composite Materials," *Journal of Engineering Materials and Technology,* ASME, Vol. 100, 77–83 (1978).

Adams, D. F. and Tsai, S. W., "The Influence of Random Filament Packing on the Transverse Stiffness of Unidirectional Composites," *J. Composite Materials, 3,* 368–381 (1969).

Adams, D. F. and Walrath, D. E., "Iosipescu Shear Properties of SMC Composite Materials," Department of Mechanical Engineering, The University of Wyoming, Larrabee, Wyoming 82738 (1981).

Ackley, R. H. and Carley, E. P., "XMC-3 Composite Material Structural Molding Compound," 34th Annual Technical Conference, Reinforced Plastics/Composites Institute, The Society of Plastics Industry, Section 21-D (1979).

Aifantis, E. C. and Gerberich, W. W., "Gaseous Diffusion in a Stressed-Thermoelastic Solid, Part II: Thermodynamic Structure and Transport Theory," *Acta Mechanica, 28,* 25–47 (1977).

Allen, N. S. and McKellar, J. F., "Photochemical Reactions in an MDI-Based Elastomeric Polyurethane," *Journal of Applied Polymer Science, 20,* 1441–1447 (1976).

Allred, R. E. and Hall, N. H., "Volume Fraction Determination of Kevlar 49/Epoxy Composites," *Polym. Engr. and Sci., 19* (13), 907–909 (1979).

Allred, R. E., Hall, N. H., and Miller, A. K., "Moisture and Post-Cure Effects on the Matrix Glass Transition Temperature of Kevlar/Epoxy Laminates," *Polym Engr. and Sci.*

Allred, R. E. and Lindrose, A. M., "The Room Temperature Moisture Kinetics of Kevlar 49 Fabric/Epoxy Laminates," *Composite Materials: Testing and Design,* ASTM STP 674, American Society for Testing and Materials, 313–323 (1979).

Aoki, H. and Yoshida, A., "The Kinetic Evaluation of Plastic Films at their Deterioration," Zairyo, Japan, Vol. 21, 309–314 (1972).

"A Report of Observed Effects on Electrical Systems of Airborne Carbon/Graphite Fibers," NASA TM 78652 (1978).

E. Artecka
    See W. Gogol, et al.

K. H. G. Ashbee
    See Farrar, N. R. and Ashbee, K. H. G.
        Sargent, J. P. and Ashbee, K. H. G.

Augl, J. M., "Research on Composite Materials," Quarterly Report to the Naval Air Systems Command (1975).

Augl, J. M., "Prediction and Verification of Moisture Effects on Carbon Fiber-Epoxy Composites," NSWC TR79-43 (March 1979).

Augl, J. M., "Moisture Sorption and Diffusion in Kevlar 49 Aramid Fiber," Naval Surface Weapons Center Report NSWC/TR-79-51, Silver Spring, MD (March 1979).

Augl, J. M. and Berger, A. E., "The Effect of Moisture on Carbon Fiber Reinforced Epoxy Composites. I. Diffusion," NSWC/WOL/TR 76-7, Naval Surface Weapons Center, White Oak, Silver Spring, Maryland (1976).

Augl, J. M. and Trabocco, R., "Environmental Degradation Studies on Carbon Fiber Reinforced Epoxies," presented at the *Workshop on Durability of Composite Materials,* held at the Batelle Memorial Institute, Columbus, Ohio (September 30–October 2, 1975).

Balasubramanian, N.
    See Rao, R. M. V. G. K., et al.

Barnes, G. J.
    See Hilado, C. J., et al.

Barrett, C. S.
    See Predecki, P. and Barrett, C. S.

Bassewitz, H. V., "Behavior of Carbon Fiber Composites Under Simulated Space Environment," Non Semiconducting Materials, Session II, International Conference on Evaluation of the Effect of the Space Environment on Materials, Toulouse, France, 525–537 (June 1974).

Beachell, H. C. and Chang, I. L., "Photodegradation of Urethane Model Systems," *Journal of Polymer Science, 10* (A-1), 503–520 (1972).

Beachell, H. C. and Ngoc Son, C. P., "Color Formation in Polyurethanes," *Journal of Applied Polymer Science, 7,* 2217–2237 (1963).

Beran, M. *Statistical Continuum Theories.* Interscience (1968).

Berger, A. E.
    See Augl, J. M. and Berger, A. E.

Blacklock, J. M.
    See Richard, R. M. and Blacklock, J. M.

Bohlmann, R. E. and Derby, E. A., "Moisture Diffusion in Graphite/Epoxy Laminates: Experimental and Predicted," Presented at the 18th Structures, Structural Dynamics and Materials Conference, and at the Conference on Aircraft Composites: The Emerging Methodology for Structural Assurance, San Diego, Calif. (March 1977), AIAA Technical Papers Volume A. (A77-25726 10–39), 219–226 (1977).

Boley, B. A. and Weiner, J. H. *Theory of Thermal Stresses.* John Wiley and Sons, Inc. (1967).

Bonniau, P. and Bunsell, A. R., "A Comparative Study of Water Absorption Theories Applied to Glass-Epoxy Composites," *J. Composite Materials, 15,* 272–93 (1981).

Breinan, E. M. and Kreider, K. G., "Axial Creep and Stress-Rupture of B-Al Composites," *Metall. Trans., 1,* 93–104 (1970).

Briley, R. B.
   See Maymon, G., et al.
      Rehfield, L. W. and Briley, R. B.

Brinson, H. F.
   See Yeow, Y. T., et al.
      Griffith, W. I., et al.

Broadway, N. J., King, R. W., and Palinchak, S., "Space Environmental Effects on Materials and Components," *Elastomeric and Plastic Materials, 1,* Appendix A-1, Battelle Memorial Institute (April 1964).

Browning, C. E.
   See Whitney, J. M. and Browning, C. E.

Browning, C. E., "The Effects of Moisture on the Properties of High Performance Structural Resins and Composites," Technical Report AFML-TR-72-94, Air Force Materials Laboratory, Air Force Systems Command, Wright-Patterson Air Force Base, Dayton, Ohio (September 1972).

Browning, C. W., "The Mechanisms of Elevated Temperature Property Losses in High Performance Structural Epoxy Resin Matrix Materials after Exposures to High Humidity Environments," *Polym Engr. and Sci., 18,* 16–24 (1978).

Browning, C. E. and Hartness, J. T., "Effects of Moisture on the Properties of High-Performance Structural Resins and Composites," *Composite Materials: Testing and Design.* (Third Conference), ASTM STP 546, American Society for Testing and Materials, 284 (1974).

Browning, C. E., Husman, G. E., and Whitney, J. M., "Moisture Effects in Epoxy Matrix Composites," *Composite Materials: Testing and Design.* (Fourth Conference), ASTM STP 617, 481–496 (1977).

Budiansky, B., "On the Elastic Moduli of Some Heterogeneous Materials," *J. Mech. Phys. Solids, 13,* 223–27 (1965).

Budiansky, B. and O'Connell, R. J., "Elastic Moduli of a Cracked Solids," *Int. J. Solids Structures, 12,* 81–97 (1976).

Bunsell, A. R.
   See Bonniau, P. and Bunsell, A. R.

Cairns, D. S. and Adams, D. F., "Moisture and Thermal Expansion of Composite Materials," Report UWME-DR-101-104-1, Department of Mechanical Engineering, University of Wyoming, Laramie, Wyoming (November 1981).

Camahort, J. L., Rennhack, E. H., and Coons, W. C., "Effects of Thermal Cycling Environment on Graphite/Epoxy Composites," *Environmental Effects on Advanced Composite Materials.* ASTM STP 602, American Society for Testing and Materials, 37 (1976).

Campbell, M. D.
   See Christian, J. L. and Campbell, M. D.
      Freeman, W. T. and Campbell, M. D.

"Carbon Fiber Electrical Resistance Modification—Its Relationship to Electrical Equipment Malfunctions," Final Report of the Office of Naval Research Carbon Fiber Study Group, Office of Naval Research, Metallurgy and Ceramic Program (1978).

Carley, E. P.
   See Ackley, R. H. and Carley, E. P.

Carlslaw, H. S. and Jaeger, J. C. *Conduction of Heat in Solids.* Oxford Press (1959).

Carpenter, J. F., "Moisture Sensitivity of Epoxy Composites and Structural Adhesives," McDonnell Aircraft Company, Report MDC A2640 (December 1973).

Carter, H. G., "Fundamental and Operational Glass Transition Temperature of Composite Resins and Adhesives," *Advanced Composite Materials—Environmental Effects,* J.R. Vinson, ed. ASTM—STP 658 (1978).

Carter, H. G.
   See Kibler, K. G., et al.

Carter, H. G. and Kibler, K. G., "Langmeir-Type Model for Anomalous Moisture Diffusion in Composite Resins," *J. Composite Materials, 12,* 118–131 (1978).

Caskey, G. R. and Pillinger, W. L., "Effect of Trapping on Hydrogen Permeation," *Metallurgical Transactions, 6A,* k67–476 (1975).

Chajes, A. *Principles of Structural Stability Theory.* Prentice-Hall (1974).

Chamis, C. C., Lark, R. J., and Sinclair, J. M., "Effect of Moisture Profiles and Laminate Configuration on the Hygro Stress in Advanced Composites," SAMPE Conference (October 1978).

Chamis, C. C. and Sinclair, J. H., "10-Degree Off-Axis Test for Shear Properties of Fiber Composites," SESA Spring Meeting, Dallas, TX (May 1977).

Chanda, M.
   See Rao, R. M. V. G. K., et al.

Chandra, M.
   See Rogers, K. F., et al.

Chang, I. L.
   See Beachell, H. C. and Chang, I. L.

Chang, S.
   See Chen, C. H. and Chang, S.

Chen, C. H. and Cheng, S., "Mechanical Properties of Fiber Reinforced Composites," *J. Composite Materials, 1,* 30–41 (1967).

Chen, I. W.
   See Cleary, M. P., et al.

Chou, T. W.
   See Ishikawa, T. and Chou, T. W.
       Nomura, S. and Chou, T. W.

Chou, T. W. and Kelly, A., "Fiber Composites," *Mat. Sci. Eng., 25,* 35 (1976).

Chou, T. W., Nomura, S., and Taya, M., "A Self-Consistent Approach to the Elastic Stiffness of Short-Fiber Composites," *J. Composite Materials, 14,* 178–188 (1980).

Christensen, R. M., "Asymptotic Modulus Results for Composites Containing Randomly Oriented Fibers," *Int. J. Solids Structures, 12,* 537 (1976).

Christensen, R. M. *Mechanics of Composite Materials.* New York: John Wiley (1979).

Christensen, R. M. and Waals, F. M., "Effective Stiffness of Randomly Oriented Fiber Composites," *J. Composite Materials, 6,* 518 (1972).

Christian, J. L. and Campbell, M. D., "Mechanical and Physical Properties of Several Advanced Metal-Matrix Composite Materials," *Cryogenic Engineering Conference, Advances in Cryogenic Engineering.* Vol. 18, New York: Plenum Press, 175–183 (1973).

Chung, T. J., "Thermomechanical Response of Inelastic Fiber Composites," *Int. Journal for Numerical Methods in Engineering, 9,* 169–185 (1975).

Chung, T. J. and Prater, J. L., "A Constitutive Theory for Anisotropic Hygrothermoelasticity with Finite Element Applications," *Journal of Thermal Stresses, 3,* 435–452 (1980).

Civelek, M. B.
   See Erdogan, F. and Civelek, M. B.

Cleary, M. P., Chen, I. W., and Lee, S. M., "Self-Consistent Techniques for Heterogeneous Media," *J. Eng. Mech. Div., ASCE, 106,* N5, 861–87 (1980).

Cohen, M. I., "Melting of a Half-Space Subjected to a Constant Heat Input," *Journal of the Franklin Institute, 283,* 271 (1967).

Coons, W. C.
    See Camahort, J. L., et al.

Corten, H. T., "Micromechanics and Fracture Behavior of Composites," *Modern Composite Materials,* L.V. Broutman and R.H. Krock, eds. Reading, Mass: Addison-Wesley (1967).

Crane, D. A. and Adams, D. F., "Finite Element Micromechanical Analysis of a Unidirectional Composite Including Longitudinal Shear Loading," Department Report UWME-DR-001-101-1, Department of Mechanical Engineering, University of Wyoming, Laramie, Wyoming (February 1981).

Crank, J. *The Mathematical Theory of Diffusion.* 2nd ed. Oxford University Press (1975).

Crossman, F. W.
    See DeRuntz, J. A. and Crossman, F. W.
        Flaggs, D. L. and Crossman, F. W.
        Mauri, R. E. and Crossman, F. W.

Crossman, F. W., Mauri, R. E., and Warren, W. J., "Moisture-Altered Viscoelastic Response of Graphite/Epoxy Composites," *Advanced Composite Materials—Environmental Effects,* ASTM STP 658, American Society for Testing and Materials, Philadelphia, Pennsylvania, 205–220 (1977).

Cushman, J. B. and McCleskey, S. F., "Design Allowables Test Program, Celion 3000/PMR-15 and Celion 6000/PMR-15 Graphite/Polyimide Composites," Boeing Aerospace Co., NASA Contract NAS1-15644, NASA CR-165840 (1982).

Dalton, T. A.
    See Hobbs, N. P., et al.

Davis, G. F.
    See Watt, J. P., et al.

Dean, G. and Turner, G., "The Elastic Properties of Carbon Fibers and Their Composites," *Composites, 4,* 174 (1973).

Decker, L. J.
    See Ward, J. R. and Decker, L. J.

Dederich, P. H.
    See Zeller, R. and Dederich, P. H.

Dederichs, P. H. and Zeller, R., "Variational Treatment of the Elastic Constants of Disordered Materials," *Z. Physik, 259, 103* (1973).

Delasi, R. and Whiteside, J. B., "Effect of Moisture on Epoxy Resins and Composites," *Advanced Composite Materials—Environmental Effects,* ASTM STP 658, J.R. Vinson, ed., American Society for Testing and Materials, 2–20 (1978).

Delasi, R., Whiteside, J., and Wolter, W., "Effects of Varying Hygrothermal Environments on Moisture Absorption in Epoxy Composites," presented at Fourth Conference on Fibrous composites in Structural Design, San Diego, CA (Nov. 1978).

Delale, F. and Erdogan, F., "Viscoelastic Analysis of Adhesively Bonded Joint," *Journal of Applied Mechanics, 48,* 331–338 (1981).

de Malherbe, M. C.
    See Hewitt, R. L. and de Malherbe, M. C.

Denton, D. L., "Mechanical Properties Characterization of an SMC-R50 Composite," 34th Annual Technical Conference, Reinforced Plastics/Composites Institute, The Society of the Plastics Industry, 1979, Section 11-F; also SAE Paper 790671 (1979).

Derby, E. A.
    See Bohlman, R. E. and Derby, E. A.

DeRuntz, J. A. and Crossman, F. W., "Time and Temperature Effects in Laminated Composites, *Nuclear Metallurgy, 20,* 1085 (1976).

de Silva, A. R. T., "A Theoretical Analysis of Creep in Fibre Reinforced Composites," *J. Mech. Phys. Solids, 16,* 169–186 (1968).

Dibenedetto, A. T.
See Mahta, B. S., et al.

DiCarlo, J. A., "Anelastic Deformation of Boron Fibers," *Scripta Met., 10,* 115–119 (1976).

DiCarlo, J. A., "Time-Temperature-Stress Dependence of Boron Fiber Deformation." *Composite Materials: Testing and Design (Fourth Conference),* ASTM STP 617, Philadelphia: American Society for Testing Materials, 443–65 (1977).

DiCarlo, J. A., "Mechanical and Physical Properties of Modern Boron Fibers," *ICCM/2, Second International Conference on Composite Materials,* New York: The Metallurgical Society of AIME, 520–538 (1978).

DiCarlo, J. A. and Maisel, J. E., "Measurement of the Time-Temperature Dependent Dynamic Mechanical Properties of Boron/Aluminum Composites," *Comosite Materials: Testing and Design (Fifth Conference),* ASTM STP 674, Philadelphia: American Society for Testing Materials, 201–229 (1979).

DiCarlo, J. A. and Williams, W., "Dynamic Modulus and Damping of Boron Silicon Carbide and Alumina Fibers," NASA TM-81422, Washington, D.C. (1980).

Do, M. H. and Springer, G. S., "Mass Loss of and Temperature Distribution in Southern Pine and Douglas Fir in the Range 100 to 800 °C," *J. of Fire Sciences, 1,* 271–284 (1983).

Do, M. H. and Springer, G. S., "Model for Predicting Changes in the Strengths and Moduli of Timber Exposed to Elevated Temperatures," *J. of Fire Sciences, 1,* 285–296 (1983).

Do, M. H. and Springer, G. S., "Failure Time of Loaded Wooden Beams During Fire," *J. of Fire Sciences, 1,* 297–303 (1983).

Donea, J., "Thermal Conductivities Based on Variational Principles," *J. Composite Materials, 6,* 262 (1972).

Doner, D. R.
See Adams, D. F. and Doner, D. R.
    Tsai, S. W. and Doner, D. R.

Donovan, J. G.
See Abbott, N. J. et al.

Dorn, W. S.
See McCracken, D. D. and Dorn, W. S.

Douglass, D. A. and Weitsman, Y., "Stresses Due to Environmental Conditioning of Cross-Ply Graphite/Epoxy Laminates." *Proc. Third International Conference on Composite Materials (ICCM3),* A.R. Bunsell, et al., editors, Vol. 1, Paris, France: Pergamon Press, 529–542 (1980).

Dvoskin, N., Jordan, V., and Wolter, W., Grunman A.C. Technical Report AC-79.5. "Collaborative BAe/GAC Environmental Fatigue Test Programme, Environmental Definition" (1979).

Eakens, W. J., "Effect of Water on Glass Fiber Resin Bonds," *Interfaces in Composites.* ASTM STP 452, Philadelphia, Pennsylvania: American Society for Testing and Materials, 137–148 (1969).

Eastham, J., "In Depth Evaluation of C.F.R.P.," BAe Warton Division, Report No. MDR 0182, Final Report on M.O.D. Contract K/LR32B/2126 (1979).

Edge, E. C., "The Implications of Laboratory Accelerated Conditioning of Carbon Fibre Composites," AGARD CP 288 (1980).

Eisenmann, J. R.
See Kibler, K. G., et al.

Engler, B. P.
See Trujillo, R. J. and Engler, B. P.

Enos, J. H., Erratt, R. L., Grancis, E., and Thomas, R. E., "Structural Performance of Vinylester Resin Compression Molded High Strength Composites," 34th Annual Technical Conference, Reinforced Plastics/Composites Institute, The Society of Plastics Industry, Section 11-E (1979).

Erdogan, F.
See Delale, F. and Erdogan, F.

Erdogan, F. and Civelek, M. B., "Contact Problem for an Elastic Reinforcement Bonded to an Elastic Plate," *Journal of Applied Mechanics, 41,* Trans. ASME, 1014–1018 (1974).

Erratt, R. L.
See Enos, J. H., et al.

Eshelby, J. D., "The Determination of the Elastic Field of an Ellipsoidal Inclusion, and Related Problems," *Proc. Roy. Soc., 241A,* 376 (1957).

Everett, R. A.
See Garber, D. P., et al.

Fanter, D. L.
See Levy, R. L., et al.

Farrar, N. R. and Ashbee, K. H. G., "Self-stress-enhanced Water Migration in Composites," ACS Symposium Resins For Aerospace, #132, 435 (1980).

Farrell, P. V.
See Pering, G. A., et al.

Findley, W. N., Lai, J. S., and Onaran, K. *Creep and Relaxation of Nonlinear Viscoelastic Materials.* North-Holland Publ. Co. (1976).

Firle, T. E., "Amplitude Dependence of Internal Friction and Shear Modulus of Boron Fibers," *J. Appl. Phys., 39,* 2839–2845 (1968).

Flaggs, D. L. and Crossman, F. W., "Viscoelastic Response of a Bonded Joint due to Transient Hygrothermal Exposure," *Modern Developments in Composite Materials and Structures,* J.R. Vinson, ed. ASME (1979).

Flaggs, D. L. and Crossman, F. W., "Analysis of Viscoelastic Response of Composite Laminates During Hygrothermal Exposure," *J. Composite Matls., 15,* 21 (1981).

Flaggs, D. L. and Vinson, J. R., "Hygrothermal Effects on the Buckling of Laminated Composite Plates," *Fibre Science and Technology, 11,* 353 (1978).

"Flexural Properties of Plastics," ASTM Designation D 790-71, American Society for Testing and Materials, Philadelphia, PA (1978).

Foster, P. K.
See McNabb, A. and Foster, P. K.

Freeman, W. T. and Campbell, M. D., "Thermal Expansion Characteristics of Graphite Reinforced Composite Materials," *Composite Materials: Testing and Design.* (Second Conference), ASTM STP 497, American Society for Testing and Materials, Philadelphia, Pennsylvania, 121–142 (1972).

Fried, N., "Degradation of Composite Materials: The Effect of Water on Glass-Reinforced Plastics," *Proc. Fifth Symp. Naval Structural Mechanics.* Philadelphia (May 8-10, 1967).

Fukuda, H. and Kawata, K., "On Young's Modulus of Short Fiber Composites," *Fibre Sci. & Tech., 7,* 207 (1974).

Furmański, P. and Gogól, W., "Investigation of Steady-state Heat Conduction in Two-dimensional Model of Composite with Symmetrically Arranged Circular Fibers," Archives of Thermodynamics and Combustion, No. 2, (Archiwum Termodynamiki i Spalania—in Polish) (1979).

Galasso, F. S.
See Veltri, R. D. and Gallasso, F. S.

Garber, D. P., Morris, D. H., and Everett, R. A., Jr., "Elastic Properties and Fracture Behavior of Graphite/Polyimide Composites at Extreme Temperatures," *Composites for Extreme Environments,* N.R. Adsit, ed. ASTM STP 768, American Society for Testing and Materials, 73–91 (1982).

Garzek, L. E. and Isenberg, L., "The Effects of Space Environment on Reinforced Plastics,"

National Symposium on the Effects of Space Environment on Materials, St. Louis, MO (May 7–9, 1962).

Gerberich, W. W.
    See Aifantis, E. C. and Gerberich, W. W.

Gerharz, J. J. and Schutz, D., "Fatigue Strength of CFRP under Combined Flight-by-Flight Loading and Flight-by-Flight Temperature Changes," AGARD CP 288 (1980).

Gibbins, M. N. and Hoffman, D. J., "Environmental Exposure Effects on Composite Materials for Commercial Aircraft," NASA Contractor Report 3502 (January 1982).

Gibson, R. F. and Plunkett, R., "A Forced Vibration Technique for Measurement of Material Damping," *Experimental Mechanics, 11,* 297–302 (1977).

Gibson, R. F. and Yau, A., "Final Report—Phase 1: Dynamic Stiffness and Internal Damping of Automotive Chopped Fiber Reinforced Plastic Materials," submitted to General Motors Manufacturing Development, University of Idaho (July 31, 1979).

Gibson, R. F. and Yau, A., "Complex Moduli of Chopped Fiber and Continuous Fiber Composites: Comparison of Measurements with Estimated Bounds," *Journal of Composite Materials, 14,* 155–167 (1980).

Gibson, R. F., Yau, A., Mende, E. W., and Osborn, W. E., "The Influence of Environmental Conditions on the Dynamic Mechanical Behavior of Automotive Fiber Reinforced Plastics," Final Report submitted to General Motors Manufacturing Development, University of Idaho (August 1980).

Gibson, R. F., Yau, A., and Riegner, D. A., "The Influence of Environmental Conditions on the Vibration Characteristics of Chopped-Fiber-Reinforced Composite Materials," Presented at AIAA/ASME/ASCE/AHS 22nd Structure, Structured Dynamics and Materials Conference (1981).

Gibson, R. F., Yau, A., and Riegner, D. A., "Vibration Characteristics of Automotive Composite Materials," presented at ASTM Symposium on Short Fiber Reinforced Composites, Minneapolis, Minnesota, April 1980, also in ASTM STP 772 (1982).

Gibson, R. F., Yau, A., and Riegner, D. A., "An Improved Forced Vibration Technique for Measurement of Material Damping," *Experimental Techniques, 6,* 10–14 (1982).

Gogól, E.
    See Gogól, W., et al.

Gogól, W.
    See Gogól, E. and Gogól, W.
    Furmański, P. and Gogól, W.

Gogól, E. and Gogol, W., "Apparatus for Measurements of Thermal Conductivity of Metals in Moderate Temperatures by the Comparison Method," (in Polish) *Biuletyn Inf. Inst. Techn. Cieplnej, 49* (1977).

Gogól, W., Gogól, E., and Artecka, E., "Thermal Conductivity Investigations of Moist Soils," (in Polish), *Biuletyn Inf. Inst. Techn. Cieplnej, 40* (1973).

Goland, M. and Reissner, E., "The Stresses in Cemented Joints," *Journal of Applied Mechanics, 1,* Trans. ASME, A17–127 (1944).

Goudreau, G. L.
    See Taylor, R. L., et al.

Grancis, E.
    See Enos, J. H., et al.

Greenwood, J. H. and Rose, P. G., "Compressive Behavior of Kevlar 49 Fibres and Composites," *J. Matl. Sci. 9,* 1809–1814 (1974).

Greever, W. L.
    See Young, H. L. and Greever, W. L.

Griffith, W. I., Morris, D. H., and Brinson, H. F., "The Accelerated Characterization of Viscoelastic Composite Materials," VPI-E-80-15, Department of Engineering Science and

Mechanics, Virginia Polytechnic Institute (April 1980).

Gurtin, M. E. and Yatomi, C., "On a Model for Two Phase Diffusion in Composite Materials," *J. Composite Materials, 13,* 126–130 (1979).

Hahn, H. T.
See Tsai, S. W. and Hahn, H. T.

Hahn, H. T. and Kim, R. Y., "Swelling of Composites Laminates," *Advanced Composite Materials—Environmental Effects,* J. R. Vinson, ed., ASTM-STP 658 (1978).

Halkias, J. E.
See McKague, E. L., Jr., et al.

Hall, N. H.
See Allred, R. E. and Hall N. H.
  Allred, R. E., et al.

Halpin, J. C.
See Shirrell, C. D. and Halpin, J. C.

Halpin, J. C. and Pagano, N. J., "The Laminate Approximation for Randomly Oriented Fibrous Composites," *J. Comp. Materials, 3,* 720 (1969).

Halpin, J. and Tsai, S. W., "Effects of Environmental Factors of Composite Materials," AFML-TR 67-423 (1969).

Hancox, N. L. and Minty, D. C., "Materials Qualification and Property Measurements of Carbon Fiber Reinforced Composites for Space Use," *British Interplanetary Society Journal, 30,* 391–399 (1977).

Hanson, M. P., "Effect of Temperature on the Tensile and Creep Characteristics of PRD 49 Fiber/Epoxy Composites," *Composite Materials in Engineering Design,* B.R. Norton, ed. Proceedings of 6th St. Louis Symposium, The American Society for Metals, 717 (May 11–12, 1972).

Hartness, J. T.
See Browning, C. E. and Hartness, J. T.

Hashin, Z., "Theory of Fiber Reinforced Materials," NASA CR-1974 (1972).

Hashin, Z., "Theory of Mechanical Behavior of Heterogeneous Media," *Appl. Mech. Rev., 17,* 1 (1964).

Hashin, Z., "Assessment of the Self-Consistent Scheme Approximation: Conductivity of Particulate Composites," *J. Comp. Mat., 2,* 284 (1968).

Hashin, Z. and Rosen, B. W., "The Effective Moduli of Fiber-Reinforced Materials," *J. Appl. Mech., 31,* 223–232 (1964).

Haskins, J. F.
See Kerr, J. R., et al.

Hasselman, D. P. H., "Effect of Cracks on Thermal Conductivity," *J. Compos. Mater., 12,* 403–07 (1978).

Heimbuch, R. A.
See Sanders, B. A. and Heimbuch, R. A.

Heimbuch, R. A. and Sanders, B. A., "Mechanical Properties of Automotive Chopped Fiber Reinforced Plastics," General Motors Manufacturing Development Report No. MD 78-032 (1978).

Heimbuch, R. A. and Sanders, B. A., "Mechanical Properties of Chopped Fiber Reinforced Plastics," in *Composite Materials in the Automotive Industry.* American Society of Mechanical Engineers, 111–139 (1978).

Herring, R. L.
See Menousek, J. F. and Herring, R. L.

Hertz, J., "Moisture Effects on the High-Temperature Strength of Fiber-Reinforced Resin Composites," *Proceedings of the 4th National SAMPE Technical Conference.* Society for the Advancement of Material and Process Engineering, Azusa, CA, 1–7 (1972).

Hertz, J., "Investigation into the High-Temperature Strength Degradation of Fiber-Reinforced Resin Composite During Ambient Aging," Materials Research Group, Convair Aerospace Division, General Dynamics Corporation Report No. GDCA-DBG71-004-3, contract NAS8-27435 (March 1972).

Hertz, J., "Investigation into the High-Temperature Strength Degradation of Fiber-Reinforced Resin Composite During Ambient Aging," Convair Aerospace Division, General Dynamics Corporation, Report No. GDCA-DBG73-005, Contract NAS8-27435 (June 1973).

Hewitt, R. L. and de Malherbe, M. C., "An Approximation for the Longitudinal Shear Modulus of Continuous Fiber Composites," *J. Composite Materials, 4,* 280 (1970).

Hilado, C. J., Barnes, G. J., Kourtides, D. A., and Parker, J. A., "The Use of High Flux Heater in the Smoke Chamber to Measure Ignitability and Smoke Evolution of Composite Panels," *J. Fire and Flammability, 8,* 324 (1977).

Hill, R., "A Self-Consistent Mechanics of Composite Materials," *J. Mech. Phys. Solids, 13,* 213–222 (1965).

Hobbs, N. P., Dalton, T. A., and Smiley, R. F., "TRAP2-A Digital Computer Program for Calculating the Response of Mechanically Loaded Structures to Laser Irradiation," KA TR-143, Kamon, Avidyne, Burlington, Mass. (October 1977).

Hoenig, A., "Electric Conductivities of a Cracked Solid," *Pageoph.,* 117 (1978/79).

Hofer, K. E., Jr., Larsen, D., and Humphreys, V. E., "Development of Engineering Data on the Mechanical and Physical Properties of Advanced Composite Materials," Technical Report AFML-TR-74-266, Air Force Materials Laboratory, Air Force Systems Command, Wright-Patterson Air Force Base, Dayton, Ohio (February 1975).

Hofer, K. E., Jr., Rao, N., and Larsen, D., "Development of Engineering Data on the Mechanical and Physical Properties of Advanced Composites Materials," Technical Report AFML-TR-72-205, Part II, Air Force Materials Laboratory, Air Force Systems Command, Wright-Patterson Air Force Base, Dayton, Ohio (February 1974).

Hoffman, D. J.
See Gibbins, M. N. and Hoffman, D. J.

Hori, M., "Statistical Theory of Effective Electrical, Thermal and Magnetic Properties of Random Heterogeneous Materials II," *J. Math. Phys., 14,* 1942 (1973).

Hori, M. and Yonezawa, F., "Statistical Theory of Effective Electrical, Thermal, and Magnetic Properties of Random Heterogeneous Materials IV," *J. Math. Phys., 16,* 352 (1975).

Howe, D. J.
See Kanury, A. M. and Howe, D. J.

Humphreys, V. E.
See Hofer, K. E., Jr., et al.

Husman, G. E.
See Browning, C. E., et al.
    Whitney, J. M. and Husman, G. E.

Husman, G. E., "Characterization of Wet Composite Laminates," Proceedings of Mechanics of Composite Review, Air Force Materials Laboratory Non-Metallic Materials Division, Bergamo Center, Dayton, Ohio (January 1976).

Isenberg, L.
See Garzek, L. E. and Isenberg, L.

Ishai, O., "Environmental Effects on Deformation, Strength, and Degradation of Unidirectional Glass-Fiber Reinforced Plastics," *Polymer Engineering and Science, 15* (1975).

Ishikawa, T., "Anti-Symmetric Elastic Properties of Composite Plates of Satin Weave Cloth," *Fib. Sci. Tech., 15,* 127 (1981).

Ishikawa, T. and Chou, T.-W., "Stiffness and Strength Properties of Woven Fabric Composites," Proceedings of ICCM 4, Tokyo, 489 (1982).

Ishikawa, T. and Chou, T.-W., "Stiffness and Strength Behavior of Woven Fabric Composites," *J. Materials Science, 17,* 3211 (1982).

Ishikawa, T. and Chou, T.-W., "Elastic Behavior of Woven Hybrid Composites," *J. Composite Materials, 16,* 2 (1982).

Ishikawa, T., Koyama, K., and Kobayashi, S., "Elastic Moduli of Carbon-Epoxy Composites and Carbon Fibers," *J. Composite Materials, 11,* 332 (1977).

Ishikawa, T., Koyama, K., and Kobayashi, S., "Thermal Expansion Coefficients of Unidirectional Composites," *J. Composite Materials, 12,* 153 (1978).

Jaeger, C. H.
  See Carslow, H. S. and Jaeger, C. H.

Jones, R. M., "Mechanics of Composite Materials," *Scripta,* Washington, D.C. (1975).

Jones, R. M., "Apparent Flexural Modulus and Strength of Multimodulus Materials," *J. Compt. Mat'ls., 10,* 342-351 (1976).

Jordan, V.
  See Dvoskin, N., et al.

Joshida, A.
  See Aoki, H. and Joshida, A.

Jost, W. *Diffusion in Solids, Liquids, Gases.* Academic Press (1960).

Judd, N. C. W. and Wright, W. W., "Voids and Their Effects on the Mechanical Properties of Composites—An Appraisal," *SAMPE J., 14,* 10-14 (1978).

Kabelka, J., "Thermal Expansion of Composites with Canvas-Type Reinforcement and Polymer Matrix," *Proceedings of ICCM 3,* Paris, 770 (1980).

Kachanov, M., "Continuum Models of Media with Cracks," *J. Eng. Mechanics Div., ASCE,* 106, EM5, 1039-51 (1980).

Kaelble, D. H., "Physical and Chemical Properties of Cured Resins," in *Epoxy Resins Chemistry and Technology,* C.A. May and Y. Tanaka, eds. NY: Marcel Dekker, 327-369 (1973).

Kaelble, D. H., "Wetometer for Measurement of Moisture in Composites," Progress in Quantitative NDE, Third Annual Report, AFML Contract F33615-74-C-5180 (1977).

Kaminski, B. E., "Effects of Specimen Geometry on the Strength of Composite Materials," *Analysis of the Test Methods for High Modulus Fibers and Composites.* ASTM STP521, 181 (1973).

Kanury, A. M., "Burning of Wood—A Pure Transient Conduction Model," *J. Fire and Flammability, 2,* 191 (1971).

Kanury, A. M. and Howe, D. J., "Response of Building Components to Heating in a Fire," *J. Heat Transfer, 101,* 365 (1979).

Kardos, J. L.
  See Mehta, B. S., et al.

Kawata, K.
  See Fukuda, H. and Kawata, K.

Kays, A. O., "Determination of Moisture Content in Composites by Dielectric Measurements," Second Interim Technical Report, AFFDL Contract F33615-78-C-3216 (December 1978).

Keller, J. B., "A Theorem on the Conductivity of a Composite Medium," *J. Mathematical Physics, 5* (4) (1964).

Kelly, A.
  See Chou, T. W. and Kelly, A.

Kennedy, J. M.
  See Poe, C. C., Jr. and Kennedy, J. M.

Kerner, E. H., "The Elastic and Thermoelastic Properties of Composite Media," *Proc. Phys. Soc., 69B,* 808 (1956).

Kerr, J. R., Haskins, J. F., and Stein, B. A., "Program Definition and Preliminary Results of a Long-Term Evaluation Program of Advanced Composites for Supersonic Cruise Aircraft Applications," *Environmental Effects on Advanced Composite Materials.* ASTM, STP 602, 3 (1975).

Kibler, K. G.
See Carter, H. G. and Kibler, K. G.

Kibler, K. G., "Effects of Moisture and Elevated Temperature on 5208/T300," in *The Effects of Relative Humidity and Elevated Temperature on Composite Structures,* J.R. Vinson, R.B. Pipes, W.J. Walker, and D.R. Ulrich, eds. AKOSR TR-7700 30, 190–211 (1977).

Kibler, K. G., "The Time Dependent Environmental Behavior of Graphite/Epoxy Laminates," Technical Report AFWAL-TR-80-4082, Materials Laboratory, Wright Patterson Air Force Base, Dayton, Ohio (May 1980).

Kibler, K. G., Carter, H. G., and Eisenmann, J. R., "Response of Graphite Composites to Laser Irradiation," TR-77-0706, Air Force Office of Scientific Research, Bolling Air Force Base, D.C. (March 1977).

Kim, R. Y.
See Hahn, H. T. and Kim, R. Y.

Kim, R. Y. and Whitney, J. M., "Effect of Environment on the Notch Strength of Laminated Composites," Presented at the Mechanics of Composites Review, Bergamo Center, Dayton, Ohio (January 1976).

Kimpara, I. and Takehana, M., "Analysis of Acoustic Emission from Internal Failure of Glass Fiber Reinforced Composites," 2nd Acoustic Emission Symp., Tokyo, Ses. 9/2-20 (1974).

King, R. W.
See Broadway, N. J., et al.

Kingston-Lee, D. M.
See Rogers, K. F., et al.
    Yates, B., et al.

Kobayashi, S.
See Ishikawa, T., et al.

Kondo, K. and Taki, T., "Transverse Moisture Diffusivity of Unidirectionally Fiber-Reinforced Composite," Proc. Japan-U.S. Conference on Composite Materials: Mechanics, Mechanical Properties, and Fabrication, Tokyo, 308–317 (1981).

Kourtides, D. A.
See Hilado, C. J., et al.

Koyama, K.
See Ishikawa, T., et al.

Kreider, K. G.
See Breinen, E. M. and Kreider, K. G.
    Prewo, K. M. and Kreider, K. G.

Kreider, K. G. and Prewo, K. M., "Boron-Reinforced Aluminum," *Composite Materials,* K.G. Kreider, ed. Vol. 4, New York: Academic Press, 399–471 (1974).

Kroner, E., "Elastic Moduli of Perfectly Disordered Composite Materials," *J. Mech. Phy. Solids, 15,* 319 (1967).

Kröner, E., "Bounds for Effective Elastic Moduli of Disordered Materials," *J. Mech. Phys. Solids, 25,* 137 (1977).

Kulkarni, S. V., Rice, J. S., and Rosen, B. W., "An Investigation of the Compressive Strength of Kevlar 49/Epoxy Composites," *Composites,* 217–225 (September 1975).

Kuze, E.
See Sato, N. and Kuze, E.

Lai, J. S.
See Findlay, W. N., et al.

Lark, R. J.
See Chamis, C. C. and Lark, R. J.

Larosa, C. N.
See Silvergleit, M., et al.

Larsen, D.
  See Hofer, K. E., Jr., et al.

Larsen, F.
  See Penn, L. and Larsen, F.

Laurities, K. N. and Sandorff, P. E., "The Effect of Environment on the Compressive Strengths of Laminated Epoxy Matrix Composites," Technical Report AFMLT-TR-79-4179, Materials Laboratory, Wright Patterson Air Force Base, Dayton, Ohio (December 1979).

Lee, E. H.
  See Morland, L. W. and Lee, E. H.
    Morland, L. W., et al.

Lee, S. M.
  See Cleary, M. P., et al.

Leisler, W. H.
  See Shirrell, C. D., et. al.

Leissa, A. W.
  See Whitney, J. M. and Leissa, A. W.

Levy, R. L., Fanter, D. L., and Summer, C. J., "Spectroscopic Evidence for Mechanochemical Effects of Moisture in Epoxy Resins," *J. Appl. Poly. Sci., 24,* 1643–1664 (1979).

Lindholm, U. S.
  See Wang, T. K., et al.

Lindrose, A. M.
  See Allred, R. E. and Lindrose, A. M.

Loos, A. C. and Springer, G. S., "Moisture Absorption of Graphite-Epoxy Composites Immersed in Liquids and in Humid Air," *Journal of Composite Materials, 13,* 131–147 (1979).

Loos, A. C. and Springer, G. S., "Effects of Thermal Spiking on Graphite-Epoxy Composites," *Journal of Composite Materials, 13,* 17–34 (1979).

Loos, A. C., Springer, G. S., Sanders, B. A., and Tung, R. W., "Moisture Absorption of Polyester-E Glass Composites," *Journal of Composite Materials, 14,* 142–154 (1980).

Lundemo, C. Y. and Thor, S. E., "Influence of Environmental Cycling on the Mechanical Properties of Composite Materials," *J. Composite Materials, 11,* 276 (1977).

Luikov, A. V. *Analytical Heat Diffusion Theory.* Academic Press (1968).

Machavariani, Z. P.
  See Romanenkov, I. G. and Machavariani, Z. P.

Maisel, J. E.
  See DiCarlo, J. A. and Maisel, J. E.

Mauri, R. E.
  See Crossman, F. W., et al.

Mauri, R. E., "Organic Materials for Structural Applications," *Space Materials Handbook,* J.B. Rittenhouse and J.B. Singletary, eds. third edition, SP-3051, 355–381 (1969).

Mauri, R. E., Crossman, F. W., and Warren, W. J., "Assessment of Moisture Altered Dimensional Stability of Structural Composites," *Proc. 23rd National SAMPE Symposium.* Anaheim, CA, 1202 (May 1978).

Maymon, G., Briley, R. P., and Rehfield, L. W., "Influence of Moisture Absorption and Elevated Temperature on the Dynamic Behavior of Resin Matrix Composites: Preliminary Results," *Advance Composite Materials—Environmental Effects.* ASTM STP 658, 221–233 (1978).

McCalla, B. A.
  See Rogers, K. F., et al.
    Yates, B., et al.

McCleskey, S. F.
  See Cushman, J. B. and McCleskey, S. F.

McFerrin, J. H. and Trulson, O. C., "Electrical Equipment Protection and Waste Disposal Practices—Carbon Fiber Manufacture," SAE paper 790034 (1979).

McGarry, F. J.
 See Sung, N. H. and McGarry, F. J.

McKague, E. L., Jr., Halkias, J. E., and Reynolds, J. D., "Moisture in Composites: The Effect of Supersonic Service on Diffusion," *J. Composite Materials, 9,* 2 (1975).

McKague, E. L., Reynolds, J. D., and Halkias, J. E., "Moisture Diffusion in Fibre Reinforced Plastics," *Transactions of ASME, Journal of Engineering Materials and Technology,* 92–95 (1976).

McKague, E. L., Jr., Reynolds, J. D., and Halkias, J. E., "Swelling and Glass Transition Relations in Epoxy Matrix Material in Humid Environments," *J. Applied Polymer Science* (1978).

McKellar, J. F.
 See Allen, N. S. and McKellar, J. F.

McNabb, A. and Foster, P. K., "A New Analysis of the Diffusion of Hydrogen in Iron and Ferritic Steels," *Transactions of the Metallurgical Society of AIME, 27,* 618–627 (1963).

McPherson, E. L.
 See Vaughan, D. J. and McPherson, E. L.

Mehta, B. S., Dibenedetto, A. T., and Kardos, J. L., "Sorption and Diffusion of Water in Glass-ribbon Reinforced Composites," *J. Applied Polymer Science, 21,* 3111–3127 (1977).

Mehta, B. S., Dibenedetto, A. T., and Kardos, J. L., "Diffusion and Permeation of Gases in Fibre Reinforced Composites," *Intern. J. Polymeric Mater., 5,* 147–161 (1976).

Mende, E. W.
 See Gibson, R. F., et al.

Menousek, J. F., Herring, R. L., and Wiggins, E. W., "Thermal Properties of Graphite Epoxy at Elevated Temperatures," MDC-IR0115, McDonnel Aircraft Company, St. Louis, Missouri (Dec. 1978).

Menousek, J. F. and Monin, D. L., "Laser Thermal Modeling of Graphite Epoxy," NWC Technical Memorandum Report 3834, Naval Weapons Center, China Lake, CA (June 1979).

Meyn, D. A., "Effect of Temperature and Strain Rate on the Tensile Properties of Boron-Aluminum and Boron-Epoxy Composites," *Composite Materials: Testing and Design* (Third Conference), ASTM STP 546, American Society for Testing Materials, Philadelphia, 225–236 (1974).

Miller, A. K.
 See Adams, D. F. and Miller, A. K.
 Allred, R. E., et al.

Miller, A. K. and Adams, D. F., "Micromechanical Aspects of the Environmental Behavior of Composite Materials," Report UWME-DR-701-111-1, Department of Mechanical Engineering, University of Wyoming, Laramie, Wyoming (January 1977).

Miller, D. B.
 See Morgan, R. J., et al.

Miller, E. *Textiles: Properties and Behavior.* (rev. edn.) London: B. T. Batsford (1976).

Minty, D. C.
 See Hancox, N. L. and Minty, D. C.

Monin, D. L.
 See Menousek, J. F. and Monin, D. L.

Morgan, R. J. and O'Neal, J. E., "The Durability of Epoxies," *Polym.-Plast. Tech. and Engr., 10,* 49–116 (1978).

Morgan, R. J., O'Neal, J. E., and Miller, D. B., "The Structure, Modes of Deformation and Failure, and Mechanical Properties of Diaminodiphenyl Sulphone-cured Tetraglycidyl 4,4′ Diaminodiphenyl Methane Epoxy," *J. Matl. Sci., 14,* 109–124 (1979).

Mori, T.
　See Tanaka, K. and Mori, T.

Mori, T. and Tanaka, K., "Average Stress in Matrix and Average Elastic Energy of Materials with Misfitting Inclusions," *Acta Metall., 21,* 571 (1973).

Morimoto, K., Suzuki, T., and Yosomiya, R., "Adhesion between Glass Fiber and Matrix of Glass Fiber Reinforced Rigid Polyurethane Foam Under Tension," *Polym. Plas. Tech. Eng.*

Morimoto, K., Suzuki, T., and Yosomiya, R., "Flexural Properties of Glass Fiber Reinforced Rigid Polyurethane Foam," *Ind. Eng. Chem. Prod. Res. Dev.*

Morimoto, K., Suzuki, T., and Yosomiya, R., "Stress Relaxation of Glass Fiber Reinforced Rigid Polyurethane Foam," *Polym. Eng. Sci.*

Morland, L. W. and Lee, E. H., "Stress Analysis for Linear Viscoelastic Materials With Temperature Variation," *Transactions Society of Rheology, IV,* 233–263 (1960).

Morris, D. H.
　See Garber, D. P., et al.
　　Griffith, W. I., et al.
　　Yeow, Y. T., et al.

Morris, J. G.
　See Silvergeit, M., et al.

Ngoc Son, C. P.
　See Beachell, H. C. and Ngoc Son, C. P.

Noma, K. and Yosomiya, R., "A Discussion on the Surface of Resin by IR Analysis," *Resin Finishing, Japan, 12,* 35–40 (1963).

Nomura, S.
　See Chou, T. W., et al.

Nomura, S. and Chou, T. W., "Bounds of Effective Thermal Conductivity of Short-Fiber Composites," *J. Composite Materials, 14,* 120–29 (1980).

Nomura, S. and Chou, T. W., "Fiber Orientation Effects on the Elastic Moduli of Short-Fiber Composites," *Fibre Sci. Tech.*

Norman, J. C.
　See Zweben, C. and Norman, J. C.

Nowacki, W., "Certain Problems of Thermodiffusion in Solids," *Arch. Mech., 23,* 731–735 (1971).

O'Connell, R. J.
　See Budiansky, and O'Connell, R. J.
　　Watt, J. P., et al.

Onaran, K.
　See Findlay, W. N., et al.

O'Neal, J. E.
　See Morgan, R. J.
　　Morgan, R. J., et al.

Osborn, W. E.
　See Gibson, R. F., et al.

Otsuka, M.
　See Wakashima, K., et al.

Overy, M. J.
　See Rogers, K. F., et al.
　　Yates, B., et al.

Pagano, N. J.
　See Halpin, J. C. and Pagano, N. J.
　　Tsai, S. W. and Pagano, N. J.

Pagano, N. J. and Pipes, R. B., "Some Observations on the Interlaminar Strength of Composite Laminates," *Int. J. Mech. Sci., 15*, 679–688 (1973).

Parker, J. A.
    See Hilado, C. J., et al.

Parker, S. F. H.
    See Rogers, K. F., et al.

Paul, B., "Prediction of Elastic Constants in Multi-phase Materials," *Trans AIME, 218*, 36 (1960).

Pears, C. D.
    See Thornburg, J. D. and Pears, C. D.

Penn, L. and Larsen, F., "Physiochemical Properties of Kevlar 49 Fiber," *J. Appl. Poly. Sci., 23*, 59–74 (1979).

Pering, G. A., Farrell, P. V., and Springer, G. S., "Degradation of Tensile and Shear Properties of Composites Exposed to Fire or High Temperature," *J. of Composite Materials, 14*, 54–68 (1980).

Phelps, H. R., "Effects of Nonionizing Space Environmental Parameters on Graphite/Epoxy Composites," Masters Thesis, George Washington University (August 1979).

Phillips, L. N.
    See Rogers, K. F., et al.
        Yates, B., et al.

Pillinger, W. L.
    See Caskey, G. R. and Pillinger, W. L.

Pipes, R. B.
    See Pagano, N. J. and Pipes, R. B.

Pipkin, A. C. *Lectures on Viscoelasticity Theory.* Springer-Verlag (1972).

Pister, K. S.
    See Taylor, R. L., et al.

Plunkett, R.
    See Gibson, R. F. and Plunkett, R.

Poe, C. C., Jr. and Kennedy, J. M., "An Assessment of Buffer Strips for Improving Damage Tolerance of Composite Laminates," *Journal of Composite Materials Supplement, 14*, 57–70 (1980).

Prater, J. L.
    See Chung, T. J. and Prater, J. L.

Prater, J. L., "Finite Element Analysis in Two-Dimensional Composites Subjected to Hygro-thermoelastic Loadings," M.S. Thesis, The University of Alabama in Huntsville (1980).

Predecki, P. and Barrett, C. S., "Residual Stresses in Resin Matrix Composites," Presented at the 28th Sagamore Army Materials Research Conference, Lake Placid, New York, (July 13–17, 1981).

Predecki, P. and Barrett, C. S., "Stress Measurement in Graphite/Epoxy Composites by X-ray Diffraction from Fillers," *J. Composite Matls., 13*, 61 (1979).

Prestwich, J. D., "Unidirectional Properties of Carbon Fibre Composites," BAe Manchester Division, Report No. HSA-MSM-R-GEN-0295 (1978).

Prewo, K. M.
    See Kreider, K. G. and Prewo, K. M.

Prewo, K. M., "Anelastic Creep of Boron Fibers," *J. Compos. Mater., 8*, 411–414 (1974).

Prewo, K. M. and Kreider, K. G., "High Strength Boron and Borsic Fiber Reinforced Aluminum Composites," *J. Compos. Mater., 6*, 338–357 (1972).

Purohit, K. S.
    See Zinsmeister, G. E. and Purohit, K. S.

Rao, N.
See Hofer, K. E., Jr., et al.

Rao, R. M. V. G. K., "Diffusion Phenomenon in Polymer Composites, Permeable and Impermeable Fibre Composites," Ph.D. Thesis, Dept. of Chemical Engineering, I.I.Sc., Bangalore (Dec. 1982).

Rao, R. M. V. G. K., Balasubramanian, N., and Chanda, M., "Moisture Absorption Phenomenon in Permeable Fibre Polymer Composites," *J. Appl. Poly. Sci.*, 26, 4069-4079 (1981).

Rehfield, L. W.
See Maymon, G., et al.

Rehfield, L. W. and Briley, R. P., "A Comparison of Environmental Effects on Dynamic Behavior of Graphite/Epoxy Composites with Aluminum Alloys," ASME Paper 78-WA/Aero-10 (1978).

Reinhart, T. J., "Air Force Materials Laboratory Perspective" in *The Effects of Relative Humidity and Elevated Temperatures on Composite Structures*, J.R. Vinson, R.B. Pipes, W.J. Walker, and D.R. Ulrich, eds. AFOSR TR-770030, 37-97 (1977).

Reissner, E.
See Goland, M. and Reissner, E.

Remakrishanan, M., "Pyrolysis and Thermal Degradation of Rigid-Urethane Foams," University of Utah, Salt Lake City (1975).

Rennhack, E. H.
See Camahort, J. L., et al.

Reynolds, J. D.
See McKague, E. L., Jr., et al.

Rice, J. S.
See Kulkarni, S. V., et al.

Richard, R. M. and Blacklock, J. M., "Finite-Element Analysis of Inelastic Structures," *AIAA Journal*, 7, 432-438 (1969).

Riegner, D. A.
See Gibson, R. F., et al.

Riegner, D. A. and Sanders, B. A., "A Characterization Study of Automotive Continuous and Random Glass Fiber Composites," General Motors Corporation, Manufacturing Development Report No. MD79-023, General Motors Technical Center, Warren, Michigan, 1979, and presented to the Society of Plastics Engineers, National Technical Conference, Detroit, Michigan (November 1979).

Rogers, K. F.
See Yates, B., et al.

Rogers, K. F., Kingston-Lee, D. M., Phillips L. N., Yates, B., Chandra, M., and Parker, S. F. H., "The Thermal Expansion of Carbon-fibre Reinforced Plastics, Part 6 The Influence of Fiber Weave in Fabric Reinforcement," *J. Materials Science*, 16, 2803 (1981).

Rogers, K. F., Phillips, L. N., Kingston-Lee, D. M., Yates, B., Overy, M. J., Sargent, J. P., and McCalla, B. A., "The Thermal Expansion of Carbon Fibre-reinforced Plastics, Part 1 The Influence of Fibre Type and Orientation," *J. Materials Science*, 12, 718 (1977).

Romanenkov, I. G. and Machavariani, Z. P., "Water Absorption of GRPS," *Soviet Plastics*, 49-51 (June 1966).

Rose, P. G.
See Greenwood, J. H. and Rose, P. G.

Rosen, B. W.
See Hashin, Z. and Rosen, B. W.
Kulkarni, S. V. and Rosen, B. W.

Rosen, B. W., "Tensile Failure of Fibrous Composites," *AIAA J.*, 2, 1985-1991 (1964).

Roylance, M.
  See Roylance, D. and Roylance, M.

Roylance, D. and Roylance, M., "Influence of Outdoor Weathering of Dynamic Mechanical Properties of Glass/Epoxy Laminates," ASTM STP 602, 85-94 (1976).

Sakai, T.
  See Sato, N. and Sakai, T.

Sanders, B. A.
  See Heimbuch, R. A. and Sanders, B. A.
      Loos, A. C. and Sanders, B. A.
      Riegner, D. A. and Sanders, B. A.
      Springer, G. S. and Sanders, B. A.
      Wang, T. K. and Sanders, B. A.

Sanders, B. A. and Heimbuch, R. A., "Engineering Properties of Automotive Fiber Reinforced Plastics," General Motors Manufacturing Development Report No. MD77-020, General Motors Technical Center, Warren, Michigan, 1977, and presented to the Conference on Environmental Degradation of Engineering Materials, Virginia Polytechnic Institute and State University, Blacksburg, Virginia (October 10, 1977) and to the Society for Experimental Stress Analysis, Fall Meeting, Philadelphia, Pennsylvania (October 11, 1977).

Sandon, F. A.
  See Shirrell, C. D., et al.

Sandorff, P. E.
  See Laurities, K. N. and Sandorff, P. E.

Sandorff, P. E. and Tajima, Y. A., "The Experimental Determination of Moisture Distribution in Carbon/Epoxy Laminates," *Composites* (January 1979).

Sargent, J. P.
  See Rogers, K. F., et al.
      Yates, B., et al.

Sargent, J. P. and Ashbee, K. H. G., "High Resolution Optical Interference Investigation of Swelling Due to Water Uptake by Model Adhesive Joints," *J. Adhesion, 11,* 175 (1980).

Sato, N., Sakai, T., and Kuze, E., "Thermal Expansion Characteristics of Glass-Epoxy Resin Composite in the Viscoelastic State (II)," *Trans. Japan Soc. Composite Materials, 4,* 5 (1978).

Schaffer, B. G. and Adams, D. F., "Nonlinear Viscoelastic Analysis of a Unidirectional Composite Material," *Journal of Applied Mechanics, 48,* 859-865 (1981).

Schapery, R. A., "Thermal Expansion Coefficients of Composite Materials Based on Energy Principles," *J. Composite Materials, 2,* 380 (1968).

Schapery, R. A., "Viscoelastic Behavior and Analysis of Composite Materials," in *Composite Materials,* Broutman and Krock, eds. Academic Press, p. 85 (1974).

Schapery, R. A., et al., "Composite Materials for Structural Design," AFOSR TR-79-0347 (March 1979).

Schoppee, M. M.
  See Abbott, N. J., et al.

Schutz, D.
  See Gerharz, J. J. and Schutz, D.

Seiffert, V. W., "Review of Recent Activities and Trends in the Field of Automobile Materials," in *Worldwide Applications of Plastics.* Society of Automotive Engineers, SP-482, 1-6 (1981).

Shen, C. H.
  See Springer, G. S. and Shen, C. H.

Shen, C. H. and Springer, G. S., "Moisture Absorption and Desorption of Composite Materials," *J. Composite Materials, 10,* 2-20 (1976).

Shen, C. H. and Springer, G. S., "Effects of Moisture and Temperature on the Tensile Strength of Composite Materials," *J. Composite Materials, 11,* 2-16 (1977).

Shen, C. H. and Springer, G. S., "Environmental Effects on the Elastic Module of Composite Materials," *J. Composite Materials, 11,* 250–264 (July 1977).

Sheppard, D. M.
See Weatherford, W. D. and Sheppard, D. M.

Shimada, J., "The Mechanism of Oxidative Degradation of ABS Resin, Part 1. The Mechanism of Thermooxidative Degradation," *Journal of Applied Polymer Sciences, 12,* 655–669 (1968).

Shirrell, C. D., "Diffusion of Water Vapor in Graphite/Epoxy Composites," *Advanced Composite Materials—Environmental Effects,* ASTM STP 658, American Society for Testing and Materials, Philadelphia, Pennsylvania, 21–42 (1977).

Shirrell, C. D. and Halpin, J., "Moisture Absorption and Desorption in Epoxy Composite Laminates," *Composite Materials: Testing and Design.* (Fourth Conference), ASTM STP 617, American Society for Testing and Materials, 514–528 (1977).

Shirrell, C. D., Leisler, W. H., and Sandon, F. A., "Moisture-Induced Surface Damage in T300/ 5208 Graphite/Epoxy Laminates," *Nondestructive Evaluation and Flaw Criticality for Composite Materials,* ASTM STP 696, American Society for Testing and Materials, 209–222 (1979).

Shollenberger, C. S. and Stewart, F. D., "Thermoplastic Urethane Structure and Ultraviolet Stability," *J. Elastoplastics, 4,* 294–331 (1972).

Silvergleit, M., Morris, J. G., and Larosa, C. N., "Flammability Characteristics of Fiber Reinforced Organic Matrix Composites," *Polymer Engineering and Science, 18,* 97 (1978).

Sinclair, J. M.
See Chamis, C. C. and Sinclair, J. M.
  Chamis, C. C., et al.

Skelton, J.
See Abbott, N. J., et al.

Smiley, R. F.
See Hobbs, N. P., et al.

Smith, R. J., "Changes in Boron Fiber Strength Due to Surface Removal by NASA TN D-8219," Washington, D.C. (1976).

Smith, W. S.
See Zweben, C., et al.

Smith, W. S., "Environmental Effects on Aramid Composites," E. I. du Pont de Nemours, Textile Fibers Dept. Report, Wilmington, DE (1980).

Springer, G. S.
See Do, M. H. and Springer, G. S.
  Loos, A. C. and Springer, G. S.
  Loos, A. C., et al.
  Pering, G. A., et al.
  Shen, C. H. and Springer, G. S.

Springer, G. S., "Moisture Contents of Composites under Transient Conditions," *J. Composite Materials, 11,* 107–122 (1977).

Springer, G. S., "Environmental Effects on Epoxy Matrix Composites," *Composite Materials: Testing and Design,* S.W. Tsai, ed. ASTM STP 647, American Society for Testing and Materials, 291–314 (1979).

Springer, G. S. *Environmental Effects on Composite Materials.* Lancaster, PA: Technomic Publishing Company (1980).

Springer, G. S., "Properties of Organic Matrix Short Fiber Composites," Materials Laboratory, Air Force Systems Command, Report AFWAL-TR-82-4004, Wright Patterson Air Force Base, Dayton, Ohio (1982).

Springer, G. S., Sanders, B. A., and Tung, R. W., "Environmental Effects on Glass Fiber Reinforced Polyester and Vinylester Composites," *J. Composite Materials, 14,* 213–232 (1980).

Springer, G. S. and Shen, C. H., "Moisture Absorption and Desorption of Composite

Materials," Technical Report AFML-TR-76-102, Air Force Materials Laboratory, Air Force Systems Command, Wright-Patterson Air Force Base, Dayton, Ohio (June 1976).

Springer, G. S. and Tsai, S. W., "Thermal Conductivities of Unidirectional Composites," *J. Composite Materials, 1,* 166–173 (1967).

"Standard Recommended Practice for Maintaining Constant Relative Humidity by Means of Aqueous Solutions," ASTM Designation: E104-51 (1971).

Stander, M.
  See Trabocco, R. E. and Stander, M.

Stein, B. A.
  See Kerr, J. R., et al.

Sternstein, S. S., "Mechanical Properties of Glassy Polymers," in *Treatise on Materials Science and Technology,* Vol. 10, J.M. Schultz, ed. Academic Press (1977).

Stewart, F. D.
  See Shollenberger, C. S. and Stewart, F. D.

Stiepanow, S. W., "Zawisimost koefficientov tieploprowodnosti uporiadocziennych dwuchfaznych sistiem ot obiemnoj koncientracji wkliuczieni," Tieplofizicziesskije swojstwa twiordych wieszczestw, Izd. "Nauka," Moskva (1971).

Stoecklin, R. L., "737 Graphite Composite Flight Spoiler Flight Service Evaluation," NASA CR-144984 (May 1976).

Summer, C. J.
  See Levy, R. L., et al.

Sung, N. H. and McGarry, F. M., "The Mechanical and Thermal Properties of Graphite Fiber Reinforced Polyphenylquinoxaline and Polyimide Composites," *Polymer Engineering and Science, 16,* 426 (1976).

Suzuki, T.
  See Morimoto, K., et al.

Symm, G. T., "The Longitudinal Shear Modulus of a Unidirectional Fibrous Composite," *J. Composite Materials, 4,* 426 (1970).

Taback, I., "Vulnerability," in *Carbon Fiber Risk Analysis.* NASA Conference Publication 2074, 109 (1978).

Tajima, Y. A.
  See Sandorff, P. E. and Tajima, Y. A.

Takehana, M.
  See Kimpara, I. and Takehana, M.

Taki, T.
  See Kondo, K. and Taki, T.

Tanaka, K.
  See Mori, T. and Tanaka, K.

Tanaka, K. and Mori, T., "Note on Volume Integrals of the Elastic Field Around an Ellipsoidal Inclusion," *J. Elasticity, 2,* 199 (1972).

Taya, M.
  See Chou, T. W., et al.

Taylor, R. L., Pister, K. S., and Goudreau, G. L., "Thermomechanical Analysis of Viscoelastic Solids," Structural Engineering Laboratory Report 68-7, Department of Civil Engineering, University of California at Berkeley (June 1968).

Tenney, D. R.
  See Unman, J. and Tenney, D. R.

"Textile Fibers for Industry," Owens-Corning Fiberglas Corp., Toledo, Ohio, 8–30 (1971).

Thomas, R. E.
  See Enos, J. H., et al.

Thor, S. E.
See Lundemo, C. Y. and Thor, S. E.

Thornburg, J. D. and Pears, C. D., "Prediction of the Thermal Conductivity of Filled and Reinforced Plastics," ASME Paper 65-WA/HT-4 (1965).

Toth, I. J., "Comparison of the Mechanical Behavior of Filamentary Reinforced Aluminum and Titanian Alloys," *Composite Materials: Testing and Design* (Third Conference), ASTM STP 546, American Society for Testing Materials, Philadelphia, 542–560 (1974).

Trabocco, R. E.
See Augl, J. M. and Trabocco, R. E.

Trabocco, R. E. and Stander, M., "Effect of Natural Weathering on the Mechanical Properties of Graphite/Epoxy Composite Materials," *Environmental Effects on Advanced Composite Materials*. ASTM STP 602, American Society for Testing and Materials, 67 (1976).

Trujillo, R. E. and Engler, B. P., "Chemical Characterization of Composite Prepreg Resins. Part I," Sandia Laboratories Report SAND 78-1504, Albuquerque, NM (December 1978).

Trulson, O. C.
See McFerrin, J. H. and Trulson, O. C.

TRW Inc., Systems Group, "Analysis of Moisture in Polymers and Composites," NASA Lewis Research Center, Contract NAS 320406.

Tsai, S. W.
See Adams, D. F. and Tsai, S. W.
Halpin, J. C. and Tsai, S. W.
Springer, G. S. and Tsai, S. W.

Tsai, S. W., "Structural Behavior of Composite Materials," NASA CR-71 (1964).

Tsai, S. W., Adams, D. F., and Doner, D. R., "Effect of Constituent Materials Properties on the Strength of Fiber Reinforced Composite Materials," Air Force Materials Laboratory Report AFML-TR-66-190 (August 1966).

Tsai, S. W. and Hahn, H. T. *Introduction to Composite Materials*. Lancaster, PA: Technomic Publ. Co. (1980).

Tsai, S. W. and Pagano, N. J., "Invariant Properties of Composite Materials," *Composite Materials Workshop*, S.W. Tsai, J.C. Halpin, and N.J. Pagano, eds. Lancaster, PA: Technomic Publishing Co. (1968).

Tung, R. W.
See Loos, A. C., et al.

Turner, G.
See Dean, G. and Turner, G.

Umekawa, S.
See Wakashima, K., et al.

Unman, J. and Tenney, D. R., "Analytical Prediction of Moisture Absorption/Desorption in Resin Matrix Composites Exposed to Aircraft Environments," AIAA Conference, San Diego (March 1977).

Updike, D. P.
See Yuceoglu, U. and Updike, D. P.

Van Amerongen, G. J., "Diffusion in Elastomers," *Rubber Chemistry and Technology, 37,* 1067–1074 (1964).

Van Der Waal, P. C. *Effects of Space Environment on Materials*. First Edition, RV-22, Fokker, Amsterdam: Royal Netherlands Aircraft Factories (1968).

Vaughan, D. J. and McPherson, E. L., "The Effects of Adverse Environmental Conditions on the Resin-Glass Interface of Epoxy Composites," *Proceedings of the 27th Annual Conference, Reinforced Plastics/Composites Institute*. New York, New York: The Society of the Plastics Industry, Inc., Section 21-C, 1–7 (1972).

Veltri, R. D. and Galasso, F. S., "High-Temperature Strength of Boron, Silicon Carbide,

Stainless Steel, and Tungsten Fibers," *J. Am. Ceram. Soc., 54,* 319-320 (1971).

Vinson, J. R.
See Flaggs, D. L. and Vinson, J. R.

Verette, R. M., "Temperature/Humidity Effects on the Strength of Graphite/Epoxy Laminates," AIAA Paper No. 75-1011, AIAA 1975 Aircraft Systems and Technology Meeting, Los Angeles, California (August 4-7, 1975).

Waals, F. M.
See Christensen, R. M. and Waals, F. M.

Wakashima, K., Otsuka, M., and Umekawa, S., "Thermal Expansion of Heterogeneous Solids Containing Aligned Ellipsoidal Inclusions," *J. Composite Materials, 8,* 391 (1974).

Walrath, D. E.
See Adams, D. F. and Walrath, D. E.

Walrath, D. E. and Adams, D. F., "Fatigue Behavior of Hercules 3501-6 Epoxy Resin," Report No. NADC-78139-60, Naval Air Development Center, Warminster, Pennsylvania (January 1980).

Wang, T. K., Sanders, B. A., and Lindholm, U. S., "A Loading Rate and Environmental Effects Study of Adhesive Bonded SMC Joints," Report GMMD80-044, General Motors Corporation, Manufacturing Development, GM Technical Center, Warren, Michigan 48090 (1980).

Ward, J. R. and Decker, L. J., "Determination of the Thermal Decomposition Kinetics of Polyurethane Foam by Guggenheim's Method," *Ind. Eng. Chem.Prod. Res. Dev., 21,* 460-461 (1982).

Wardle, M. W.
See Zweben, C. and Wardle, M. W.
    Zweben, C., et al.

Warren, W. J.
See Crossman, F. W., et al.
    Mauri, R. E., et al.

Watanabe, T., "Weathering Test of Fiberglass Reinforced Plastic Under Flexural Load," *Kobunshi Ronbunshu, 39,* Japan, 1-6 (1982).

Watt, J. P., Davies, G. E., and O'Connell, R. J., "The Elastic Properties of Composite Materials," *Rev. Geophys. Space Phys., 14,* 541-63 (1976).

Weatherford, W. D. and Sheppard, D. M., "Basic Studies of the Mechanism of Ignition of Cellulosic Materials," Tenth Symposium (International) on Combustion, The Combustion Institute, 897 (1965).

Weiner, J. H.
See Boley, B. A. and Weiner, J. H.

Weitsman, Y.
See Douglass, D. A. and Weitsman, Y.

Weitsman, Y., "Diffusion With Time-Varying Diffusivity, With Application to Moisture-Sorption in Composites," *Journal Compt. Mat., 10,* 193-204 (1976).

Weitsman, Y., "Hygrothermal Viscoelastic Analysis of a Resin Slab Under Time-Varying Moisture and Temperature," AIAA Structures, Dynamics and Materials Conference, San Diego (March 1977).

Weitsman, Y., "Residual Thermal Stresses Due to Cool-Down of Epoxy-Resin Composites," *J. of Applied Mechanics, 46,* 563 (1979).

Whiteside, J. B.
See DeIasi, R. and Whiteside, J. B.
    DeIasi, R., et al.

Whitney, J. M.
See Browning, C. E., et al.
    Kim, R. Y. and Whitney, J. M.

Whitney, J. M. and Browning, C. E., "Some Anomalies Associated with Moisture Diffusion in Epoxy Matrix Composite Materials," *Advanced Composite Materials—Environmental Effects,* J.R. Vinson, ed. ASTM STP 658, American Society for Testing and Materials, 43 (1978).

Whitney, J. M. and Husman, G. E., "Use of the Flexure Test for Determining Environmental Behavior of Fibrous Composites," *Experimental Mechanics, 18,* 185-190 (1978).

Whitney, J. M. and Leissa, A. W., "Analysis of Heterogeneous Anisotropic Plate," *J. Applied Mech., 36,* 261 (1969).

Wiggins, E. W.
See Menousek, J. F., et al.

Williams, J. H., Jr., "Stresses in Adhesive Between Dissimilar Adherends," *Journal of Adhesion, 7,* 97-107 (1975).

Williams, W.
See DiCarlo, J. A. and Williams, W.

Willis, J. R., "Bounds and Self-Consistent Estimates for the Overall Properties of Anisotropic Composites," *J. Mech. Phys. Solids, 25,* 185-202 (1977).

Wills, J. L.
See Wright, M. A. and Wills, J. L.

Wolter, W.
See Delasi, R., et al.
Dvoskin, N., et al.

Wright, M. A. and Welch, D., "Failure of Centre Notched Specimens of 6061 Aluminum Reinforced with Unidirectional Boron Fibers," *Fibre Sci. Technol., 11,* 447-461 (1978).

Wright, M. A. and Wills, J. L., "The Tensile Failure Modes of Metal Matrix Composite Materials," *J. Mech. Phys. Solids, 22,* 161-175 (1974).

Wright, W. W.
See Judd, N. C. W. and Wright, W. W.

Yates, B.
See Rogers, K. F., et al.

Yates, B., Overy, M. J., Sargent, J. P., McCalla, B. A., Kingston-Lee, D. M., Phillips, L. N., and Rogers, K. F., "The Thermal Expansion of Carbon Fibre-Reinforced Plastics, Part 2. The Influence of Fiber Volume Fraction," *J. Materials Science ,13,* 433 (1978).

Yatomi, C.
See Gurtin, M. E. and Yatomi, C.

Yau, A.
See Gibson, R. F. and Yau, A.
Gibson, R. F., et al.

Yau, A., "Experimental Techniques for Measuring Dynamic Mechanical Behavior of Composite Materials," M.S. Thesis, Mechanical Engineering Department, University of Idaho (July 1980).

Yeow, Y. T., Morris, D. H., and Brinson, H. F., "Time-Temperature Behavior of a Unidirectional Graphite/Epoxy Composite," in *Composite Materials: Testing and Design,* 5th Conference, S.W. Tsai, ed. ASTM STP 674, 263-281 (1979).

Yonezawa, F.
See Hori, M. and Yonezawa, F.

Yosomiya, R.
See Noma, K. and Yosomiya, R.
Morimato, K., et al.

Young, H. L. and Greever, W. L., "High Temperature Strength Degradation of Composites During Aging in the Ambient Atmosphere," in *Composite Materials in Engineering Design,* B.R. Noton, ed. Metals Park, OH: American Society for Metals, 695-715 (1972).

Yuceoglu, U. and Updike, D. P., "Stress Analysis of Bonded Plates and Joints," *Journal of the Engineering Mechanics Division, ASCE, 106,* 37-56 (1980).

Zeller, R.
  See Dederich, P. H. and Zeller, R.

Zeller, R. and Dederichs, P H., "Elastic Constants of Polycrystals," *Phys. Stat. Sol., b55,* 831 (1973).

Zinsmeister, G. E. and Purohit, K. S., "Comments on Springer and Tsai's Method of Predicting Effective Thermal Conductivities of Unidirectional Materials," *J. Composite Materials, 4,* 278 (1970).

Zweben, C., "The Flexural Strength of Aramid Fiber Composites," *J. Comp. Mat'ls., 12,* 422–430 (1978).

Zweben, C. and Norman, J. C., "Kevlar 49/Thornel 300 Hybrid Fabric Composites for Aerospace Applications," *SAMPE Quarterly,* 1 (July 1976).

Zweben, C., Smith, W. S., and Wardle, M. W., "Test Methods for Fiber Tensile Strength, Composite Flexural Modulus and Properties of Fabric-Reinforced Laminates," *Composite Materials: Testing and Design.* ASTM STP 674, American Society for Testing and Materials, 228–262 (1979).

Zweben, C. and Wardle, M. W., "Flexural Fatigue of Marine Laminates Reinforced with Woven Rovings of E-Glass and of Kevlar 49 Aramid," *Proceedings of 34th Annual Conference Reinforced Plastics/Composites Institute.* Society of the Plastics Industry, New Orleans, LA, Section 4C, 1–6 (1979).

Index